Immanuel Velikovsky

AGES IN CHAOS
Volume I
From the Exodus to King Akhnaton

Abacus edition published in 1973
by Sphere Books Ltd
30/32 Grays Inn Road, London WC1X 8JL
Reprinted 1974

First published in Great Britain
by Sidgwick & Jackson Ltd 1953
Copyright Immanuel Velikovsky 1952

ISBN 0 349 13575 4

Set in Intertype Times

*Printed in Great Britain by Cox & Wyman Ltd,
London, Reading and Fakenham*

Foreword

Ages in Chaos was conceived in the spring of 1940. It was then that I realized that the Exodus had occurred in the midst of a natural upheaval and that this catastrophe might prove to be the connecting link between the Israelite and Egyptian histories, if ancient Egyptian texts were found to contain references to a similar event. I found such references and before long had worked out a plan of reconstruction of ancient history from the Exodus to the conquest of the East by Alexander the Great. Already by October of the same year I had come to understand the nature and extent of that catastrophe. For a decade after that I worked simultaneously on *Ages in Chaos* and *Worlds in Collision*, the present work requiring the lion's share of the toil.

Ages in Chaos covers largely the period dealt with in *Worlds in Collision*—the eight hundred years from the Exodus of the Israelites from Egypt to the invasion of Palestine by Sennacherib in 687 before the present era, and the additional three and a half centuries to Alexander of Macedonia, altogether twelve hundred years of the history of the ancient East. But whereas the first work concentrated on the description of the physical history of the period, the present work deals with its political and cultural aspects. The occurrence of a widespread natural catastrophe serves here only as the point of departure for constructing a revised chronology of the times and lands under consideration.

I searched the records of one land after another and went from one generation to another, taking from everywhere hints and clues, evidence and proof. Because I had to discover and to collate them, this book is written like a detective story. It is well known that in detective work unexpected associations are often built on minute details: a fingerprint on a bar of metal, a hair on a window sill, a burnt-out match in the bushes. Some details of an archaeological, chronological, or paleographic nature may seem minor matters, but they are the fingerprints of an investigation in which the history of many nations in many generations is vitally involved. Such details are not included to make the reading difficult; they are necessary to establish the main thesis of this work. Therefore, any attempt to read this book cursorily will prove to be a fruitless undertaking.

Correct strategy requires that once a bridgehead is established it should be fortified. Is it good judgment, instead, to open a second front against a new adversary?

After the publication of *Worlds in Collision,* a volume describing two acts of a celestial and terrestrial drama, reconstructed from the collective memory of the human race, a wise and proper move would have been to strengthen my position by following with a volume of geological and paleontological evidence of the same dramatic events in the life of the earth. And since this material from the realm of stones and bones is not rare but abundant, such an undertaking would seemingly not be difficult. It was therefore a great temptation for me to continue from where I had left off in *Worlds in Collision,* to prove again and again, from new angles, that catastrophes did take place and did disrupt slow-moving evolution in inanimate as well as animate nature. And in fact, since the publication of *Worlds in Collision,* I have devoted myself to organizing the evidence from geology and prehistory to supplement the literary and historical evidence of cosmic catastrophism, and to writing *Earth in Upheaval,* only little concerned with the storm aroused by my first book. But I found that the arguments presented in that book were not given a careful hearing, or even reading, particularly by those who protested the loudest. Would it help to produce in haste still more evidence? In my inner council on strategy, I decided to tarry no longer with *Ages in Chaos,* my opus magnum.

I call *Ages in Chaos* the second front because, after having disrupted the complacent peace of mind of a powerful group of astronomers and other textbook writers, I offer here major battle to the historians. The two volumes of the present work will be as disturbing to the historians as *Worlds in Collision* was to the astronomers. It is quite conceivable that historians will have even greater psychological difficulties in revising their views and in accepting the sequence of ancient history as established in *Ages in Chaos* than the astronomers had in accepting the story of cosmic catastrophes in the solar system in historical times. Indeed, a distinguished scholar who has followed this work from the completion of the first draft in

1942, expressed this very idea. He said that he knows of no valid argument against the reconstruction of history presented here, but that psychologically it is almost impossible to change views acquired in the course of decades of reading, writing, and teaching.

The attempt to reconstruct radically the history of the ancient world, twelve hundred years in the life of many nations and kingdoms, unprecedented as it is, will meet severe censure from those who, in their teaching and writing, have already deeply committed themselves to the old concept of history. And many of those who look to acknowledged authorities for guidance will express their disbelief that a truth could have remained undiscovered so long, from which they will deduce that it cannot be a truth.

Should I have heeded the abuse with which a group of scientists condemned *Worlds in Collision* and its author? Unable to prove the book or any part of it wrong or any quoted document spurious, the members of that group indulged in outbursts of unscientific fury. They suppressed the book in the hands of its first publisher by the threat of a boycott of all the company's textbooks, despite the fact that when the book was already on the presses the publisher agreed to submit it to the censorship of three prominent scientists and it passed that censorship. When a new publisher took the book over, this group tried to suppress it there, too, by threats. They forced the dismissal of a scientist and an editor who openly took an objective stand, and thus drove many members of academic faculties into clandestine reading of *Worlds in Collision* and correspondence with its author. The guardians of dogma were, and still are, alert to stamp out the new teaching by exorcism and not by argument, degrading the learned guild in the eyes of the broad public, which does not believe that censorship and suppression are necessary to defend the truth. And here is a rule by which to know whether or not a book is spurious: Never in the history of science has a spurious book aroused a storm of anger among members of scientific bodies. But there has been a storm every time a leaf in the book of knowledge has been turned over. "We are most likely to get angry and excited in our opposition to some idea when we ourselves are not quite certain of our own position, and are inwardly tempted to take the other side."[1]

A scientific approach requires, first, reading, then thinking and investigation, and, lastly, the expression of an opinion. In the case of *Worlds in Collision* the procedure was repeatedly reversed. Scientific rejection demands invalidation of the evidence presented. Nothing of the kind was done with *Worlds in Collision*. The few arguments offered—whatever could be gathered from numerous reviews—I answered point by point in a debate with Professor J. Q. Stewart, astronomer at Princeton University, published in the June 1951 issue of *Harper*'s magazine, fourteen months after the publication of the book. No argument was left unanswered, and no new one has been presented since then, though emotional outbursts have not ceased. Finally, a new strategy was employed: the views expounded in *Worlds in Collision* were appropriated piecemeal by those who first opposed them, though not with frankness and candor, but rather under the guise of showing how wrong the author of that heretical book is. At present no chapter of *Worlds in Collision* needs to be rewritten and no thesis revoked.

Great are the changes in the political history of the ancient East offered in *Ages in Chaos*. I claim the right to fallibility in details and I eagerly welcome constructive criticism. However, before proclaiming that the entire structure must collapse because an argument can be made against this or that point, the critic should carefully weigh his argument against the whole scheme, complete with all its evidence. The historian who permits his attention to be monopolized by an argument directed against some detail, to the extent of overlooking the work as a whole and the manifold proofs on which it stands, will only demonstrate the narrowness of his approach to history. He will be like that "conscientious scientist," Professor Twist, in Ogden Nash's verse, who went on an expedition to the jungles, taking his bride with him. When, one day, the guide brought the tidings to him that an alligator had eaten her, the professor could not smile. "You mean," he said, "a crocodile?"

I believe that the evidence collated in *Ages in Chaos* warrants the reconstruction offered. Sooner or later, and it may be any day, some new archaeological discovery will verify the main thesis of this book, and then it will become apparent even to the indolent, for whom only a fulfilled prophecy is an argument.

The recent discovery of two-language texts – old Hebrew and "Hittite" pictographs – and thus of a clue to the undeciphered pictographs of Asia Minor and Syria, gives promise of revealing facts of unmeasured significance.

For this reason, too, I should delay no longer the present publication. Is it not the case that at first a new idea is regarded as not true, and later, when accepted, as not being new?

IMMANUEL VELIKOVSKY

February 1952

1. Thomas Mann, *Buddenbrooks*.

Dedication

This work is dedicated to my late father. I want to say in a few sentences who Simon Yeheil Velikovsky was.

From the day when, at the age of thirteen, he left the home of his parents and went on foot to one of the old centers of talmudic learning in Russia, to the day when, in December 1937, at the age of seventy-eight, he ended his years in the land of Israel, he devoted his life, his fortune, his peace of mind, all that he had, to the realization of what was once an idea, the renaissance of the Jewish people in its ancient land. He contributed to the revival of the language of the Bible and the development of modern Hebrew by publishing (with Dr. J. Klausner as editor) collective works on Hebrew philology, and to the revival of Jewish scientific thought by publishing, through his foundation, Scripta Universitatis, *to which scientists of many countries contributed and thus laid the groundwork for the Hebrew University at Jerusalem. He was first to redeem the land in Negeb, the home of the patriarchs, and he organized a co-operative settlement there which he called Ruhama; today it is the largest agricultural development in northern Negeb. I do not know whom I have to thank for intellectual preparedness for this reconstruction of ancient history if not my late father, Simon.*

Acknowledgments

In composing this reconstruction, I incurred a debt to archaeologists, who for more than a century have toiled in excavating numerous places in the lands of the ancient East; to generations of philologists, who have read the ancient texts; and to those among the scholars who have made easier the work of research by collecting and classifying the material.

I am grateful to Dr. Walter Federn, of the Asia Institute, New York, who has always been ready to help me with his incomparable knowledge of Egyptological literature. I feel my obligation to him all the more because he has never committed himself to my thesis. It took him more than six years to concede that conventional history is not built on unshakable foundations. His arguments have been a steady incentive for me to collect more and more proof, to collate more and more historical material, until the book attained its present form. His criticism has always been constructive.

I am also indebted to Dr. Robert H. Pfeiffer, outstanding authority on the Bible. Director of the Harvard excavation at Nuzi, curator of the Semitic Museum at Harvard University, professor of ancient history at Boston University, editor of the *Journal of Biblical Literature* (1943–47), and author of a distinguished standard work on the Old Testament, he is eminently qualified to pass judgment. In the summer of 1942, when the manuscript was still in its first draft, he read *Ages in Chaos* and suggested that I try to prove my thesis on archaeological art material. I followed his advice, and the second volume of this work carries chapters on "Ceramics and Chronology" and "Metallurgy and Chronology", in addition to a number of scattered sections dealing with the problems of ancient art, paleography, and stratigraphical archaeology. He read later drafts, too, and showed a great interest in the progress of my work. Neither subscribing to my thesis nor rejecting it, he kept an open mind, believing that only objective and free discussion could clarify the issue.

Neither he nor Dr. Federn nor anyone else shares any responsibility whatsoever for any statement in this book.

Professor J. Garstang, excavator of Jericho, read an early draft of the first chapter. It was his opinion that the Egyptian record of the plagues, as set forth in this book, and the biblical passages dealing with the plagues are so similar that they must have had a common origin.

Dr. I. J. Gelb and the late Dr. S. I. Feigin, both of the Oriental Institute of the University of Chicago, graciously answered questions put to them without being informed as to the thesis of my work. Dr. C. H. Gordon of Dropsie College also was kind enough to answer a number of questions in his field. I express to them my appreciation.

Dr. Horace M. Kallen, professor and sometime dean of the Graduate Faculty of the New School for Social Research, New York, a humanist and humanitarian, gave me his unfailing moral support during all these years, because he knew the odds against which I had to work and the opposition I would meet.

I was fortunate to have had the help of Miss Marion Kuhn, who with great care went over the entire manuscript more than once and offered numerous improvements in style. Mrs. Kathryn Tebbel of the Doubleday staff proved to be a copy editor of great keenness, well versed in the Scriptures.

I should not omit to say that I have had every possible consideration from Mr. Walter Bradbury, managing editor of Doubleday and Company, who made me feel that all the facilities of the great publishing house were at my disposal.

Contents

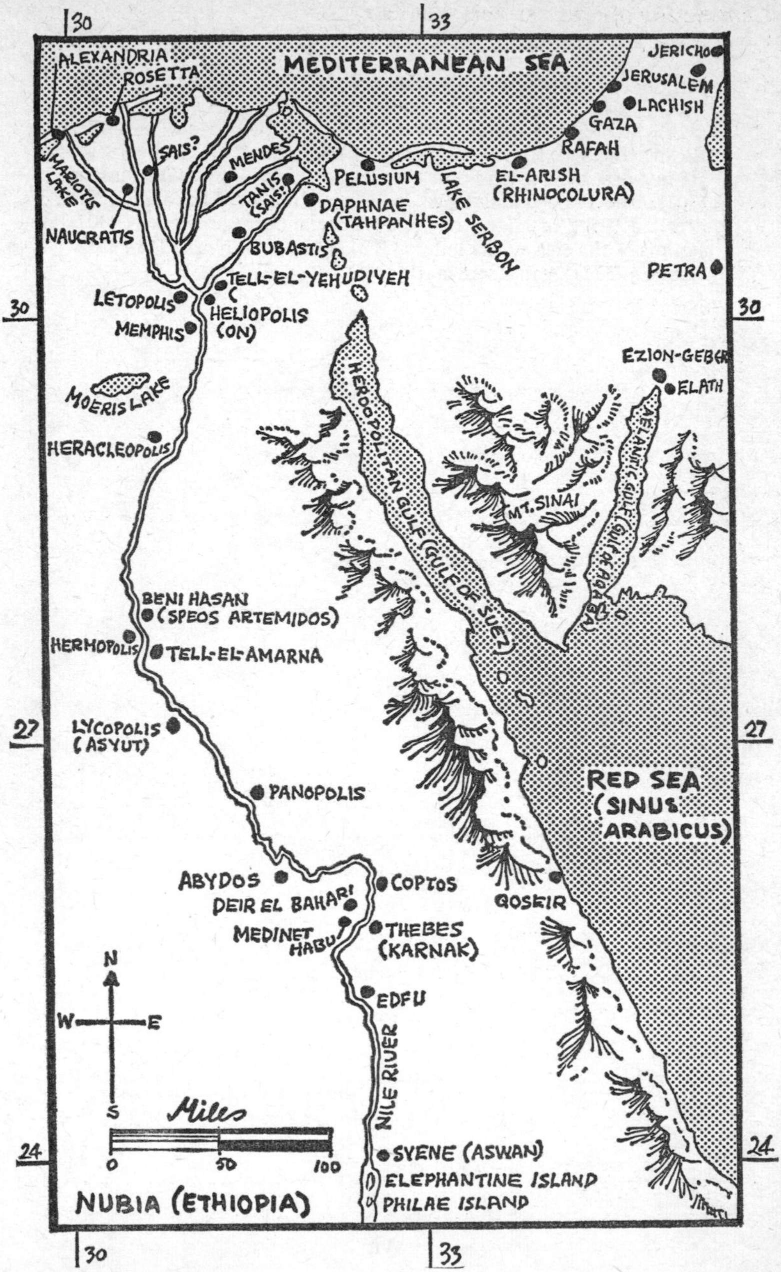

30
33
MEDITERRANEAN SEA
ALEXANDRIA
ROSETTA
MAREOTIS LAKE
SAIS?
NAUCRATIS
MENDES
TANIS (SAIS)
PELUSIUM
LAKE SERBON
EL-ARISH (RHINOCOLURA)
JERICHO
JERUSALEM
LACHISH
GAZA
RAFAH
DAPHNAE (TAHPANHES)
BUBASTIS
TELL-EL-YEHUDIYEH
LETOPOLIS
HELIOPOLIS (ON)
MEMPHIS
PETRA
30
30
EZION-GEBER
ELATH
MOERIS LAKE
HEROOPOLITAN GULF (GULF OF SUEZ)
AELANITIC GULF (GULF OF AQABA)
MT. SINAI
HERACLEOPOLIS
BENI HASAN (SPEOS ARTEMIDOS)
HERMOPOLIS
TELL-EL-AMARNA
27
27
LYCOPOLIS (ASYUT)
RED SEA (SINUS ARABICUS)
PANOPOLIS
ABYDOS
COPTOS
DEIR EL BAHARI
QOSEIR
MEDINET HABU
THEBES (KARNAK)
N
W
E
S
EDFU
NILE RIVER
Miles
0
50
100
24
24
SYENE (ASWAN)
ELEPHANTINE ISLAND
PHILAE ISLAND
NUBIA (ETHIOPIA)
30
33

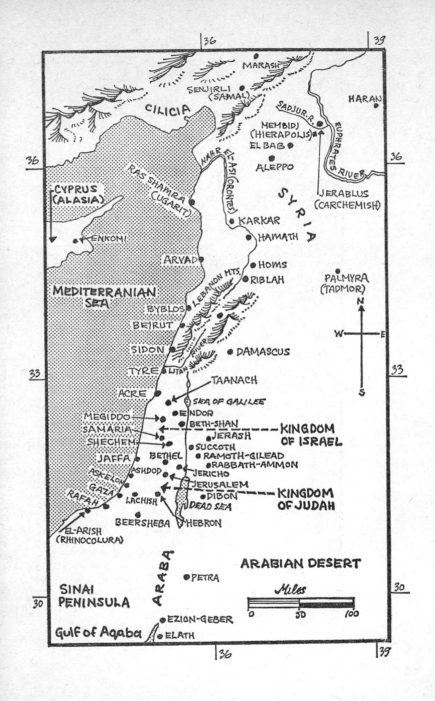

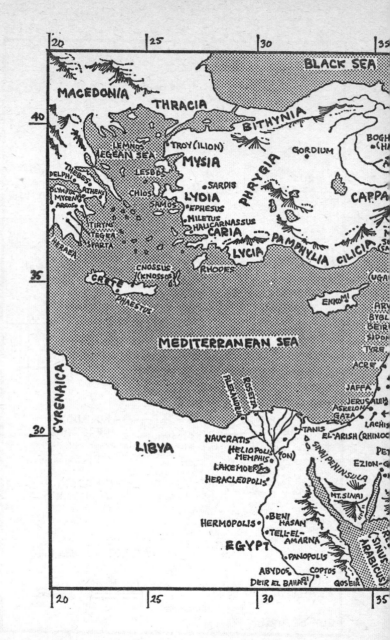

Introduction

This is not descriptive history in the usual meaning of the term. It is a sequence of chapters, each of which is like a court hearing, where witnesses are brought to the stand to disprove the validity of the old and to attest to the accuracy of a new concept of ancient history. The old story of mankind, never before disputed, is here assailed as distorted, and a reconstruction is presented. The period under investigation covers over one thousand years, ending with the advent of Alexander the Great.

Many wondrous things happen when historical perspective is distorted. In order to understand the scope of the displacements in the history of the ancient world, one must try to conceive of the chaos which would result if a survey of Europe and America were written in which the history of the British Isles were some six hundred years out of line, so that in Europe and America the year would be 1941 while in Britain it would be 1341.

As Columbus discovered America in 1492, the Churchill of 1341 could not have visited this country, but must have visited some other land—the scholars would be divided in their opinion as to the whereabouts of that land—and met its chief. Another chief, not Franklin Delano Roosevelt of Washington, would live in history as cosigner of a charter with Churchill of Britain in 1341.

But as American records would speak of Churchill who crossed the ocean in the early forties of the twentieth century, British history would also have a Churchill II, six hundred years after the first one. Cromwell would also be doubled by the same process. He would have to live three hundred years before Churchill I and also three hundred years after him, or three hundred years before Churchill II.

The First World War would be fought twice, as would the Second. The First World War, in its second variant, would follow the Second

World War, in its first variant, by five and three quarter centuries.

By the same token, the development of the Constitution, the cultural life, the progress of technology and the arts, would appear in chaotic distortion.

Newton in England would become an early forerunner of Copernicus instead of following him. Joan of Arc would revive the old traditions of the suffragettes of the post-Victorian days; she would be burned twice with an interval of six hundred years between; or, with the growing confusion of history, she would have to return to the stake a few centuries from today to suffer her death again.

In the case presented, not only the history of the British Isles would be doubled and distorted, but also the history of the entire world. Difficulties would, of course, arise, but they would be swept away as oddities. Complicated theories would be proposed and discussed, and if accepted, they would establish themselves as new, strong obstacles to a correct perception of past history.

Ancient history is distorted in this very manner. Because of the disruption of synchronism, many figures on the historical scene are "ghosts" or "halves" and "doubles". Events are often duplicates: battles are shadows; many speeches are echoes; many treaties are copies; even some empires are phantoms.

The primary error can be found in Egyptian history; because of retardation, the history of Egypt was taken out of real contact with the histories of other peoples. Events in which the people of Egypt and the people of Assyria or Babylonia or Media were involved were recorded in the histories of these peoples from the Egyptian annals; the same events were then described for the second time in the history of Egypt, the annals of these other peoples, participants in the events, being the source.

Thus the histories of Assyria, Babylonia, and Media are disrupted and spoiled; the history of the "Hittite Empire" is entirely invented; the Greek history of the Mycenaean period is displaced and that of the pre-Alexander period is lacerated and Spartan and Athenian warriors, even those with well-known names, appear once more on the pages of history as archaic intruders out of the gloom of the past.

The process of levelling out the histories of the peoples of the ancient world to an exact synchronism will provide some exciting moments. We shall see in a new light many historical documents which had been misinterpreted when presented in an incorrect historical perspective.

We shall read the story of the plagues of the days of the Exodus, written by an Egyptian eyewitness and preserved on papyrus. We shall be able to establish the identity of the mysterious Hyksos, and also to indicate the site of their stronghold Auaris, probably the greatest fortress of ancient times. We shall read the Queen of Sheba's record of her journey to Jerusalem in the days of Solomon, and shall see illustrations depicting this voyage and showing the inhabitants, animals, and plants of Palestine of that time. We shall have before our eyes photographs of the vessels and furniture and utensils of the Temple of Solomon as cut in stone bas-reliefs by a contemporaneous artist. Then will follow texts of letters written by the Jewish kings Jehoshaphat of Jerusalem and Ahab, the sinner of Jezreel, and also by their military chiefs, signed with names we know from the Scriptures.

Of still greater scope are the effects of the revision for Egyptian, Assyrian, and Babylonian histories, and also for the concept of the Greek past. The rectification of chronology, without altering the sequence of the Hebrew past, enriches its records bountifully. The history of Egypt, and following it the histories of Babylonia, Assyria, Media, Phoenicia, Crete, and Greece, change their length. The architectonics of the world past, when redesigned, shows its structure properly joined in time and space. It shows that kings were made their own great-great-grandchildren. Imaginary empires were described, and museum halls have been opened to display the arts of empires which did not exist: the art objects are products of other centuries, even of another millennium. This is the case with the Hittite Empire and its art. This is the case with the Hurrian people and their language.

Through the laborious efforts of scholars, achievements have been recorded without knowledge of their real nature. The Chaldean language was deciphered, but its decipherers did not know that they were reading Chaldean; handbooks about the Carian language were written, but the industrious philologists did not know that it was Carian.

It is not possible in a short introduction to point out all the things which appear in a new light when correctly placed in time. When the hinges of world history are lifted to an adequate height, facts about

peoples and countries, their art and religion, their battles and treatises pour down as if out of a horn of plenty. Certainly more than one
fact and more than one parallel must have been overlooked in this
book, but this is a shortcoming from which a pioneer work is seldom
free.

CHAPTER ONE

In Search of a Link Between Egyptian and Israelite Histories

Two Lands and Their Past

Palestine, one of the westernmost lands in Asia, and Egypt, in the northeast corner of Africa, are neighbouring countries. The history of Egypt reaches back to hoary antiquity; the Jewish people has a history that claims to describe the very beginning of this nation's march through the centuries. At the dawn of their history the Israelites, an unsettled tribe, came from Canaan to Egypt. There they grew to be a people; there, also, they bore the yoke of bondage. Their eventful departure from Egypt is the most treasured recollection of their past, and their traditions tell its story numberless times.

The annals of Egypt, we are told, did not preserve any record of this sojourn of the Israelites or of their departure. It is not known when the Exodus occurred, if it happened at all. A few scholars have expressed the opinion that the sojourn of the Israelites, their bondage, and their departure are mythological motifs; the absence of direct reference to these events on Egyptian monuments and papyri seems to corroborate this view. However, it has been argued that no people would invent legends about bondage which were not calculated to enhance the dignity of the nation, and therefore, it was insisted, there must be a historical basis for the story.

Historians disagree as to the date of the Exodus, and many hypotheses have been proposed. But for more than two thousand years they have agreed that the Exodus took place during the period called in present terminology "the New Kingdom" of Egypt.

Egypt's past is divided into the following periods:

1. The predynastic period belonging mainly to the Neolithic or the Late Stone Age.

2. The Old Kingdom, when most of the pyramids were built: the

Fourth Dynasty, that of Cheops, and the Sixth, that of Phiops, are the best known.

3. The first interregnum, when the land fell into chaos: central authority was abolished in this dark age. Of the dynasties from the Seventh to the Tenth almost nothing is known.

4. The Middle Kingdom, comprising the Eleventh, Twelfth, and Thirteenth Dynasties: feudal Egypt was united under the Twelfth Dynasty, and Egyptian literature reached a height never again to be attained.

5. Another period of chaos, exploited by certain invaders known as Amu in Egyptian and called Hyksos by authors writing in Greek.[1] The Hyksos kings were pharaohs of the Fourteenth to the Seventeenth Dynasties and they ruled over Egypt without mercy;[2] It is not known of what race they were.

6. The New Kingdom. The Hyksos were expelled at the time of Ahmose (Amasis I), who founded the Eighteenth Dynasty, the most renowned of all, the dynasty of Thutmose I; the famous Queen Hatshepsut; Thutmose III, the greatest of all Egyptian conquerors; Amenhotep II; Thutmose IV; Amenhotep III, the builder of magnificent temples at Luxor and Karnak; and Amenhotep IV, who called himself Akhnaton, the great heretic. The epigoni followed; among them the young king Tutankhamen is best known, not because of the distinction of his reign, which is obscure, but because of the riches in his tomb, discovered early in the twenties of this century, and because of the mystery that has surrounded his burial place.

The Eighteenth Dynasty declined under conditions that are not sufficiently known, and history records that the Nineteenth Dynasty, that of Seti the Great, Ramses II (the Great), and Merneptah, followed.

The period of transition from the Nineteenth to the Twentieth Dynasties is obscure.

Among the kings of the Twentieth Dynasty Ramses III was the most prominent; he was the last great emperor of ancient Egypt.

7. The pharaohs from the Twenty-first to the Thirtieth Dynasties were small kings, who left no important records, and their age is called the "late period". Some of these pharaohs, we are told, led armies against Palestine and Babylonia; however, the sources are usually not Egyptian, but for the most part scriptural. Some of these dynasties were Libyan or Ethiopian; later ones (from —525) were

under Persian supremacy; still later ones rebelled against the Persians. The last native king was removed by the Persians in −342. In −332 Egypt was conquered by Alexander the Great.

8. The Ptolemaic Dynasty of descendants of Ptolemy, a general of Alexander, expired with Cleopatra in −40.

The present work covers the time from the end of the Middle Kingdom to the conquest of Egypt by Alexander (the periods marked above as 5, 6, and 7), over a thousand years in the history of the ancient East.

It is useful to remark here that the division into "kingdoms" is modern;[3] the divisions into dynasties comes from Manetho, an Egyptian priest of the third century before the Christian era, who wrote in Greek; the designation of kings as "first", "second", "third", is an arrangement of modern scholars.

The beginning of the New Kingdom is established to have been about −1580 (the expulsion of the Hyksos in the time of Kamose and Ahmose); Akhnaton must have reigned from −1375 to −1358; Ramses II of the Nineteenth Dynasty from −1300 to −1234; and Merneptah from this last year on. Ramses III of the Twentieth Dynasty began to reign in −1200 or a few years later. These dates are regarded as of importance for establishing the time of the Exodus.

The history of Israel from the days of the Exodus on is composed of the years of wandering in the desert, according to tradition forty years, of the time of Joshua and Judges and the first king, Saul, about four hundred years, and of the time of the kings of the House of David. David established his kingship about − 1000. For only approximately one hundred years, during the reigns of Saul, David, and Solomon, was the kingdom undivided. In the days of Solomon's heir it was split in two—Israel in the north and Judah in the south. About −722 the Ten Tribes of Israel, after the capture of their capital, Samaria, by Sargon II of Assyria, went into exile, from which they did not return. In −587 or −586 Judah, after the destruction of its capital, Jerusalem, by Nebuchadnezzar, went into Babylonian exile, from whence small groups of the nation came back after Babylon was captured (in −538) by Cyrus the Persian. Additional groups returned to Palestine during the following century.

Alexander the Great conquered Palestine on his way to Egypt in −333.

Although Egypt and Palestine are closely neighboring countries,

3

"the truth is that there is in Egypt singularly little evidence which bears directly on the Bible narrative".[4] The Scriptures tell of the sojourn of Israel in Egypt and of the Exodus; but no documents referring to these events have been found. There is no scriptural mention of Egypt during the time of the Judges. However, in the days of the Kings, Palestine repeatedly came into contact with Egypt—mostly through being attacked by armies of the pharaohs, campaigns which the pharaohs of the tenth to the sixth centuries usually forgot to mention.

It is strange that there is no real link between the histories of Egypt and Palestine for a period of many hundreds of years. At least the Exodus of the Israelites from Egypt was an event that should belong to both histories and thus supply a connecting link. We shall therefore try to determine during what period of Egyptian history the Exodus took place.

Whether it occurred only one or two hundred years before David, or three or four or five hundred years before him, depending on the length of the periods of the wandering and of the Judges—in other words, whether the Israelites left Egypt in the sixteenth, fifteenth, fourteenth, thirteenth, or twelfth century—this event occurred during the New Kingdom. There has never been any doubt on this point; where scholars have differed is concerning the king of the New Kingdom to whose reign the Exodus is to be ascribed. Although, as noted above, no definite statement bearing directly on the Exodus has been found in Egyptian historical documents, certain details do appear which invited discussion.

What Is the Historical Time of the Exodus?

The oldest theory places the Exodus at the earliest date: the Israelites were identified with the Hyksos, and the Exodus was identified with the expulsion of the Hyksos. Manetho, the priest previously mentioned, wrote that the Hyksos, when expelled from Egypt, went to Syria and there built Jerusalem.[5] Josephus Flavius, the Jewish historian of the first century, polemized against Apion, the grammarian, and against Manetho, his source, but accepted and supported the view that the Israelites were the Hyksos. Julius Africanus, one of the Fathers of the Church, wrote on the authority of Apion that in the

4

days of Ahmose the Jews revolted under Moses.[6] Eusebius, another Father of the Church, in his Canon wrote a gloss to the name Cencheres of one of the later kings of the Eighteenth Dynasty (his identity is not known): "About this time Moses led the Jews in their march out of Egypt."[7]

This divergence of opinion has not been settled after nineteen centuries, though modern scholars have probably not always been aware that they repeated an old controversy. The neglect of early Christian sources seemed justifiable: did not Augustine make Moses and Prometheus contemporaries?[8]

The identification of the Israelites with the Hyksos was many times accepted[9] and as often rejected. Even today a group among the historians maintains that the Exodus took place at the very beginning of the Eighteenth Dynasty and that the story of the Exodus is but an echo of the expulsion of the Hyksos.[10] However, in view of the bondage of the Israelites in Egypt and the bondage of Egypt under the Hyksos, the identity of martyred slaves and cruel tyrants must be regarded as a very strained hypothesis. Therefore a variant has been proposed, according to which the Israelite nation never sojourned in Egypt; the Hyksos sojourned there and then departed; the Israelites, hearing of the traditions of a strange people, adapted them to the stories of their own past.

Apart from the incongruity of identifying the Hyksos with the Israelites, the tyrants with the oppressed, there is a further difficulty in the fact that during the time of the successors to Ahmose there was no likely moment for an invasion of Palestine by Israelite refugees from Egypt. The pharaohs who followed Ahmose were strong kings, and it is regarded as established that Palestine was under their domination.

The same argument was employed to defend the theory that the Exodus occurred in — 1580, the time of the expulsion of the Hyksos. "If the expulsion of the Hyksos (c. 1580 B.C.) is too early for the Exodus, where in the history of the powerful Eighteenth Dynasty can we find a probable place for an event which, like the Exodus of tradition, presupposes internal trouble and weakness in Egypt, until the reign of Akhenaten [Akhnaton]?"[11]

In the days of strong pharaohs the Israelites were unable to enter Palestine; but how were they able to put off the yoke of bondage in the days of equally strong pharaohs?

5

A large group of scholars regard another moment as providing the clue for determining the time of the Exodus. In the 1880s, in the Nile Valley, at a place to which archaeologists gave the name of Tell el-Amarna, a correspondence on clay tablets was found which dated from the time of Amenhotep III and his son Akhnaton. Some of them were anxious letters written from Jerusalem (Urusalim), warning the pharaoh of an invasion by the "Habiru [Khabiru]",[12] approaching from Trans-Jordan. Granting that the Habiru were identical with the Hebrews, the Exodus must have taken place one or two generations earlier.[13] The scriptural statement (I Kings 6: 1)that the Temple of Solomon was built four hundred and eighty years after the Exodus would point to the middle of the fifteenth century, and computations have been made which indicate −1447 as the year of the Exodus. This year would fall in the reign of Amenhotep II; and the invasion of Palestine in −1407 would coincide with the time of the el-Amarna letters. The view that the Habiru were invading Hebrews was corroborated by the results of excavations of Jericho, where in the walls of the ancient city were found indications of earthquake and signs of fire, which the excavator referred to −1407 or thereabouts—the time of the el-Amarna correspondence.[14] This earthquake might have been the cause of the fall of the walls of Jericho when the Israelites, after crossing the Jordan, besieged the city.

A combination of the first and second views has also been offered: Israel left Egypt at the time of the expulsion of the Hyksos and reached Palestine as the Habiru in the reign of Akhnaton. But this hypothesis would entail more than two hundred years of wandering in the desert, instead of the scriptural forty, and it is therefore regarded as improbable.[15] An exodus in the days of Amenhotep II, on the other hand, does not present this difficulty and seems to agree with the chronological figures of the Bible. However, in the view of students of Egyptology, the time of Amenhotep II hardly seems to have been suitable for such a venture. "Of all theories, to place the Exodus, say, in the reign of Amenhotep II, in order to agree with traditional dates, seems to the historian of Egypt the least probable."[16]

Stress has also been laid on the fact that Palestine was under Egyptian rule as late as the disturbances of −1358, which put an end to the reign of Akhnaton. "Joshua did not find any such Egyptian hold during his conquest."[17] The end of Akhnaton's reign and the close of the Eighteenth Dynasty in the days of Tutankhamen and Aye was a time favorable for rebellion and the withdrawal of the slaves from Egypt. No reference has been found that could be interpreted as even hinting at an exodus during the interregnum between the Eighteenth and Nineteenth Dynasties, and only the fact that the situation was such as to make an exodus possible favors this hypothesis. This idea found its way into the work of a psychologist who, following in the footsteps of certain historians,[18] tried to show that Moses was an Egyptian prince, a pupil of Akhnaton; that Akhnaton was the founder of monotheistic idealism; that when Akhnaton ceased to rule and his schism fell into disfavor, Moses preserved his teachings by bringing them to the slaves, with whom he left Egypt.

The next theory reduces the age of the Exodus further: it has for its cornerstone a stele of Merneptah, in which this king of the Nineteenth Dynasty says that Palestine "is a widow" and that "the seed of Israel is destroyed". This is regarded as the earliest mention of Israel in an Egyptian document. Merneptah did not perish in the sea, nor did he suffer a debacle; he obviously inflicted a defeat on Israel and ravaged Palestine. The circumstances do not correspond with the pronounced tradition of Israel, but since it is the first mention of Israel, Merneptah is regarded by many as the Pharaoh of the Exodus (about −1220), and Ramses II, his predecessor, as the Pharaoh of Oppression.[19] Other scholars, however, consider the mention of Israel in Palestine in the days of Merneptah not as a corroboration, but as a refutation of the theory that Merneptah was the Pharaoh of the Exodus. They argue that if he found Israel already in Palestine, he could not have been the Pharaoh of the Exodus.[20]

A further obstacle to placing the Exodus in the reign of Merneptah has also been emphasized. If he really was the Pharaoh of the Exodus, then the Israelites must have entered Palestine at least a generation later, about −1190 to −1180; on this theory there remains only a century for the events of Judges. "The attribution of the exodus to the reign of Meneptah [Merneptah] (c. 1220 B.C.), hitherto generally accepted as a probable guess, has always suffered from the reproach of being almost impossibly late."[21]

Some scholars assumed that the Exodus occurred in successive waves.[22] A combination of the "Habiru theory" and the "Merneptah theory" puts events into the following order: "When the

7

Hebrews were entering Canaan, the Israelites were still in Egypt ... All Israelites were Hebrews, but not all Hebrews were Israelites. Thus while the Israelites or Jacob tribes were in Egypt, other Hebrew tribes were knocking at the door of Canaan."[23] The conciliators among the scholars proposed the following solution: "Some of the Hebrews remained in Egypt after the Exodus of the main body."[24]

Still later, Ramses III of the Twentieth Dynasty carried on a war against the Pereset or Peleset in Palestine. These have been identified as Philistines. Inasmuch as in the detailed reports of this war no mention is made of the Israelites, it is supposed by many scholars that they had not yet reached Palestine. They are believed to have left Egypt in the days of Merneptah (though his stele mentions Israel as already in Canaan), but they did not appear in Palestine until after the invading Philistines, with whom Ramses III battled.[25] Accordingly, the invasion of Palestine by the Philistines is put some fifty years after the Exodus and a few years before the conquest of Canaan by Israel.

The arrival of the Israelites in Palestine in the days of Merneptah, and still less in the days of Ramses III after his campaign there in −1186, leaves no room for the events of the Judges who guided the people for four centuries prior to Saul and David (−1000); but a school of historians argued in favor of that theory: "The entry ... could not be till after the last war of the Egyptians there by Rameses [Ramses] III, 1186 B.C. ... There is no free play of uncertainty left."[26] Archaeological considerations were presented to support this view. The excavation of Bethel in Palestine, it was claimed, "shows continuous native occupation until the break after −1200 due to Israelite conquest". Consequently the conclusion was drawn that for the invasion of Palestine by Israel "no earlier date is possible".[27]

The divergence of opinion is even greater. We have been told that an Exodus in the days of Merneptah is "almost impossibly late", but a scholar challenged all other opinions by bringing the Israelites, not *from,* but *into,* Egypt in the days of Merneptah.[28] During his reign Asiatics crossed the frontier and were registered by authorities there as immigrants.

Expulsion of the Hyksos, invasion of the Habiru, defeat of Israel in the days of Merneptah—these are the three events on which the

8

various schools of historians base their respective theories. It is hopeless to try to reconcile the irreconcilable. Each group points to the distortions in which its rivals indulge. Two hundred years of wandering in the desert destroys one theory; one hundred years for the period of the Judges undermines another, and so on. All of them have one and the same obstacle to surmount: "Under any chronological system which can reasonably be advanced, the date of Israel's invasion and settlement falls within the period (1500–1100 before the present era) when the country was ruled by Egypt as an essential portion of its Syrian Empire."[29] But if this is so, how could the Israelites have left Egypt, and, having left Egypt, how could they have entered Palestine? Moreover, why do the Books of Joshua and Judges, which cover four hundred years, ignore the rule of Egypt and, indeed, fail to mention Egypt at all.

An explanation was found to account for the fact that Israel left Egypt in the days of the strong pharaohs, but none to account for the strange silence of the Books of Joshua and Judges. The pharaohs were very strong, and the Exodus was only the daily passage of Bedouins across the Egyptian border. When the Israelites came to the frontier in a year of drought, they were admitted, but they had to do some work of benefit to the state to pay for the hospitality they and their herds enjoyed. When they left Egypt, an officer gave them a permit for departure, and it may be that he noted their leaving, but it was too trifling and stereotyped an event to become the subject of a monumental inscription. "The Exodus from Egypt was apparently a minor occurrence in the history of that time, so minor, indeed, that the nation most concerned in it next to the Jews themselves, the Egyptians, never took the trouble to record it."[30] "One merely has to bear in mind what this event meant, or rather, what it did not mean to Egypt."[31]

If this point of view is correct, then the archaeologists can have little hope of finding in Egypt a parallel to the Book of Exodus, and historians have no basis on which to decide the time of an event without significance.

If the people of Egypt did not care to notice the Exodus of the Israelites, this search for what passed unnoticed by contemporaries may be only a waste of time and effort.

9

The biblical story does not present the departure from Egypt as an everyday occurrence, but rather as an event accompanied by violent upheavals of nature.

Grave and ominous signs preceded the Exodus: clouds of dust and smoke darkened the sky and colored the water they fell upon with a bloody hue. The dust tore wounds in the skin of man and beast; in the torrid glow vermin and reptiles bred and filled air and earth; wild beasts, plagued by sand and ashes, came from the ravines of the wasteland to the abodes of men. A terrible torrent of hailstones fell, and a wild fire ran upon the ground; a gust of wind brought swarms of locusts, which obscured the light; blasts of cinders blew in wave after wave, day and night, night and day, and the gloom grew to a prolonged night, and blackness extinguished every ray of light. Then came the tenth and most mysterious plague: the Angel of the Lord "passed over the houses of the children of Israel ... when he smote the Egyptians, and delivered our houses" (Exodus 12:27). The slaves, spared by the angel of destruction, were implored amid groaning and weeping to leave the land the same night. In the ash-gray dawn the multitude moved, leaving behind scorched fields and ruins where a few hours before had been urban and rural habitations.

There are two scholarly approaches to this story of the plagues as told in Exodus, Chapters 7–11.[32] One holds it to be a fairy tale.[33] The story was taken apart and analyzed, and it was found that originally the legend had told of the death of the crown prince; then the death of one person was expanded into a plague that was inflicted on all the firstborn.[34] Later on, one plague was increased to three; but the storytellers, still not satisfied, continued to spin out their tale until they had a story in ten episodes. With precision the authorship of "Elohist" and "Yawist" was discerned.

"Neither group of legends has any historical truth at its source. The plagues are a later substitution for older miracles. However, miracles have never occurred anywhere."[35] "And since neither the plagues nor the miracles are historical, no conclusion can be drawn about the time of the Exodus."[36]

And when a purely realistic analysis was applied, the lever technique of the narrators was exposed: "These are scholarly reflections: by the hail only flax and barley were destroyed, because they already were ripe, and the wheat with the rye were spared because they used to ripen later. This gloss was added in order that at the next plague the locusts should have something to devour [*etwas zu fressen haben*]."[37] Sometimes the tale-spinners' self-control failed, as can be seen in the story of the boils: "Pest boils don't fly like ashes in the air, and still . . . Moses was made to sprinkle the ashes of the furnace 'toward the heaven'."[38]

The other approach tried to find a natural explanation for the plagues. In Egypt the sirocco blows in the fall and in the spring; the hot wind is also called *khamsin*, meaning "fifty", because for fifty days in the year this breath from the desert brings clouds of dust. Pictures were produced to show the darkened sky on a day when the *khamsin* blew. The desert wind may bring clouds of locusts; they cover the sky like a screen so that during their passage the sun's disk is obscured. The brownish color of the waters of the Nile, especially before it overflows, is well known to all tourists, and certain special observations near the cataracts of the Nile were described in detail.[39]

The lice, fleas, and frogs of the Egypt of today have been the subject of attentive study by reverend authors. It has been repeatedly pointed out that the order of the plagues as described in the Book of Exodus is exactly the order of the annual discomforts caused by the climate and insects of Egypt under Turkish rule, and is largely the same today.

This approach to the problem of the plagues makes of them a year-in, year-out occurrence. Little wonder, then, that they impressed the Egyptians in the same measure as the year-in, year-out entry and departure of some Bedouins with their cattle.

For hundreds of years thousands of scholars have paid tribute to the story of the plagues. If pious, they have not asked questions; if enlightened, they have defended the narrative, proving that wonders are but trivial phenomena; if critical, they have rejected the story, explaining it as a myth of relatively late origin.

The Book of Exodus then proceeds to tell how the Israelites were pursued by the army of the king, who regretted their escape. They were trapped between mountains and sea. The night was frightful. A

11

heavy cloud darkened the sky, which was rent by incessant lightning. A hurricane raged the whole night, and at dawn the sea was cleft, the waters torn by a double tide of gigantic force. The slaves passed through; the pursuers followed in chariots; but the waters returned, and the Egyptians with their king, fleeing against them, met their death in the waves.

Attempts were made to explain this story as a natural phenomenon. It seemed difficult to give credence to its miraculous element; but the vivid description of the night, the hurricane, and the mountainous waves suggested that some event had actually taken place, the memory of which was later clothed in fantastic elaborations.[40] The constant return, through the centuries, of Jewish thought to some experience by the sea also suggested that not the whole story had been invented. Historians agree that the most precious tradition of the people was born on the shores of the Jam-Suf, generally translated as "Red Sea".

A river or sea cleft in two is a frequent motif in folklore. The pursuers probably experienced some catastrophe, not because of a sea rent in twain, but because of a tide swollen by the storm.

But an explanation based on ebb and flood tides is obviously invalid. Whether the Sea of Passage was the Gulf of Suez or the Gulf of Aqaba on the Red Sea, or Lake Sirbonis (Serbon)[41] connected with the Mediterranean, or some other lake—the Crocodile Lake, the Salt Lake, through the waters of which ships pass today from the Mediterranean into the Red Sea—there are no perceptible movements of flood and ebb on any of these water surfaces—either the Mediterranean or the Red Sea, of course, the inland seas (lakes).

A more plausible explanation would therefore omit the tides and content itself with the storm. Some of the chariots of the pursuing Egyptians sank in the sea when its billows broke over the shore. Then the Israelites sang their song of deliverance, or received the inspiration out of which the exaggerated picture of the catastrophe was later born. How could it be otherwise than exaggerated when the annals of Egypt know nothing about the sea engulfing a king and his chariots, while the descendants of the fugitives from royal bondage glorified themselves with a story of a miraculous storm unwitnessed by the Egyptians?

Is there then any use in endeavoring to show that a strong east wind, blowing from evening till dawn, could force the sea to retreat, and that a change in the direction of the wind could overwhelm an army marching on land? Strange, indeed, is the persistence with which the Jewish people have clung to this story, making it the beginning and at the same time the most dramatic episode of their history as a nation.

The fugitives, after their escape, entered a desert, a desolate wasteland. The Book of Exodus relates that a pillar of smoke went before them by day and a pillar of fire by night. A simple explanation of this portent has been found: at the head of wandering caravans a torch is usually carried lifted high to show the way to the moving train. Because of the heat of the day, caravans prefer to move at night, and the burning and smoking pitch is intended to prevent anyone from being lost and to frighten the beasts of the desert.[42]

Although this explanation is the one that is accepted and found in numerous Bible dictionaries, it is too simple. The pillar of smoke and fire deeply impressed the Israelites; it was said to be the Angel of God. Did the Israelites not know the manner and custom of caravans journeying in the desert, and were they so impressed by common things and so anxious for wonders that the torch in the hand of the leader became for them an angel?

But the pillar of cloud and of fire could have been less than illusion and only the invention of storytellers.

In the last century an Englishman named Charles Beke, a man of more than one strange idea, published a pamphlet entitled *Mount Sinai a Volcano*.[43] On the title page he placed an epigraph of two sentences, one from the Book of Exodus, the other from the Greek poet Pindar. The verse from Exodus 13:21 reads: "... by day in a pillar of cloud, to lead them the way; and by night in a pillar of fire, to give them light." The lines from Pindar (Odes *Pythia*, I, 22–23) describe Etna: "By day a burning stream of smoke; but by night a ruddy eddying flame." Beginning with this parallel and going over to the biblical description of the day of the lawgiving, Beke came to the startling conclusion expressed in the title of his pamphlet. The day of the lawgiving is described in these words: "And it came to pass on the third day in the morning, that there were thunders and lightnings, and a thick cloud upon the mount, and the voice of the trumpet exceeding loud; so that all the people that was in the camp trembled ... And mount Sinai was altogether on a smoke ... and the smoke

13

thereof ascended as the smoke of a furnace, and the whole mount quaked greatly ... And all the people saw the thunderings, and the lightnings, and the noise of the trumpet, and the mountain smoking: and when the people saw it, they removed, and stood afar off."[44]

Beke explained the pillar of smoke and fire as the ignited column of ashes and vapors erupted by the volcano. He cited instances from volcanic regions showing that volcanic eruptions can produce black clouds of ashes that darken the sky and are sometimes swept over great distances. Eruptions are usually accompanied by rumblings in the bowels of the earth; earthquakes and eruptions are often concurrent phenomena; earthquakes that shake the bottom of the sea create tidal waves that may retreat from the shore and then return to engulf the land, leaving destruction in their wake. At the Sea of Passage, according to this explanation, an earthquake created havoc; and the reference to chariots that were unable to move (Exodus 14:25) has its parallel in the description of the earthquake that accompanied the eruption of Vesuvius in the year 79 when Pompeii and Herculaneum perished, a description preserved in a letter from Pliny the Younger to Tacitus (*Epistles*, VI, 20): "We stood still, in the midst of a most dangerous and dreadful scene. The chariots, which we had ordered to be drawn out, were so agitated backwards and forwards, though upon the most level ground, that we could not keep them steady, even by supporting them with large stones. The sea seemed to roll back upon itself, and to be driven from its shores by the convulsive motion of the earth."

The interpretation of the wondrous events at the Sea of Passage and at Mount Sinai as seismic and volcanic phenomena of nature met most vigorous opposition and derision from high ecclesiastics. "It is well known that no volcanic phenomena exist in the desert to account for these appearances. In fact, all the expressions used in the sacred writers are those which are usually employed in the Hebrew Scriptures to describe a thunderstorm."[45]

Beke did not regard the mountainous heights of the Sinai Peninsula as the Mount Sinai of the Scriptures. He had previously published a work arguing for the fallacious notion that Mizraim of the Scriptures was not Egypt, but some vanished kingdom on the Sinai Peninsula; the Israelites, when leaving the country, crossed the tip of the Gulf of Aqaba and came to the Arabian shore of that gulf. He announced that he was staking his reputation as a traveler and bib-

14

lical scholar, and that, granted public assistance, he would locate Mount Sinai in Harra Radjla, "which was formerly in activity, but has now been extinct during many ages". An old man, he followed, as he thought, in the footsteps of the prophet Elijah, who also made his pilgrimage to the mountain in the desert.

When he returned, declaring that Mount Sinai is Har-Nur (Mount of Fire), east of Ghor, a peak that, like a number of others, has a halo around it but does not appear to have been a volcano, he wrote that he was "egregiously mistaken with respect to the volcanic character of Mount Sinai".[46] His confession was published posthumously in a gilt-edged volume, very different in appearance from *Mount Sinai a Volcano*. It gave an account of the voyage: "I am therefore bound to confess that I was in error, as regards the physical character of Mount Sinai, and that the appearances mentioned in Scripture were as little volcanic as they were tempestuous."[47]

Thirty years after the theory of the volcanic character of Sinai was enunciated, it was adopted by one scholar,[48] ten years later by a few others,[49] and recently the idea that Yahweh was a local deity of a volcano has become an oft-repeated notion; its acceptance, however, has not prevented some scholars from denying the historicity of the visit of wandering Israelites to Mount Sinai.[50]

Upheaval

If we do not limit ourselves to the few passages from the Book of Exodus cited in support of the idea that Mount Sinai was a volcano, the activity of which impressed the Israelites, but turn our attention to the many other passages in the various books of the Scriptures referring to the Exodus, we soon feel bound to make the unusual admission that, if the words mean what they say, the scope of the catastrophe must have exceeded by far the extent of the disturbance that could be caused by one active volcano. Volcanic activity spread far and wide, Mount Sinai was but one furnace in a great plain of smoking furnaces.

Earth, sea, and sky participated in the upheaval. The sea overflowed the land, lava gushed out of the riven ground. The Scriptures thus describe the uproar of the unchained elements:

15

... the earth shook and trembled; the foundations ... of ... hills moved and were shaken ...

Smoke ... and fire ... coals were kindled ... the channels of waters were seen, and the foundations of the world were discovered.[51]

In a great geologic catastrophe the bottom of the sea fell, and the waters rushed into the chasms. The earth trembled, the volcanoes threw smoke and fire out of the interior of the earth, cliffs were torn away, molten rock ran along the valleys, the dry earth became sea, the bowels of the mountains groaned, and the skies thundered unceasingly.

His lightnings enlightened the world: the earth saw, and trembled. The hills melted like wax ...[52]

Tectonic strata collapsed. Formations changed their profile in major displacements.

[He] removeth the mountains. ... [He] overturneth them in his anger, [He] shaketh the earth out of her place.[53]

This seismic and volcanic activity is constantly referred to the time when the Israelites went from Egypt.

... the earth trembled ... The mountains melted ... even that Sinai ...'[54]

The last quotation is from the Song of Deborah, one of the oldest fragments inserted in the Scriptures.

The pious imagination conceives these utterances to be only metaphoric. Critical analysis likewise sees in them but the expression of an effusive ecstasy. Was there no real experience of any kind to which the metaphors could apply? Is the following a description of flood and ebb in the salt marshes in Egypt: "... the channels of waters were seen, and the foundations of the world were discovered"?[55] Folklore does not work in such an indiscriminating manner. These narratives of geologic changes, persistently repeated in the above and many other similar passages in connection with the time of the Exodus, must have had some underlying experience that folklore molded and remolded. The experience, according to the Scriptures, was so majestic and terrible that even after a long line of succeeding generations it could not be forgotten.

16

I call to remembrance my song in the night: ... Will the Lord cast off for ever? ... Hath God forgotten to be gracious? ... I will remember ... thy wonders of old. ... The depths also were troubled. ... The lightnings lightened the world: the earth trembled and shook.... Thou leddest thy people like a flock by the hand of Moses and Aaron.[56]

The poet of this psalm was visited at night by a vision of the past, when wondrous things were performed at the sea and in the wilderness in the sight of a folk escaped from the house of bondage.

The turbulence and uproar of nature stirred the fugitives in the desert to a state of exaltation:

Thou hast made the earth to tremble; thou hast broken it: heal the breaches thereof; for it shaketh.

Thou hast shewed thy people hard things: thou hast made us to drink the wine of astonishment.[57]

The nights under the raging sky of the wilderness torn by unceasing lightnings, when flaming lava flowed and hills melted, were unforgettable. During the long years in which the Israelites lived in their land, they never forgot the convulsions of the desert, the explosion of the burning mountain, the fury of the waters. The events of these weeks or months, when the surface of the earth underwent violent changes in its tectonic structure, became the most important tradition of this nation.

The scriptural tradition persists that before the Israelites left Egypt this land was visited by plagues, forerunners of a great holocaust caused by frenzied elements. When the Israelites departed from the country they witnessed gigantic tidal waves on the sea; farther off, in the desert, they experienced spasmodic movements of the earth's surface and volcanic activity on a great scale, with lava gushing out of the cleft ground, suddenly yawning chasms,[58] and springs disappearing or becoming bitter.[59]

The logical question to be asked here is: Is this testimony entirely false? And if it is not a collection of misleading inventions, can it be that nothing of this was noticed by the Egyptians? If their land did suffer from the disaster, are we not on the right track in our search for a synchronic moment in the Jewish and Egyptian histories? A desert close to Egypt was convulsed by earth tremors. Were these seismic disturbances of great magnitude confined to a comparatively

17

small area? Is any earthquake at all mentioned in Egyptian records?

The standard works on Egyptian history contain no mention of an earthquake, and none of plagues. Nevertheless, we should like to persist with our question. In this persistence we are guided by the consideration that something great may be at stake. If we could help this witness on the stand—the annals of ancient Egypt—to remember some vast catastrophe, we might perchance obtain a precious clue to an obscure problem, at whose door all disputes, now more than two thousand years old, have remained without decision.

An Egyptian Eyewitness Testifies to the Plagues

In this trial of history the judgment will depend on the following statement and its probing by cross-examination.

There was a great natural catastrophe, the after effects of which lasted for years. The impression it made endured, and its story was handed down from generation to generation and echoed and re-echoed in the Scriptures and in other writings. Is no reference to it to be found in Egyptian documents?

Or was the Exodus really an obscure and insignificant passage through the control of the collectors of royal revenues at the boundaries of the state? If so, how is it that it became the most exciting memory of generations of the Jewish people? Whence came the visions of an upheaval that rent earth and sea? Is the turmoil that visited the land and its river, the sea and the desert, really not to be found in Egyptian writings? Did every recollection of it vanish?

In view of the failure of works on the history of Egypt to mention any natural catastrophe, we should investigate the ancient sources.

It is not known under what circumstances the papyrus containing the words of Ipuwer was found. According to its first possessor (Anastasi), it was found in "Memphis", by which is probably meant the neighborhood of the pyramids of Saqqara. In 1828 the papyrus was acquired by the Museum of Leiden in the Netherlands and is listed in the catalogue as Leiden 344.

The papyrus is written on both sides. The face (recto) and the back (verso) are differentiated by the direction of the fiber tissues; the story of Ipuwer is written on the face, on the back is a hymn to a deity. A fascimile copy of both texts was published by the authorities of the museum together with other Egyptian documents.[60] The text of Ipuwer is now folded into a book of seventeen pages, most of them containing fourteen lines of hieratic signs (a flowing writing used by the scribes, quite different from pictorial hieroglyphics). Of the first page only a third—the left or last part of eleven lines—is preserved; pages 9 to 16 are in very bad condition—there are but a few lines at the top and bottom of the pages—and of the seventeenth page only the beginning of the first two lines remains.

The first interpretation of the text of Ipuwer was presented in the introduction to the facsimile.[61] It was explained that eight pages of the recto were proverbs or axioms, and the following pages were supposed to be a chapter out of a philosophic work.

The author of the next attempt to translate the text (only the first nine pages) understood it as a collection of proverbs and examples of sayings brought together for didactic use.[62] Another scholar[63] called the papyrus a collection of riddles.

At the beginning of this century an effort was made to translate the entire Ipuwer text.[64] The words of Ipuwer were interpreted as prophetic in character: a time of evil was foretold for the people of Egypt. The prophet might have been inspired by some similar political situation in the past, before the inauguration of the Twelfth Dynasty.

In 1909 the text, translated anew, was published by Alan H. Gardiner under the title, *The Admonitions of an Egyptian Sage from a Hieratic Papyrus in Leiden.*[65] Gardiner argued that all the internal evidence of the text points to the historical character of the situation. Egypt was in distress; the social system had become disorganized; violence filled the land. Invaders preyed upon the defenceless population; the rich were stripped of everything and slept in the open, and the poor took their possessions. "It is no merely local disturbance that is here described, but a great and overwhelming national disaster."[66]

Gardiner, following Lange, interprets the text as though the words of a sage named Ipuwer were directed to some king, blaming him for inactivity which has brought confusion, insecurity, and suffering to the people. "The Almighty", to whom Ipuwer directs his words, is a customary appellation of great gods.[67] Because the introductory

19

passages of the papyrus, where the author and his listeners would be likely to be mentioned, are missing, the presence of the king listening to the sage is assumed on the basis of the preferred form of certain other literary examples of the Middle Kingdom. In accordance with this interpretation, the papyrus containing the words of Ipuwer is called, in the Gardiner edition, *Admonitions of an Egyptian Sage.*

Egypt in Upheaval

The Papyrus Ipuwer is not a collection of proverbs (Lauth, Chabas) or riddles (Brugsch); no more is it a literary prophecy (Lange) or an admonition concerning profound social changes (Gardiner, Sethe). It is the Egyptian version of a great catastrophe.

The papyrus is a script of lamentations, a description of ruin and horror.

> PAPYRUS 2:8 Forsooth, the land turns round as does a potter's wheel.
> 2:11 The towns are destroyed. Upper Egypt has become dry (wastes?).
> 3:13 All is ruin!
> 7:4 The residence is overturned in a minute.
> 4:2 . . . Years of noise. There is no end to noise.

What do "noise" and "years of noise" denote? The translator wrote: "There is clearly some play upon the word *hrw* (noise) here, the point of which is to us obscure." Does it mean "earthquake" and "years of earthquake"? In Hebrew the word *raash* signifies "noise", "commotion", as well as "earthquake".[68] Earthquakes are often accompanied by loud sounds, subterranean rumbling and roaring, and this acoustic phenomenon gives the name to the upheaval itself.

Apparently the shaking returned again and again, and the country was reduced to ruins, the state went into sudden decline, and life became unbearable.

Ipuwer says:

> PAPYRUS 6:1 Oh, that the earth would cease from noise, and tumult (uproar) be no more.

The noise and the tumult were produced by the earth. The royal residence could be overthrown "in a minute" and left in ruins only

20

by a mighty earthquake. The upheaval seems to have wrought havoc on the high seas, where ships were thrown into whirlpools; in the passage where "the towns are destroyed", it is also said that ships were set adrift.

The papyrus of Ipuwer contains evidence of some natural cataclysm accompanied by earthquakes and bears witness to the appearance of things as they happened at that time.

I shall compare some passages from the Book of Exodus and from the papyrus. As, prior to the publication of *Worlds in Collision* and *Ages in Chaos*, no parallels had been drawn between the Bible and the text of the Papyrus Ipuwer, the translator of the papyrus could not have been influenced by a desire to make his translation resemble the biblical text.[69]

> PAPYRUS 2:5–6 Plague is throughout the land. Blood is everywhere.

> EXODUS 7:21 . . . there was blood throughout all the land of Egypt.

This was the first plague.

> PAPYRUS 2:10 The river is blood.

> EXODUS 7:20 . . . all the waters that were in the river were turned to blood.

This water was loathsome, and the people could not drink it.

> PAPYRUS 2:10 Men shrink from tasting —— human beings, and thirst after water.

> EXODUS 7:24 And all the Egyptians digged round about the river for water to drink; for they could not drink of the water of the river.

The fish in the lakes and the river died, and worms, insects, and reptiles bred prolifically.

> EXODUS 7:21 . . . and the river stank.

> PAPYRUS 3:10–13 That is our water! That is our happiness! What shall we do in respect thereof? All is ruin!

21

The destruction in the fields is related in these words:

EXODUS 9:25 . . . and the hail smote every herb of the field, and brake every tree of the field.[70]

PAPYRUS 4:14 Trees are destroyed.
6:1 No fruit nor herbs are found . .

This portent was accompanied by consuming fire. Fire spread all over the land.

EXODUS 9:23–24 . . . the fire ran along the ground.
. . . there was hail, and fire mingled with the hail, very grievous.

PAPYRUS 2:10 Forsooth, gates, columns and walls are consumed by fire.

The fire which consumed the land was not spread by human hand but fell from the skies.[71]
By this torrent of destruction, according to Exodus,

EXODUS 9:31–32. . . . the flax and the barley was smitten: for the barley was in the ear, and the flax was boiled.
But the wheat and the rye were not smitten: for they were not grown up.

It was after the next plague that the fields became utterly barren. Like the Book of Exodus (9:31–32 and 10:15), the papyrus relates that no duty could be rendered to the crown for wheat and barley; and as in Exodus 7:21 ("And the fish that was in the river died"), there was no fish for the royal storehouse.

PAPYRUS 10:3–6 Lower Egypt weeps. . . . The entire palace is without its revenues. To it belong (by right) wheat and barley, geese and fish.

The fields were entirely devastated.

EXODUS 10:15 . . . there remained not any green thing in the trees, or in the herbs of the fields, through all the land of Egypt.

PAPYRUS 6:3 Forsooth, grain has perished on every side.

22

5:12 Forsooth, that has perished which yesterday was seen. The land is left over to its weariness like the cutting of flax.

The statement that the crops of the fields were destroyed in a single day ("which yesterday was seen") excludes drought, the usual cause of a bad harvest; only hail, fire, or locusts could have left the fields as though after "the cutting of flax". The plague is described in Psalms 105:34–35 in these words: ". . . the locusts came, and caterpillars, and that without number. And did eat up all the herbs in their land, and devoured the fruit of their ground."

PAPYRUS 6:1 No fruit nor herbs are found . . . hunger.

The cattle were in a pitiful condition.

EXODUS 9:3 . . . the hand of the Lord is upon thy cattle which is in the field . . . there shall be a very grievous murrain.

PAPYRUS 5:5 All animals, their hearts weep. Cattle moan. . . .

Hail and fire made the frightened cattle flee.

EXODUS 9:19 . . . gather thy cattle, and all that thou hast in the field . . .
21 And he that regarded not the word of the Lord left his servants and his cattle in the field.

PAPYRUS 9:2–3 Behold, cattle are left to stray, and there is none to gather them together. Each man fetches for himself those that are branded with his name.

The ninth plague, according to the Book of Exodus, covered Egypt with profound darkness.

EXODUS 10:22 . . . and there was a thick darkness in all the land of Egypt.

PAPYRUS 9:11 The land is not light. . . .

"Not light" is in Egyptian equivalent to "without light" or "dark". But there is some question as to whether the two sentences are entirely parallel. The years of wandering in the desert are described as spent in gloom under a cover of thick clouds.[72] The Jewish written tradition persists that for a number of years after the

Exodus the light of the sun was dimmed by clouds. It is rather this biblical "shadow of death" to which the quotation of the papyrus seems to be parallel. The Egyptian parallel to the plague of impenetrable darkness will be found on a subsequent page; the "shadow of death" will also have additional parallels.

The Last Night before the Exodus

According to the Book of Exodus, the last night the Israelites were in Egypt was a night in which death struck instantly and took victims from every Egyptian home. The death of so many in a single night, even at the same hour of midnight, cannot be explained by a pestilence, which would last more than a single hour. The story of the last plague does seem like a myth; it is a stranger in the sequence of the other plagues, which can be explained as natural phenomena.

The plagues have been described here as forerunners of the catastrophe which reached its climax at the Jam-Suf (the Sea of Passage); the phenomena in the desert were the subsequent spasms of the earth's crust. Testimony from Egyptian sources about an earthquake was sought, with the purpose of establishing a synchronic moment in Egyptian and Jewish history. The evidence, when found, brought forth more analogies and showed greater resemblance to the scriptural narrative than I had expected. Apparently we have before us the testimony of an Egyptian witness of the plagues.

On careful reading of the papyrus, it appeared that the slaves were still in Egypt when at least one great shock occurred, ruining houses and destroying life and fortune. It precipitated a general flight of the population from the cities, while the other plagues probably drove them from the country into the cities.

The biblical testimony was reread. It became evident that it had not neglected this most conspicuous event: it was the tenth plague.

In the papyrus it is said: "The residence is overturned in a minute."[73] On a previous page it was stressed that only an earthquake could have overturned and ruined the royal residence in a minute. Sudden and simultaneous death could be inflicted on many only by a natural catastrophe.

EXODUS 12:30 And Pharaoh rose up in the night, he, and all his

24

servants, and all the Egyptians; and there was a great cry in Egypt: for there was not a house where there was not one dead.

A great part of the people lost their lives in one violent shock. Houses were struck a furious blow.

EXODUS 12:27 [The Angel of the Lord] passed over the houses of the children of Israel in Egypt, when he smote the Egyptians, and delivered our houses.

The word *nogaf* for "smote" is used for a violent blow, e.g. for thrusting with his horns by an ox.[74]

The residence of the king and the palaces of the rich were tossed to the ground, and with them the houses of the common people and the dungeons of captives.

EXODUS 12:29 And it came to pass, that at midnight the Lord smote all the firstborn in the land of Egypt, from the firstborn of Pharaoh that sat on his throne unto the firstborn of the captive that was in the dungeon.

PAPYRUS 4:3, and 5:6 Forsooth, the children of princes are dashed against the walls.

6:12 Forsooth, the children of princes are cast out in the streets.

The sight of the children of princes smashed on the pavement of the dark streets, injured and dead amid the ruins, moved the heart of the Egyptian eyewitness. No one saw the agony in the dungeon, a pit in the ground where prisoners were locked in, when it was filled by landslides.

PAPYRUS 6:3 The prison is ruined.[75]

Why is this unreasonable "firstborn" inserted in the Hebrew text? The explanation will follow later.

In the papyrus (2:13) it is written:

He who places his brother in the ground is everywhere.

To it corresponds Exodus (12:30):

. . . there was not a house where there was not one dead.

25

In Exodus (12:30) it is written:

> . . . there was a great cry in Egypt.

To it corresponds the papyrus (3:14):

> It is groaning that is throughout the land, mingled with lamentations.

The statues of the gods fell and broke in pieces:[76] "this night . . . against all the gods of Egypt I will execute judgment" (Exodus 12:12).

A book by Artapanus, no longer extant, which quoted some unknown ancient source and which in its turn was quoted by Eusebius, tells of "hail and earthquake by night [of the last plague], so that those who fled from the earthquake were killed by the hail, and those who sought shelter from the hail were destroyed by the earthquake. And at that time all the houses fell in, and most of the temples."[77]

The earth was equally pitiless towards the dead in their graves: the sepulchers opened, and the buried were disentombed.

> PAPYRUS 4:4, also 6:14 Forsooth, those who were in the place of embalmment are laid on the high ground.

A legend is preserved in the Haggada:[78] in the last night, when the land of Egypt was smitten, the coffin of Joseph was found lying upon the ground, lifted from its grave.

Similar effects of powerful earthquakes have occasionally been observed in modern times.[79]

The unborn, Ipuwer mourned, entered into everlasting life ere they had seen the light of the world.

In Midrash Rabba on Exodus it is written:

> Even pregnant women about to give birth miscarried and then themselves died; because the Destroyer stalked abroad and destroyed all he found.[80]

"Firstborn" or "Chosen"

The biblical story of the last plague has a distinctly supernatural quality in that all the firstborn and only the firstborn were killed on the night of the plague.[81] An earthquake that destroys only the firstborn is inconceivable, because events can never attain that degree of coincidence. No credit should be given to such a record.

Either the story of the last plague, in its canonized form, is a fiction, or it conceals a corruption of the text. Before proclaiming the whole a strange tale interpolated later, it would be wise to inquire whether or not the incredible part alone is corrupted. It may be that "the firstborn" stands for some other word.

> ISAIAH 43:16 Thus saith the Lord, which maketh a way in the sea and a path in the mighty waters;
> 20 . . . I give waters in the wilderness, and rivers in the desert, to give drink to my people, my chosen.

In the Book of Exodus it is said that Moses was commanded:

> EXODUS 4:22–23 And thou shalt say unto Pharaoh, Thus saith the Lord, Israel is my son, even my firstborn.
> . . . and if thou refuse to let him go, behold I will slay thy son, even thy firstborn.

The "chosen" are here called "firstborn". If Israel was the firstborn, revenge was to be taken against Egypt by the death of its firstborn. But if Israel was the chosen, then revenge was to be taken against Egypt by the death of its chosen.

"Israel, my chosen," is *Israel bechiri, or bechori.*

"Israel, my firstborn," is *Israel bekhori.*[82]

It is the first root which was supposed to determine the relation between God and his people. Therefore: "at midnight the Lord smote all the firstborn in the land of Egypt" (Exodus 12:29) must be read "all the select of Egypt," as one would say, "all the flower of Egypt" or "all the strength of Egypt". "Israel is my chosen: I shall let fall all the chosen of Egypt."

Natural death would usually choose the weak, the sick, the old. The earthquake is different; the walls fall upon the strong and the weak alike. Actually the Midrashim say that "as many as nine tenths of the inhabitants have perished".[83]

In Psalms 135 my idea is illustrated by the use of both roots where two words of the same root would have been expected.

> For the Lord hath chosen Jacob unto himself, and Israel for his peculiar treasure. . . . Who smote the firstborn of Egypt.

27

In Psalms 78 the history of Exodus is told once more:

PSALMS 78:43 How he had wrought his signs in Egypt . . .
51 And smote all the firstborn in Egypt . . .
52 But made his own people to go forth . . .
56 Yet they tempted and provoked the most high God . . .
31 The wrath of God came upon them, and slew the fattest of them, and smote down the chosen men of Israel. . . .

Were the firstborn destroyed when the wrath was turned against Egypt, and were the chosen destroyed when the wrath was turned against Israel?

AMOS 4:10 I have sent among you the pestilence [plague] after the manner of Egypt: your young men [chosen] have I slain.

In the days of *raash* (commotion) during the reign of Uzziah, the select and the flower of the Jewish people shall perish as perished the chosen, the strength of Egypt, was the prophecy of Amos.

It is possible that the king's firstborn died on the night of the upheaval. The death of the prince could have been an outward reason for changing the text. The intrinsic reason lies in the same source that interrupted the story of the Exodus at the most exciting place—after the houses of the Egyptians had crumbled—with these sentences:

EXODUS 13:2 Sanctify unto me all the firstborn, whatsoever openeth the womb among the children of Israel, both of man and of beast: it is mine.
13 . . . and all the firstborn of man among thy children shalt thou redeem.

Jeremiah testifies to the fact that burnt offerings and sacrifices were not ordered on the day Israel left Egypt.

JEREMIAH 7:22 For I spake not unto your fathers, nor commanded them in the day that I brought them out of the land of Egypt, concerning burnt offerings or sacrifices.

This is in contradiction to the text of Exodus 12:43 to 13:16. To free the people from this bondage is the task of Amos, Isaiah, and Jeremiah.

AMOS 5:22 Though ye offer me burnt offerings and your meat

offerings, I will not accept them: neither will I regard the peace offerings of your fat beasts.

24 But let judgment run down as waters, and righteousness as a mighty stream.

25 Have ye offered unto me sacrifices and offerings in the wilderness forty years, O house of Israel?

Revolt and Flight

My endeavor has been to find in Egyptian sources some mention of a natural catastrophe. The description of disturbances in the Papyrus Ipuwer, when compared with the scriptural narrative, gives a strong impression that both sources relate the very same events. It is therefore only natural to look for mention of revolt among the population, of a flight of wretched slaves from this country visited by disaster, and of a cataclysm in which the pharaoh perished.

Although in the mutilated papyrus there is no explicit reference to the Israelites or their leaders, three facts are clearly described as consequences of the upheaval: the population revolted; the wretched or the poor men fled; the king perished under unusual circumstances.

If, in addition to the closely parallel description of the plagues, I should try to extract more than these three facts from the papyrus, I should expose myself to the charge that I use the defective condition of the document to support preconceived ideas. But the references to the catastrophe and to the population that rebelled and fled are not ambiguous; their meaning is clear and not open to misunderstanding. Consequently, when, in the next few paragraphs, I try to uncover additional parallels in certain passages, I do so with restraint. The papyrus is damaged and obscure in many places; if one or another comparison is incomplete or arbitrary, it may add nothing to, but neither does it detract anything from, the fact established here, that a sequence of earthquakes and other natural phenomena occurred in Eygpt, accompanied by plagues bringing destruction to men, animals, plants, and sources of water.

The first omens of an approaching catastrophe brought unrest to the land, and the captives longed to escape to freedom. The papyrus narrates that "men ventured to rebel against the Uraeus" (the emblem of royal authority) and that magical spells connected with the serpent are divulged (6:6–7 to 7:5–6), that gold and jewels "are fastened on the neck of female slaves" (3:2–3; compare with Exodus 11:2: "and let every man borrow of his neighbor, and every woman of her neighbor, jewels of silver, and jewels of gold . . .").

The collapse of stone structures, the dead and wounded in the debris, the fall of many statues of the gods, inspired dread and horror; all these were looked upon as the acts of the God of the slaves.

> EXODUS 12:33 And the Egyptians were urgent upon the people, that they might send them out of the land in haste; for they said, We be all dead men.

The Egyptians were to use still bitterer words; when the calamity was repeated, they no longer expressed anguish and fear of death but invoked it in passages such as these:

> PAPYRUS 4:2 Forsooth, great and small say: I wish I might die.
> 5:14f. Would that there might be an end of men, no conception, no birth! Oh, that the earth would cease from noise, and tumult be no more!

The following lines speak of a population escaping a disaster. "Men flee. . . . Tents are what they make like the dwellers of the hills" (Papyrus 10:2). In the Book of Exodus it is said that the Israelites left the country "in haste" (Exodus 12:33) and "could not tarry" (12:39). No doubt flight and living in makeshift tents was shared by the majority of the survivors, as has happened many times since then whenever a violent shock has occurred, devastating cities; a new shock is feared by those who have escaped with their lives.

A "mixed multitude" of Egyptians joined the Israelite slaves, and with them hastily made toward the desert (Exodus 12:38). Their first brief stop was at Succoth (13:20)—which in Hebrew means "huts".

The escaped slaves hurried across the border of the country. By day a column of smoke went before them in the sky; by night it was a pillar of fire.

EXODUS 13:21 ... by day in a pillar of a cloud, to lead them the way; and by night in a pillar of fire, to give them light; to go by day and night.

PAPYRUS 7:1 Behold, the fire has mounted up on high. Its burning goes forth against the enemies of the land.

The translator added this remark: "Here the 'fire' is regarded as something disastrous."

After the first manifestations of the protracted cataclysm the Egyptians tried to bring order into the land. They traced the route of the escaped slaves. The wanderers became "entangled in the land, the wilderness hath shut them in" (Exodus 14:3). They turned to the sea, they stood at Pi-ha-Khiroth. "The Egyptians pursued after them. The Egyptians marched after them." A hurricane blew all the night and the sea fled.

In a great avalanche of water "the sea returned to his strength", and "the Egyptians fled against it". The sea engulfed the chariots and the horsemen, the pharaoh and all his host.

The Papyrus Ipuwer (7:1–2) records only that the pharaoh was lost under unusual circumstances "that have never happened before". The Egyptian wrote his lamentations, and even in the broken lines they are perceptible:

... weep ... the earth is ... on every side ... weep ...

The Hyksos Invade Egypt

There was no longer any royal power in Egypt. In the following weeks the cities turned into scenes of looting. Justice ceased to function. The mob dug in the debris and in the public records, where contracts, notes and pledges, and deeds to real property were filed. The plunderers searched among the wreckage of the royal storehouses.

PAPYRUS 6:9 Forsooth, the laws of the judgment-hall are cast forth. Men walk upon [them] in the public places.
10:3 The storehouse of the king is the common property of everyone.

The papyrus furnishes information as to what happened afterwards. The earth's crust repeatedly contracted in violent spasms ("years of noise"). The roads became impassable—"dragged" and "flooded" (Papyrus 12:11f.). The realm was depopulated, and Ipuwer bewails the "lack of people". The residence of the pharaoh was a heap of ruins. Governmental authority was completely shattered. "Behold, the chiefs of the land flee" (8:14); "Behold, no offices are in their (right) place, like a frightened herd without a herdsman" (9:2). The "poor men" who ran away roamed the desert. Slaves who remained in Egypt raised their heads.

> PAPYRUS 6:7 Forsooth, public offices are opened and their census-lists are taken away.

Then invaders approached out of the gloom of the desert; they crossed the borders and entered the shattered land.

> PAPYRUS 3:1 Forsooth, the Desert is throughout the land. The nomes are laid waste. A foreign tribe from abroad has come to Egypt.

The catastrophe that rendered Egypt defenseless was a signal to the tribes of the Arabian desert.

> PAPYRUS 15:1 What has happened? —— through it is to cause the Asiatics to know the condition of the land
> 14:11 Men —— They have come to an end for themselves. There are none found to stand and protect themselves.

Prostrated by the appalling blows of nature, the Egyptians did not defend themselves. It is not clear whether "a million of people" in the next sentence is the number of the perished or of the intruders.

> PAPYRUS 12:6ff. Today fear —— more than a million of people. Not seen —— enemies —— enter into the temples —— weep.

To all the previous plagues this was added: pillagers completed the destruction, killing and raping.

The double catastrophe—caused by nature and by the invasion—destroyed all class distinction and brought about a social revolution. "Behold, noble ladies go hungry." "Behold, he who slept without a wife through want finds precious things." "He who passed the night in squalor" raised his head. "She who looked at her face in the water is possessor of a mirror."

It was anarchy. Nobody worked. "Behold, no craftsmen work." "A man strikes his brother, the son of his mother." "Men sit [behind] the bushes until the benighted (traveler) comes, in order to plunder his burden." "Behold, one uses violence against another. . . . If three men journey upon a road, they are found to be two men; the greater number slay the less. . . . The land is as a weed that destroys men."

"How terrible it is. What am I to do?" laments Ipuwer. "Woe is me because of the misery in this time!"

Several expressions of Ipuwer indicate that the text of the papyrus was composed shortly after the major catastrophe; the aftermath and the subsequent disturbances of nature were not yet at their end.

The starting point of this research was this: the Exodus from Egypt took place at the time of a great natural catastrophe. In order to find the time of the Exodus in Egyptian history, we had to search for some record of catastrophe in the physical world. This record is contained in the Papyrus Ipuwer.

Many parts of the papyrus are missing. The beginning and the end, doubtless containing details, possibly names, are destroyed. But what is preserved is sufficient to impress us with this fact: before us is not merely the story of a catastrophe, but an Egyptian version of the plagues.

It was surprising to find in the papyrus, in addition to the story of "dwellers in marshes" and "poor men" who fled the land scourged by plagues, laments about the invaders, who came from the desert of Asia, preyed on the disorganized country, and became its violent oppressors. Amu or the Hykos were the invaders who ruled Egypt during the centuries that separate the period of the Middle Kingdom from that of the New Kingdom. In a subsequent section the divergent views on the date of origin of the papyrus of Ipuwer will be presented.

Pi-ha-Khiroth

In the sixties of the last century, in el-Arish, a town on the border between Egypt and Palestine, the attention of a traveler[84] was attracted to a shrine (naos) of black granite inscribed with hieroglyphics over all its surfaces. It was used by the Arabs of the locality as a cattle trough. An account of this shrine and a partial translation of the text were published in 1890;[85] the shrine was still being used as a trough. Sometime during the present century the stone was brought to the Museum of Ismailia and a new attempt to translate the text was undertaken.[86]

Since its discovery, the monolith of el-Arish has been mentioned only infrequently, and its strange text has been regarded as rather mythological, though kings, residences, and geographical places are named and an invasion of foreigners described. The names of deities appearing in the text are royal cognomens. The inscription is of the Ptolemaic or Hellenistic age, but the events related are of a much earlier period, that of King Thom and his successor. In this inscription the name of King Thom is written in a royal cartouche, a fact that points to the historical background of the text.

The text, as will be demonstrated, deserves attentive study and a new definitive translation. Not even the sequence of the text is conclusively established.

In the mutilated text there are these lines.

> The land was in great affliction. Evil fell on this earth. . . . It was a great upheaval in the residence. . . . Nobody left the palace during nine days, and during these nine days of upheaval there was such a tempest that neither the men nor the gods could see the faces of their next.

Similar mention of a darkness that lasted a number of days, that confined everyone to his place, that was accompanied by a great upheaval, and that was so complete that no one could discern the face of his neighbor, is found in the Book of Exodus in the story of the ninth plague.

> EXODUS 10:22–23 . . . and there was a thick darkness in all the land of Egypt three days. They saw not one another, neither rose any from his place for three days.

The Egyptian text differs from the Hebrew in that it numbers nine days of darkness.[87] A strong wind is mentioned in the Scriptures in connection with the removal of the preceding plague—the locusts. They were brought by the "east wind" and "covered the face of the whole earth, so that the land was darkened" (Exodus 10:13ff.), and they were removed by "a mighty strong west wind":

EXODUS 10:19 And the Lord turned a mighty strong west wind, which took away the locusts, and cast them into the Red Sea.

Immediately thereafter came the ninth plague—the thick darkness.

In the story of the darkness in Egypt, as told in old Midrash books, additional details are given. The plague of darkness endured seven days. During the first three days one could still change his position; during the next three days one could not stir from his place. The rabbinical sources so describing the calamity of darkness are numerous.[88] Josephus Flavius[89] and Philo the Alexandrian[90] of the first century of the present era also belong to what may be called rabbinical sources. The collation of this material presents the following picture:

> ... exceeding strong west wind ... endured seven days. All the time the land was enveloped in darkness. ... The darkness was of such a nature that it could not be dispelled by artificial means. The light of the fire ... was either extinguished by the violence of the storm, or else it was made invisible and swallowed up in the density of the darkness. ...

> None was able to speak or to hear, nor could anyone venture to take food, but they lay themselves down ... their outward senses in a trance. Thus they remained, overwhelmed by the affliction. ...[91]

The last, the seventh day of darkness, overtook the land when the Israelites were at the Sea of Passage.[92] In the Book of Exodus it is said that "it was a cloud and darkness" so that the camp of the Egyptians came not near the camp of the Israelites "all the night". "And the Lord caused the sea to go back by a strong east wind all that night."

Tradition puts the time from the tenth plague, which followed immediately the plague of darkness, until the passage of the sea at six days and a few hours. The Exodus that followed the night of the tenth plague is commemorated on the first day of Passover (fifteenth day of Nisan), and the passage of the sea on the last, the seventh day of Passover (twenty-first day of Nisan).

The Hebrew version of the plague of darkness is not unlike the Egyptian version; it was not the darkness of a quiet night—a violent storm rushed in fine ashes. "The darkness came from the hell and could be felt."[93] In the tempest and darkness that lasted for days and nights, it was difficult to measure time; men, overwhelmed by misfortune, were bereft of exact judgment of the passage of time. Under such circumstances the slight discrepancy between the text on the stone (nine days[94] of darkness) and the tradition of the Midrashim (seven days of darkness) is negligible.

The Hebrew sources tell that cities were devastated in the darkness and that many Israelites were among the dead from the ninth plague.[95] The land fell into distress and ruin.

"Evil fell upon this earth . . . the earth is in a great affliction . . . a great disturbance in the residence," are passages on the shrine describing the nine days of darkness and tempest when nobody could see anything and none could leave the palace.

In the midst of the savageries of nature "his majesty of Shou" assembled his hosts and ordered them to follow him to regions where, he promised, they would again see light: "We shall see our father Ra-Harakhti in the luminous region of Bakhit."

Under cover of darkness intruders from the desert approached the border of Egypt. ". . . his majesty of Shou went to battle against the companions of Apopi" Apopi was the fierce god of darkness. The king and his hosts never returned; they perished.

> Now when the majesty of Ra-Harmachis [Harakhti?] fought with the evil-doers in this pool, the Place of the Whirlpool, the evil-doers prevailed not over his majesty. His majesty leapt into the so-called Place of the Whirlpool.[96]

And in the Book of Exodus it is related:

> EXODUS 14:27–28 . . . the sea returned to his strength . . . and the Egyptians fled against it; and the Lord overthrew the Egyptians in the midst of the sea.
>
> And the waters returned, and covered the chariots, and the horsemen, and all the host of Pharaoh that came into the sea after them.

Pharaoh himself perished too:

> EXODUS 15:19 For the horse of Pharaoh went in with his chariots and with his horsemen into the sea, and the Lord brought again the waters of the sea upon them.

The story of the darkness in Egypt as told in Hebrew and Egyptian sources is very similar. The death of the pharaoh in the whirling waters is also similar in both Hebrew and Egyptian sources, and the value of this similarity is enhanced by the fact that in both versions the pharaoh perished in a whirlpool during or after the days of the great darkness and violent hurricane.

And yet even a striking similarity is not identity. The subject of the two records should be regarded as identical only if some detail can be found in both versions, the Hebrew and the Egyptian, that cannot be attributed to chance.

The march of the pharaoh with his hosts is related amidst the description of the great upheaval in the residence and the tempest that made the land dark. He arrived at a place designated by name:

> His Majesty —— [here words are missing] finds on this place called Pi-Kharoti.

A few lines later it is said that he was thrown by a great force. He was thrown by the whirlpool high in the air. He departed to heaven. He was no longer alive.

The explanation of the translator of the text concerning this geographical designation "Pi-Kharoti" is: ". . . is not known except in this example."[97]

Here attempted identification of the subject of the two versions, the Hebrew and the Egyptian, is not incorrect if the locality where the pharaoh perished was a place by the Sea of Passage.

> EXODUS 14:9 But the Egyptians pursued after them, all the horses and chariots of Pharaoh . . . and overtook them encamping by the sea, beside Pi-ha-hiroth [Khiroth].

Pi-Kharoti is Pi-Khiroth of the Hebrew text.[98] It is the same place. It is the same pursuit. It is erroneous to say that the name is met nowhere else except on the shrine.

The inscription on the shrine relates that after a time a son of the pharaoh, "his majesty Geb", set out himself. "He asks information. . . ." The eyewitnesses from neighboring abodes "give him the information about all that happened to Ra in Yat Nebes, the combats of the king Thoum".

All who accompanied the prince were killed by a terrible blast, and the prince, "his majesty Geb", sustained burns before he returned from his expedition to seek his father, who had perished.

Invaders approaching by way of Yat Nebes came in the gloom and overpowered Egypt. "The children of Apopi, the rebels that are at Ousherou [not identified, has a sign 'dry', meaning 'desert'] and in the desert, they approached by way of Yat Nebes, and fell upon Egypt at the fall of darkness. They conquered only to destroy. . . . These rebels, they came from the mountains of the Orient by all the ways of Yat Nebes."[99]

The prince retreated before the invaders. He did not return to Heliopolis: "He did not go to On with [or 'like'] the companions of the thieves of the scepter"; he was robbed of his heritage. He secluded himself in the provincial residence of Hy-Taoui "in the land of the plants *henou*". From there he made an attempt, entirely unsuccessful, to communicate with "the foreigners and the Amu", that they leave the country. In his helplessness he recalled how his father, who succumbed in the whirlpool, in better days had battled all the rebels and "massacred the children of Apopi".

After a time "the air cooled off, and the countries dried".

It is not known what happened to the unhappy prince. His end was certainly sad. Egypt was devastated by the tempest and scorched by fire. The residence was seized by the Amu.

The inscription on the shrine at el-Arish says that the name of the pharoah who perished in the whirlpool was Thom or Thoum. It is of interest that Pi-Thom means "the abode of Thom". Pithom was one of the two cities built by the Israelite slaves for the Pharaoh of Oppression.[100] In Manetho, the pharaoh in whose days the "blast of heavenly displeasure" fell upon Egypt, preceding the invasion of the Hyksos, is called Tutimaeus or Timaios.[101]

The question, centuries or even millennia old, as to where the Sea of Passage was, can be solved with the help of the inscription on the shrine. On the basis of certain indications in the text, Pi-ha-Khiroth, where the events took place, was on the way from Memphis to Pisoped.[102]

The Ermitage Papyrus

A papyrus text preserved in the Ermitage in Leningrad and listed in the catalogue of that museum as "number 1116b recto",[103] is a literary echo of fateful days when the empire of Egypt perished and the land fell prey to invading nomads. In this papyrus the same story is related that we now know from the Papyrus Ipuwer, but in a different way. The upheavals of nature and the subsequent subjugation of Egypt by the desert tribes are recounted, not as events in the past or present, but as things that are to come. Obviously this indicates only a preference for the literary form of foretelling.

A sage by the name of Neferrohu asks his royal listener whether he would like to hear about things past or things future. "Said his majesty: Nay, of things future." The seer "was brooding over what should come to pass in the land and conjuring up the condition of the East, when the Asiatics [Amu] approach in their might and their hearts rage. . . . And he said: 'Up my heart and bewail this land thou art sprung'."

"The land is utterly perished and nought remains. Perished is this land. . . . The sun is veiled and shines not in the sight of men. None can live when the sun is veiled by clouds. . . ." "The river is dry (even the river) of Egypt." "The South Wind shall blow against the North Wind." "The earth is fallen into misery . . . Bedouins pervade the land. For foes are in the East [side of sunrising] and Asiatics shall descend into Egypt." "The beasts of the desert shall drink from the rivers of Egypt. . . . This land shall be in perturbation. . . ." "I show thee the land upside down, happened that which never (yet) had happened. . . ." "Men laugh with the laughter of pain. None there is who weepeth because of death." "None knoweth that midday is there; his [sun's] shadow is not discerned. Not dazzled is the sight when he is beheld. . . . He is in the sky like the moon. . . ."

From the description of the changes in nature we can recognize them as belonging to the period when the Israelites roamed in the desert, under a cloudy sky, in "the land of shadow, shadow of death".[104] Jeremiah centuries later complained: "Neither said they, Where is the Lord that brought us up out of the land of Egypt, that led us through the wilderness, through a land of deserts and of pits, through a land of drought, and of the shadow of death . . . ?" (2:6.) In numerous other passages in the Scriptures this "shadow of death" is mentioned: during the years of wandering in the desert the sky was veiled, clouds hung over the desert;[105] all life processes were impaired, and for this reason the gloom was called "shadow of death". The plague of darkness, of which, I maintain, the "shadow of death" was a lasting remainder, is dealt with in *Worlds in Collision*, which investigates the physical side of the catastrophe.

After giving this picture of natural disaster combined with the political subjugation of Egypt by the Amu, the seer Neferrohu prophesied the liberation by a king who would be born of a Nubian woman and called Ameny—"the Amu shall fall by his sword". Thereafter "there shall be built the 'Wall of the Prince' so as not to allow the Amu to go down into Egypt".

It is questionable whether Ameny is a historical personality. Because of some surmise as to his identity, the papyrus was supposed to have been written during the Old Kingdom or shortly after its end; it is also apparent that this text has much in common with the text of the papyrus in Leiden. However, the Hyksos (Amu) period in Egypt followed the end of the Middle Kingdom. The name Ameny may refer to Amenhotep I, one of the first kings of Egypt after it was freed from the Hyksos; at the time of liberation he was a prince. He is usually pictured as black,[106] which would also conform with the words "born of a Nubian woman". He was highly revered in later times.

A literary remnant that closely resembles the Ermitage papyrus 1116b recto is a prophecy of a potter under King Amenophis [Amenhotep]. "The waterless Nile will be filled, the displaced winter will come in its own season. The sun will resume its course and the winds will be restrained. For in the Typhon time the sun is veiled." This prophecy is preserved in a papyrus written in Greek, being a translation of an older Egyptian text.[107] The pharaoh's name points to one of the Amenhoteps of the New Kingdom[108] and, I assume, refers to the same Ameny or Amenhotep I.

Two Questions

There are two questions that demand an answer.

The first is: What were the nature and dimensions of this catastrophe, or this series of catastrophes, accompanied by plagues, about which we have now very similar scriptural and Egyptian testimony? In the next chapter we shall also have the corroborating

40

autochthonous tradition of the Arabian peninsula. A reply to this question involves a study, not only of history, but also of many other fields. A work comprising an investigation into the nature of great catastrophes of the past preceded this volume.

Leaving aside here the problem of the extent and character of the catastrophe, we turn to the other question: When did the upheaval occur? In Jewish history the answer is at hand: in the days of the Exodus. As for Egyptian history, we must first find out when the text of the Papyrus Ipuwer originated.

Scholars who have studied this papyrus agree that the document is a copy of a still older papyrus: "The scribe used a manuscript a few centuries older."[109] The copy was made sometime during the Nineteenth Dynasty, but "The spelling is, on the whole, that of a literary text of the Middle Kingdom, if this term be interpreted in a very liberal way."[110]

The question, When did the text originate? grows in importance in view of the parallels with the Book of Exodus presented here.

It was understood that the question of the age of the text "is inextricably bound up with the problem as to the historical situation that the author [Ipuwer] had in his mind".[111] "The text tells both of civil war and of an Asiatic occupation of the Delta." "There are two periods which might possibly answer the requirement of the case: the one is the dark age that separates the sixth from the eleventh dynasty [or the Old Kingdom from the Middle Kingdom]; the other is the Hyksos period [between the Middle and the New Kingdoms]." The opinions of the papyrologists (Gardiner and Sethe) were divided on the question, To which of these periods does the text of the papyrus relate? There is no definite knowledge of any invasion by Asiatics (Amu) during the first period—that between the Old Kingdom and the Middle Kingdom—and to conform with the historical background of the papyrus, such an invasion at that time must be first postulated with the help of the papyrus.[112] "There is no such difficulty in the view preferred by Sethe," who maintained that the time described is that of the invasion of the Hyksos, Gardiner conceded and added: "The view that our Leiden papyrus contains allusion to the Hyksos has the better support from the historical standpoint." But a philological consideration "makes us wish to put back the date of the composition as far as possible". The language was found beyond doubt to be not of the New Kingdom but of an earlier time. The text also contains some references to the establishment of "Great Houses" (law courts), which became obsolete "in or soon after the Middle Kingdom".

We should remember that these Great Houses are described in the papyrus as fallen down and trodden upon by the throngs who dug in the ruins. This mention, it seems to me, points even more precisely to the time when the Middle Kingdom collapsed, and the papyrus should not, on the basis of it, be interpreted as a literary document composed in pre-Hyksos times. From the standpoint of style and language, Gardiner admitted, "it is of course possible that our text may have been composed while the Hyksos were still in the land".

The discussion of whether the text describes the period between the Old and Middle Kingdoms, or between the Middle and New Kingdoms, was closed by the supporter of the first view with this remark: "It is doubtless wisest to leave the question open for the present."

I take up the question. The historical background is that of the invasion of the Hyksos (Sethe). Philological considerations show that the text has all the signs of a literary product of the Middle Kingdom (Gardiner).[113] When the historical and philological proofs are combined, all point to the end of the Middle Kingdom and the very beginning of the invasion of the Hyksos. The style would still be, of course, that of the Middle Kingdom, because in the few months since the end of this great age no change in the language and form of poetic works could have occurred. In the centuries of domination by the Hyksos literary activity ceased in Egypt. Besides, it is obvious that Ipuwer bewails the tragedy of his own time and not of a past age.

In the dispute between Gardiner and Sethe, Gardiner is right in his philological arguments that the latest period from which the text could have originated is the time of the Hyksos. However, he is mistaken in assuming that the text describes events of the dark period between the Old and Middle Kingdoms, and Sethe is right in the historical argument that the events described are those of the invasion of the Hyksos after the fall of the Middle Kingdom. On the other hand, Sethe is mistaken when he ascribes the composition of the text to the New Kingdom, at the very beginning of the Hyksos period. The historical arguments as well as the philological arguments are in harmony with this solution.

Those who have tried to place the Exodus in the sequence of Egyptian history have not dared to set it as far back as between the Middle and New Kingdoms (time of the Hyksos), and of course not as far back as between the Old and the Middle Kingdoms.

Not only is the invasion of the Amu (Hyksos) the historical background of the *Admonitions*; so are the physical catastrophe and the plagues, and these are analogous to those of the time of the Exodus. On this basis, too, there can be no doubt that the earlier date—between the Old and Middle Kingdoms—is out of the question; and even the later one is embarrassing enough, for it appears to be too early for the Exodus.

Is there any physical evidence known which would indicate that some major change occurred in the geological structure of Egypt at some point of time following the Middle Kingdom?

Lepsius has observed the remarkable fact that the Nilometers at Semneh dating from the Middle Kingdom show an average rise in the waters of the Nile at that place, where the river is channeled in rock, twenty-two feet higher than the highest level of today.[114] Theoretically the dropping of the water level by twenty-two feet must be ascribed either to a change in the quantity of water in the Nile since the Middle Kingdom or to a change in the rock structure of Egypt.[115] If the Nile had contained so much more water before the catastrophe, many residences and temples would have been regularly covered with water. Apparently the Nilometers at Semneh indicate that some violent changes in the rock formation of Egypt took place at the end of the Middle Kingdom or later.

The Middle Minoan II Age, which is the period of the Cretan culture, coeval with the Middle Kingdom in Egypt, also came to a close in a terrible natural catastrophe, as the excavations at Knossos reveal.[116]

There exists an important Egyptian inscription of the Queen Hatshepsut, who came to power two or three generations after the expulsion of the Hyksos, in which it is written:

> The abode of the Mistress of Qes was fallen in ruin, the earth has swallowed her beautiful sanctuary and children played over

her temple. . . . I cleared and rebuilt it anew. . . . I restored that which was in ruins, and I completed that which was left unfinished. For there had been Amu in the midst of the Delta and in Hauar (Auaris), and the foreign hordes of their number had destroyed the ancient works; they reigned ignorant of the god Ra.[117]

These lines contain an inference that temples were swallowed in the ground[118] and that the Hyksos (Amu), who took possession of the country, did not care to restore the ruins and even added to the destruction. The Hyksos destroyed, but they did not bury buildings in the ground. "Does this mean that the temple disappeared in an earthquake?"[119]

In all three Egyptian documents quoted—the papyrus of Leiden, the writings on the shrine of el-Arish, and the papyrus of Ermitage—and also in the inscription of Hatshepsut, the natural catastrophe and the invasion of the Amu are described as one following the other; and as the natural catastrophe consisted of a series of upheavals and disturbances, the invasion of the people from Asia occurred before the elements calmed.

At various times since the days of antiquity several authors have identified the Israelites with the Hyksos (Amu); other authors have placed the arrival of the Israelites in the period of the Hyksos rule in Egypt and the Exodus of the Israelites in the reign of one of the kings of the Eighteenth Dynasty. The majority of scholars, however, think that the sojourn of the Israelites took place in a still later period. They relate the coming of the Israelites to some point of time during the Eighteenth Dynasty, and place the period of oppression in the days of Ramses II of the Nineteenth Dynasty and the Exodus in the days of his successor Merneptah.

I have arrived at a very different result. The Israelites left Egypt amid the outbreak of a great natural catastrophe. The Amu, who invaded Egypt and became the masters of the land immediately afterward, were obviously not the Israelites. The tradition of the Israelites definitely connects their departure from Egypt with the days when earth, sky, and sea excelled in wrath and destruction, but knows nothing of their arrival in Egypt in the days of a cataclysm.

"The dwellers of the marshes" or "the poor men" left the country under these very circumstances; they must have been the Israelites and the multitude of the Egyptians who accompanied them in the Exodus.[120] The Amu-Hyksos reached the land of Egypt, upon which they came to prey, a short time after the catastrophe.

If the above parallels and these conclusions are correct, then: the Exodus of the Israelites preceded by a few days or weeks the invasion of the Hyksos.

NOTES

1. The name Hyksos as "rulers of foreign countries" is found in the Egyptian text of the Turino Papyrus and on a few scarabs.

2. The Seventeenth Dynasty is generally regarded as the native dynasty of princes in submission to, and then in revolt against, the last kings of the Sixteenth, the Hyksos, Dynasty. But in Manetho's list, as given by Julius Africanus and Eusebius, the Seventeenth Dynasy is the last of the Hyksos.

3. The division into kingdoms is modern, but the Egyptians themselves had similar concepts of their past. Compare H. Ranke in *Chronique d'Egypte*, VI (1931), 277, 86.

4. T. E. Peet, *Egypt and the Old Testament* (Liverpool, 1922), p. 7.

5. Manetho, though making the Hyksos expelled from Egypt the builders of Jerusalem, told another story, that he assigned to a later epoch, in which he related that lepers, segregated in Auaris on the eastern border of Egypt, usurped the power in Egypt with the help of the Solymites (the people of Jerusalem) and were utterly cruel, and their chief, Osarsiph, adopted the name of Moses and led them to Palestine when they were expelled. Josephus did not separate the two Manetho stories.

6. Julius Africanus, "Chronography", in *The Ante-Nicene Fathers*, ed. A. Roberts and J. Donaldson (New York, 1896), VI, 134. There he confused Ahmose I, the first king of the New Kingdom, with Ahmose II (Amasis of Herodotus), the last king before the conquest of Egypt by Cambyses, the Persian.

   But in his Canon condensing the list of Dynasties of Manetho, he added this remark to the list of the kings of the Eighteenth Dynasty: "The first of these was Amos [Ahmose], in whose reign Moses went forth from Egypt, as I have declared; but, according to the convincing evidence of the present calculation it follows that in this reign Moses was still young." *Manetho* (trans. W. G. Waddell; Loeb Classical Library, Cambridge, Mass., 1941), p. 111.

7. Georgius Syncellus, a Byzantine chronographer, who copied Eusebius added: "Eusebius alone places in this reign the Exodus of Israel under Moses, although no argument supports him, but all his predecessors hold a contrary view—as he testifies."

8. Augustine, *The City of God*, Bk. 18, Chap. 8.

9. Cf., for instance, A. T. Olmstead, *History of Palestine and Syria* (New York, 1931), p. 128.

10. H. R. Hall, "Israel and the Surrounding Nations", in *The People and the Book*, ed. A. S. Peake (Oxford, 1925), p. 3; Sir E. A. W. Budge, *Egypt* (New York, 1925), p. 110; A. H. Gardiner, in *Etudes Champollion*, 1922, pp. 205ff.; *Journal of Egyptian Archeology*, X (1924), 88.

11. Hall, in *The People and the Book*, ed. Peake, p. 7.

12. Scholars writing in English have no unified method of transliterating the guttural letters in Semitic languages. The *Cambridge Ancient History* acknowledges the inconsistency, in some cases following the established spelling of names in English, in others preferring the closest phonetic equivalents.

13. Eduard Meyer, *Geschichte des Altertums*, Vol. 2, Pt. II (2nd ed.; Stuttgart, 1931), p. 214.

14. John Garstang, *The Foundations of Bible History* (New York, 1931): "The Israelite invasion . . . corresponds with a period of apathy under Amenhotep III."

15. Peet, *Egypt and the Old Testament*, pp. 74–75.

16. Hall, in *The People and the Book*, ed. Peake, p. 7.

17. Sir W. M. Flinders Petrie, *Palestine and Israel* (London, 1934), p. 56.

18. S. Freud, *Moses and Monotheism* (New York, 1939). Compare Strabo, *The Geography*, XVI, 2, 35.

19. This view is found in R. Lepsius, "Extracts from the Chronology of the Egyptians", in his *Letters from Egypt, Ethiopia and the Peninsula of Sinai* (London, 1853), p. 449. Even before the discovery of the Merneptah stele, he was identified by not a few scholars as the Pharaoh of the Exodus, because his predecessor, Ramses II, was thought to be the Pharaoh of Oppression. This role was ascribed to Ramses II because of the mention of the city of Ramses in the Book of Exodus. The adherents of the Habiru theory do not regard this as a weighty argument. "*Plusieurs historiens remarquant que ces villes* [Ramses and Pithom] *sont antérieures à Ramsès II estiment que les travaux en question ont pu être ordonnés par un roi de la XVIII Dynastie.*" P. Montet, *Le Drame d'Avaris* (Paris, 1941), p. 144.
Under the statue of Merneptah in the hall of the Metropolitan Museum of Art in New York, until recently a sign by a modern hand read, "Pharaoh of Exodus", and under that of Ramses II, "Pharaoh of Oppression". See H. E. Winlock, *The Pharaoh of the Exodus*, Metropolitan Museum Bulletin 17 (New York, 1922), pp. 226–34.

20. "If Israel did not leave Egypt until the reign of Merneptah, and if they spent about forty years en route to Palestine, how could Merneptah have defeated them in Palestine in the third year of his reign?" S. A. B. Mercer, *Tutankhamen and Egyptology* (Milwaukee, 1923), pp. 48ff.

21. Hall, in *The People and the Book*, ed. Peake, p. 7.

22. In an inscription of Ramses II, and also in one of his predecessor Seti, there is mention of Asher in Palestine, which is the name of one of the Twelve

46

Tribes. This reference and other similar instances led scholars to suppose that the Exodus took place in successive waves.

23. S. A. B. Mercer, *Extra-Biblical Sources for Hebrew and Jewish History* (New York, 1913). He identifies the Habiru as Hebrews, and the Pharaoh of Oppression as Ramses II, one hundred years later.

24. Peet, *Egypt and the Old Testament*, p. 124, referring to the theory of Driver and others.

25. Cf. W. F. Albright, *The Archaeology of Palestine and the Bible* (New York, 1932), p. 144, ascribing Exodus to the early thirteenth century. However, Albright advocates the sojourn of the Israelites in Egypt in the days of the Hyksos.

26. Petrie, *Palestine and Israel*, p. 58.

27. Albright, quoted by Petrie, *Palestine and Israel*, p. 57. Bethel fell "sometime about the first half of the thirteenth century, in Albright's opinion" – thus Wright, "Epic of Conquest", *Biblical Archaeologist*, III (1940), p. 36.

28. B. D. Eerdmans, *Alttestamentliche Studien* (Giessen, 1908), II, 67.

29. Garstang, *The Foundations of Bible History*, p. 51.

30. S. W. Baron, *A Social and Religious History of the Jews* (New York, 1937), I, 16.

31. Hugo Winckler, *Kritische Schriften* (Berlin, 1901–7), I, 27. Cf. also Peet, *Egypt and the Old Testament*, p. 21: "The sojourn may well have been on so small a scale that the Egyptians never thought it worthy of recording."

32. Variations, with somewhat differing sequences of the plagues, are found in Psalms 78 and 105.

33. The details of the story ought to be regarded as no less mythical than the details of creation as recorded in Genesis." A. H. Gardiner, in *Etudes Champollion*, 1922, p. 205.

34. Eduard Meyer says that the only plague, in the early version of the legend, was that of the locusts (*Die Israeliten und ihre Nachbarstämme* [Halle, 1906], p. 30). He says also, "There is no folkloristic tradition in the tale of the plagues. They are the creation of the narrator" (ibid., p. 31).

35. H. Gressmann, *Mose und seine Zeit* (Göttingen, 1913), p. 107.

36. Ibid., p. 108.

37. Ibid., p. 73.

38. Ibid., p. 92.

39. Vansleb (1677) observed that water in the Nile changed its color from green to ocherous red. "When the Nile first begins to rise, toward the end of June, the red marl brought from the mountains of Abyssinia stains it to a dark colour, which glistens like blood in the light of the setting sun." A. H. Sayce, *The Early History of the Hebrews* (London, 1897), p. 168.

40. Gressmann, *Mose und seine Zeit*, p. 117: "The picture is drawn so graphically that every detail is clear before the eyes and one would almost think of a realistic description of historical events, but for the miracles. Thus the vividness of description is also a mark of a saga."

47

41. So A. H. Gardiner, *Etudes Champollion*, 1922, pp. 205ff.; *Journal of Egyptian Archaeology*, X (1924), 82f.

42. See, for example, S. R. Driver, *The Book of Exodus in the Revised Version* (Cambridge, England, 1911), p. 113: ". . . the variously attested custom of a brazier filled with burning wood being borne at the head of a caravan of pilgrims."

43. London, 1873.

44. Exodus 19: 16, 18; 20: 18.

45. Dean Arthur P. Stanley, *Lectures on the History of the Jewish Church* (New York, 1863–76), I, 167.

46. Charles Beke, *Discoveries of Sinai in Arabia and of Midian* (London, 1878), p. 561.

47. Ibid., p. 436.

48. H. Gunkel, *Deutsche Literaturzeitung*, 24 (1903), col. 3058f.

49. Meyer, *Die Israeliten und ihre Nachbarstämme*, pp. 69ff.; H. Gressmann, *Der Ursprung der israelitisch-jüdischen Eschatologie* (Göttingen, 1905), pp. 31ff.; also Gressmann, *Mose und seine Zeit*, pp. 417ff. A Musil identified Mount Sinai with the extinct volcano al-Bedr.

50. Meyer, *Geschichte des Altertums*, Vol. II, Pt. 2 (2nd ed.), p. 210: "*So kann kein Zweifel bestehen, dass der Sinai in einem der zahlreichen jetzt erloschenen Vulkane der Harra's zu suchen ist*"; compare ibid., p. 205: "It is very possible that the saga [of the Sinai experience] belonged first to some tribe of the Sinai Peninsula, and then was taken over by the Israelites as a great act of Yahwe."

   Gressmann (*Mose und seine Zeit*, p. 418) also denied the visit of the Israelites to Mount Sinai.

51. Psalms 18: 7–8, 15.

52. Psalms 97: 4–5.

53. Job 9: 5–6.

54. Judges 5: 4–5.

55. Psalms 18: 15.

56. Psalms 77.

57. Psalms 60: 2–3.

58. Numbers 16: 32.

59. Exodus 15: 23; Psalms 107: 33–35.

60. C. Leemans, *Aegyptische Monumenten van het Nederlandsche Museum van Oudheden te Leyden* (Leiden, 1846), Pt. 2, Face: Plates 105–13.

61. By F. Chabas, reprinted in *Bibliothèque égyptologique*, X (Paris, 1902), 133ff., especially 139–40.

62. F. J. Lauth, "Altaegyptische Lehrsprüche", *Sitzungsberichte der Bayerischen Akademie der Wissenschaften, Philosphisch-philologische und historische Classe* (1872).

63. H. K. Brugsch, cited by Lange (see note 64).

64. H. O. Lange, "Prophezeiungen eines aegyptischen Weisen", *Sitzungsberichte der Preussischen Akademie der Wissenschaften*, 1903, pp. 601–610.

65. Published in Leipzig.

66. Gardiner, *Admonitions*, note to 1: 8.

67. "*Er steht vor dem Allherscher, was sonst ein Epitheton der grossen Goetter ist, hier aber wohl den König bezeichnet.*" Lange, *Sitzungsberichte der Preussischen Akademie der Wissenschaften*, 1903, p. 602.

68. The other Hebrew word for "noise", *shaon*, also means "earthquake". See S. Krauss, "Earthquake", The Jewish Encyclopedia (New York, 1901–1906).

69. The Bible quotations are from the King James version; the quotations from the text of the papyrus are from the translation by A. H. Gardiner.

70. In Psalms 105: 33 this plague is described: "He smote [with hail] their vines also and their fig trees; and brake the trees of their coasts."

71. See the Notes to the text of Gardiner, *Admonitions*, with a reference to Papyrus Leiden 345 recto, 1.3.3.

72. Jeremiah 2:6.

73. Gardiner accompanies the translation of the word "to overturn" with an explanatory example: "To overthrow a wall".

74. J. Levy, *Wörterbuch über die Talmudim und Midrashim* (Vienna, 1924).

75. In his notes to another passage Gardiner translates "storehouse" as "prison".

76. Eusebius, *Preparation for the Gospel* (trans. E. H. Gifford; Oxford, 1903); Book IX, Chap. xxvii.

77. Ibid.

78. Cf. Louis Ginzberg, *Legends of the Jews* (1925), III, 5–6.

79. Compare C. S. Osborn, *The Earth Upsets* (Baltimore, 1927), p. 127, on the earthquake at Valparaiso, Chile, the night of August 15, 1906: "I visited the scene as soon as I could get there. Untombed coffins protruded from the graves in hillside cemeteries that had shaken open."

80. *Midrash Rabbah* (English trans. edited by H. Freedman and M. Simon; London, 1939), 10 vols.

81. According to the haggadic tradition, not only the firstborn but the majority of the population in Egypt was killed during the tenth plague.

82. *Bechor*, to choose, select, prefer; *bachur*, a young man, is of the same root. *Bekhor*, to be early, produce first fruits, to be first in ripening. Levy, *Wörterbuch über die Talmudim und Midrashim*.

83. Ginzberg, *Legends*, II, 369.

84. V. Guérin, *Judée* (Paris, 1869), II, 241.

85. F. L. Griffith, *The Antiquities of Tell el Yahudiyeh and Miscellaneous Work in Lower Egypt during the Years 1887–1888* (London, 1890) (published with Naville, *The Mound of the Jew and the City of Onias*).

86. Georges Goyon, "Les Travaux de Chou et les tribulations de Geb d'aprés le Naos 2248 d'Ismailia", *Kêmi, Revue de philogie et d'archéologie égyptiennes et coptes*, *VI* (1936), 1–42.

49

87. In A. S. Yahuda, *The Accuracy of the Bible* (London, 1934), on p. 84, we find the following passage: "In the 'Myth of the God-Kings' which is as old as Egypt itself it is said that the world was filled with darkness and the text proceeds literally, 'and no one of the men and the gods could see the face of the other *eight* days'. The Hebrew author was less fantastic and excessive than his Egyptian predecessor and therefore reduced the 8 days to only 3." With this remark the author of *The Accuracy of the Bible* contented himself.

88. Ginzberg, *Legends*, II, 359–60; V, 431–39. Among the sources are Midrash Shemoth Raba, Midrash Shir Hashirim Raba, Targum Yerushalmi, Midrash Tanhuma Hakadom Hajashan, Sefer Hajashar, Sefer Mekhilta Divre Ishmael.

89. *Jewish Antiquities*, II, 14, 5.

90. *Vita Mosis*, I, 21.

91. Ginzberg, *Legends*, II, 359–60.

92. Ibid.

93. Ibid.

94. See the reading of A. S. Yahuda, note 87, above.

95. Josephus, *Jewish Antiquities*, II, 14, 5. Ginzberg, *Legends*, II, 345.

96. Griffith, *The Antiquities of Tell el Yahudiyeh*, p. 73.

97. "*N'est connu que par cet exemple; sans doute peu éloigné de Saft el Henneh on sur la route de Memphis à Pisoped.*" Goyon. *Kêmi*, VI (1936), 31, note 4.

98. "Ha" in Pi-ha-Khiroth is the Hebrew definite article. It belongs between Pi and Khiroth. The vowels in the translation of the Egyptian text are a conjecture of the translator: the name can also be read Pi-Khirot.

99. Goyon, *Kêmi*, VI (1936), II (text), and 27 (translation).

100. The treasury city of Pithom was discovered by E. Naville in 1885 at Tell el Maskhuta, and identified with the help of an inscription.

The name of the other city, Ramses, was largely the reason why Ramses II of the Nineteenth Dynasty was identified as the Pharaoh of Oppression. It is well to remember that "second" is our modern reckoning of kings, and Ramses of the Nineteenth Dynasty may have had some predecessors of the same name in pre-Hyksos dynasties. Ramses could also be a city named for divinity. It is also possible that the name of the city Ramses (Exodus 12:37) is a later name of the place; similarly we call by the name Tell el-Amarna the historical Akhet-Aton. The argument of the Ramses city was sometimes raised against the identification of the Habiru with the Israelites.

101. Gutschmidt and Reinach read the name Τίμαιος. See Josephus, *Against Apion* (trans. H. St. Thackeray; London, New York, 1926), I, 75, note.

102. See note 97.

103. A. H. Gardiner, "New Literary Works from Ancient Egypt", *Journal of Egyptian Archaeology*, I (1914), 100–106.

104. Compare Psalms 23:4; 44:19; 107:10, 14; Isaiah 9:2; 51:16; Jeremiah 13:16; Amos 5:8; Job 24:17; 28:3; 34:22, etc.

105. A thick veil of clouds over the desert is mentioned repeatedly in the scriptures and in the Talmud and Midrashim.

106. However, his being pictured as black may refer to his being worshiped as a deceased saint.

107. Literature on this prophecy is found in G. Manteuffel, *De opusculis graecis Aegypti* . . . (Warsaw, 1930); *Mélanges Maspero*, II (1934), 119–27.

108. H. Ranke in Gressmann, *Altorientalische Texte* (Tübingen, 1909), pp. 207–208: "*Der Name Amenophis weist jedenfalls auf einen der Amenhotep der XVIII Dynastie.*"

109. Gardiner, *Admonitions*, p. 3.

110. Ibid., p. 2.

111. Ibid., p. 17.

112. In this case "Amu" would designate not only the Hyksos people, but Asiatics generally.

113. Both Sethe and Gardiner regarded the text as not contemporaneous with the events described, but in discussing the age of the text, Sethe saw in it a description of the events of the Hyksos period and considered the beginning of the New Kingdom as the time of the composition; Gardiner thought the Middle Kingdom, or perhaps the Hyksos period, was the time of the composition.

114. Lepsius, *Letters from Egypt, Ethiopia and the Peninsula of Sinai*, pp. 19–20: "Semneh. The Nile is here compressed within a breadth of only about 1150 feet between high rocky shores. . . . We found a considerable number of inscriptions from the Twelfth and Thirteenth Manethonic Dynasties. . . . Many of them were intended to indicate the highest rising of the Nile during a series of years, especially in the reigns of the Kings Amenemhet III and Sebekhotep I, and by comparing them, we obtained the remarkable result, that about 4000 years ago the Nile used to rise at that point, on an average, twenty-two feet higher than it does at present."

But compare L. Borchardt, *Altägyptische Festungen an der zweiten Nilschwelle* (Leipzig, 1923), p. 15, and S. Clarke, *Journal of Egyptian Archaeology*, III (1916), 169; also Borchardt, "Nilmesser und Nilstandsmarken", *Anhang* of the *Abhandlungen der Preussischen Akademie der Wissenschaften*, 1906, pp. 1–5, and *Sitzungsberichte*, 1934, pp. 194–202.

115. The Nile was low at least temporarily after the catastrophe, as "references to foreign invaders, to the scanty Nile and to a veiled or eclipsed sun" are "much of the characteristic stock-in-trade of the Egyptian prophet". Gardiner, *Journal of Egyptian Archaeology*, I (1914), 101.

116. Sir Arthur J. Evans, *The Palace of Minos* (London, 1921–35), III, 14.

117. Inscription at Speos Artemidos. W. M. Flinders Petrie, *A History of Egypt: During the Seventeenth and Eighteenth Dynasties* (7th ed.; London, 1924), II, 19. Breasted, *Records*, II, 300ff., differs in translation. A new translation was published by Gardiner, *Journal of Egyptian Archaeology*, XXXII (1946), 46f.

118. Old Midrash sources narrate that the walls of Pithom and Ramses fell and were partly swallowed by the earth, and that many Israelites perished on that

occasion. If the place Edouard Naville identified as Pithom (*The Store-City of Pithom and the Route of the Exodus* [2nd ed.; London 1885]) is the site mentioned in the Book of Exodus, excavation into deeper strata (Naville explored the level of the Nineteenth Dynasty, as he referred the Exodus to that time) may show whether or not this Midrash account is legendary.

In general, the swallowing up of cities and villages in earthquakes is authenticated.

119. "It is not easy to understand what the queen means. ... I translate, as Golenischeff does, 'the land which had swallowed up the sanctuary'. Does this mean that the temple disappeared in an earthquake?" Edouard Naville, "The Life and Monuments of Hatshopsitu" in *The Tomb of Hatshopsitu* by Theodore M. Davis (London, 1906), p. 69.

120. Exodus 12:38.

CHAPTER TWO

The Hyksos

Who Were the Hyksos?

The Egyptian historian Manetho, who lived in the Ptolemaic age, is our main source of information on the invasion of the Hyksos. His history of Egypt is not extant, but some passages dealing with that invasion are preserved by Josephus Flavius, Eusebius and Sextus Julius Africanus.[1]

Josephus, in his pamphlet, *Against Apion*, preserved more of the second book of Manetho's history of Egypt than Eusebius and Africanus did. Here is the first passage:

> I will quote his [Manetho's] own words, just as if I had produced the man himself in the witness box:
> "Tutimaeus. In his reign, I know not why, a blast of God's displeasure broke upon us.
> "A people of ignoble origin from the east, whose coming was unforeseen, had the audacity to invade the country, which they mastered by main force without difficulty or even a battle."[2]

From where did the Hyksos come? The theories of the scholars differ. Some thought that the Hyksos were Mitannians, of the Aryan race;[3] others, that they were Scythians.[4] Still others, for two millennia, have supposed that they were the Israelites, whose sojourn in Egypt is preserved in quite a different version in the Bible. And, finally, even the very fact of the invasion of Egypt by a tribe called Hyksos was disputed by a scholar who expended more effort than others in the investigation of the remnants of the so-called Hyksos dynasties in Egypt.[5] However, he found no followers.

Already in the days of Manetho the land of their origin was not known with certainty; but this he did know: "Some say that they were Arabians."[6]

Manetho, who wrote in Greek, explained their being named Hyksos:

> Their race bore the generic name of Hycsos [Hyksos], which means "king-shepherds". For *Hyc* in the sacred language denotes "king", and *sos* in the common dialect means "shepherd" or "shepherds"; the combined words form "Hycsos".[7]

As already noted on a previous page, in the remnants of Egyptian literature the Hyksos are called "Amu".

The Hyksos were a people imbued to the core with a spirit of destruction. As far as is known, no monuments of any historical or artistic value were erected under their rule, no literary works survived their dominion in Egypt, with the exception of lamentations, such as those contained in the Ipuwer papyrus. The memory of the wickedness of these nomads is preserved by Manetho-Josephus.[8]

The Scriptures furnish no information about what happened in Egypt after the Israelites departed. In the midst of a natural upheaval and a holocaust they left.

The papyrus of Ipuwer completes the records: it tells of invaders that vexed and tormented the land of Egypt. The tenth plague was not the last; yet one more was to follow. A cruel conqueror invaded the once mighty realm, now overthrown and prostrate; he subjugated it without meeting resistance; he desecrated its sacred places; he raped its surviving women, enslaved the decimated population. He utterly destroyed the temples, if any remained erect, robbed the sepulchers of the dead, and mutilated the victims who remained alive.

We read the complaint of one of those who escaped alive the fury of the trembling earth and who witnessed the misery inflicted by a fierce foe. By comparing the evidence of the Book of Exodus and that of the Ipuwer papyrus it becomes clear that this eleventh plague followed the preceding ones when the Israelites were already out of the country, but only a very short time after their departure. The Israelites left the land that lay devastated under the blows from heaven. Another people—called Amu by the Egyptians—overran the country and turned its exhaustion to their advantage. The upheaval was not yet over, was still in progress, when Egypt was invaded by the Amu.

The invaders came from Asia; this is stated in the Ipuwer papyrus. The Israelites went in the direction of Asia. The two continents, Asia and Africa, are connected by a small, triangular piece of land. There was a good chance that the Israelites, moving toward Asia, would meet the Amu invaders moving toward the Egyptian frontier. Did they meet each other?

They really met.

The Israelites Meet the Hyksos

Even before the Israelites reached Mount Sinai they met the multitudes of Amalek. At Meriba, which is at the foot of Horeb, where the people thirsted for water, in the bed of a stony valley "came Amalek, and fought with Israel in Rephidim" (Exodus 17:8). Moses, Aaron, and Hur went up to the top of the hill and prayed there when Joshua fought with Amalek. At some stages of the battle Amalek had the upper hand, but

> EXODUS 17:13–16 ... Joshua discomfited Amalek and his people with the edge of the sword.
>
> And the Lord said unto Moses, Write this for a memorial in a book ... for I will utterly put out the remembrance of Amalek from under heaven.
>
> And Moses built an altar, and called the name of it Jehovah-nissi: For he said, Because the Lord hath sworn that the Lord will have war with Amalek from generation to generation.

The rabbinic tradition says that in this encounter Joshua faced four hundred thousand Amalekite warriors.[9]

The victory at Rephidim was costly: in that battle the Israelites prevailed only after they had been hard pressed and were close to defeat.

This was but one of the battles with the Amalekites. Soon the Israelites, escaping from Egypt, came upon them in every direction they chose to go. The tribes of the desert, migrating in large bands toward Syria and Egypt, continually involved the wanderers in minor skirmishes, night raids, and irregular engagements. Suffering from lack of water in a desert covered with dust and ashes, the Israelites were plagued by the merciless despoilers, who pillaged and plundered like pirates.

The upheaval at the sea, says the Haggada, had created alarm among the heathen, and none would oppose the Israelites. But this fear vanished when the Amalekites attacked them.[10]

The southern approach to Canaan was closed to the Israelites. "The Amalekites dwell in the land of the south" (Numbers 13:29). The people in the wilderness heard this message from the twelve men when they returned from their mission to investigate Canaan.

"To intensify to the utmost," says the Midrash,[11] "their fear of the inhabitants of Palestine, the scouts said: 'The Amalekites dwell in the land of the south.' The statement concerning Amalek was founded on fact, for although southern Palestine had not originally been their home, still they had recently settled there."

The people to whom this message was brought by the twelve men "wept that night" (Numbers 14:1): "Would God that we had died in the land of Egypt! or would God we had died in this wilderness!" (14:2.)

To inspire this boundless fear, the Amelekites must have been not mere bands of Bedouin robbers but a force superior to other peoples in the area.

At first Moses charged the Israelites to try to break through into Canaan across its southern border, but they were afraid and threatened to stone their leader. Then came the condemnation: "Surely they shall not see the land which I sware unto their fathers," and they were doomed to wander forty years in the wilderness. And it was said to them:

> NUMBERS 14:25 Now the Amalekites and the Canaanites dwelt in the valley. Tomorrow turn you, and get you into the wilderness by the way of the Red Sea.

The desert with its dreadful experiences—earthquakes, cleavage of the earth, outbursts of flame, disappearance of springs—terrified the Israelites. A plague killed thousands, the flesh of wild fowl in their teeth. These were quail, put to flight like the Israelites who fled from Egypt, the Amalekites from Arabia, and the wild beasts that fled to

56

the domicile of man, and the locusts that were flung by the wind into the Red Sea. The human abode and the lair of the wild beast and the nest of the fowl became unsafe, and a mighty roving instinct awoke simultaneously in all of them. The Israelites wandered toward Canaan, their ancestral home, and were disheartened when faced with the choice of battling the Amalekites or going into the wilderness. A life as wanderers in that desolate region terrified them, and in desperation they decided to battle their way through. And now Moses said:

> NUMBERS 14:42–45 Go not up, for the Lord is not among you. . . . For the Amalekites and the Canaanites are there before you, and ye shall fall by the sword. . . . But they presumed to go up. . . .
> Then the Amalekites came down, and the Canaanites which dwelt in that hill [-country], and smote them, and discomfited them, even unto Hormah.

It was the second battle between the Israelites and the Amalekites.

The above remark in the Midrash that southern Palestine was not originally the home of the Amalekites and that they had only recently occupied that area draws our attention. The Amalekites evidently had conquered the south of Palestine only a short time before, because the arrivals from Egypt were not aware of their presence in that region. In the course of their migration the Amalekites apparently divided and turned simultaneously toward Egypt and toward the south of Palestine.

Because of the Amalekites the Israelites were compelled to roam the desert for a whole generation.

At this point, before introducing the entire body of evidence from Hebrew and Egyptian sources to establish the cardinal point of the identity of the Hyksos and the Amalekites, I shall ask a question. If the Hyksos really came from Arabia, might not some evidence of this fact be found in Arabian sources also?

The old Arabian writings, when investigated, provided the desired evidence. I shall therefore proceed to compare the three sources: the Egyptian, the Hebrew, and the Arabian.

The Amalekites were an old Arabian tribe who, from ancient

57

days, dominated Arabia. In the Book of Genesis, in a genealogical table, Amalek is said to have been an offspring of Eliphaz, son of Esau, Isaac's son.[12] But obviously this statement does not refer to the Amalek who was father of the tribe. The Book of Genesis also has another record: as early as before the destruction of Sodom the Amalekites were at war with the kings of the Two-Stream Land, a mighty coalition. The Amalekites who participated in these battles in the days of Abraham could not have been descendants of Amalek, descendant of Esau, himself a descendant of Abraham. The Amalekites were thus of an older clan and no kinsmen of the Twelve Tribes.

The Islamic historians consider Amalek as one of the most ancient of the Arab tribes. Abulfeda, an Arab scholar of the thirteenth century, wrote: "Shem [son of Noah] had several sons, among them Laud, to whom were born Pharis, Djordjan, Tasm, and Amalek,"[13] thus ascribing to these tribes a primeval existence. But there are other Moslem historians who declare this Arabian tribe to have been of Hamite stock, and give its ancestral line correspondingly.[14]

The Amalekites ruled in Mecca and from their central position on the great peninsula dominated other Arabian tribes. All parts of Arabia Felix, Arabia Petraea, and Arabia Deserta alike were within reach of their bows. And then came the upheaval.

The Upheaval in Arabia

There was a flood, an immense wave. People were swept away by a blast. The earth quaked violently. The catastrophe was preceded by plagues.

The tradition is thus handed down by Abu'l Faradj (c. 897–967) in *Kitab-Alaghaniy* (*Book of Songs*):

> The tradition reports that the Amalekites violated the privileges of the sacred territory and that the Almighty God sent against them ants of the smallest variety which forced them to desert Mecca.
>
> Afterwards the Lord sent drought and famine and showed them the clouded[15] sky at the horizon. They marched without rest toward those clouds which they saw near them, but were not

58

able to reach them; they were pursued by the drought which was always at their heels.

The Lord led them to their native land, where He sent against them "toufan"—a deluge.[16]

Saba (Sheba) in the south of Arabia, Mecca, and all the thousand miles of the Tehama coast were shattered. All the tribes on the peninsula suffered similar horrifying experiences.

Masudi[17] (d. about 956) also relates the tradition of this catastrophe and tells of "swift clouds, ants, and other signs of the Lord's rage", when many perished in Mecca. A turbulent torrent overwhelmed the land of Djohainah, and the whole population drowned in a single night.

"The scene of this catastrophe is known by the name of 'Idam' (Fury)."[18] "Omeyah son of Abu-Salt of the tribe of Takif alluded to this event in a verse worded thus: 'In days of yore, the Djorhomites settled in Tehama, and a violent flood carried all of them away.' "

The Amalekites were put to flight by plagues that fell upon them in Arabia, and in their escape they followed swift clouds. Meanwhile Mecca was destroyed in a single night filled with a terrible din. The land became a desert.

> MASUDI: From el-Hadjoun up to Safa all became desert; in Mecca the nights are silent, no voice of pleasant talk. We dwelt there, but in a most tumultuous night in the most terrible of devastations we were destroyed.[19]

In tumult and disorder, fleeing the ominous signs and plagues and driving their herds of animals infuriated by earthquakes and evil portents, the fugitive bands of Amalekites reached the shores of the Red Sea.

Plagues of insects, drought, earthquake in the night "the most terrible devastation", clouds sweeping the ground, a tidal flood carrying away entire tribes—these disturbances and upheavals were experienced in Arabia and Egypt alike.

This succession of phenomena helps us to recognize that they occurred at the time of the Israelites' escape from Egypt, also visited by plagues. They also witnessed the destructive flood at the Sea of Passage, at Pi-ha-Khiroth, shortly before they met the Amalekites. The

Israelites met the Amalekites for the first time a few days after they had crossed the sea.

Not only the Egyptians but also many Amalekites perished at the sea. Other tribes, too—Djorhomites and Katan (Yaktan)—were swept away by the flood and perished in great numbers.[20] The thick clouds covering the desert are repeatedly mentioned in the Scriptures and in the Midrashim. The Midrashim narrate that the Israelites encountered the Amalekites in a thick veil of clouds.

The Arab historians were not conscious of any link between their story of the flood at the shores of the Red Sea and the events of the Exodus, and did not connect them; if they had, they would have been suspected of having handed down a passage of the Bible in arbitrary form; but they were unaware of the significance of their report.

The Arabian Traditions about the Amalekite Pharaohs

Many ancient Arab writers recorded the invasion of Egypt by the Amalekites. There is a great deal of fancy in some of these stories. In several instances they are spoiled by the clumsy attempts of these authors to adapt their Arab traditions to the traditions of the Hebrews, but not to the correct ones. So it happens that Joseph was sold into Egypt when an Amalekite was the pharaoh, or that Moses left Egypt when an Amalekite was the pharaoh. An Arab author admitted that there was no concurrence as to the race of the pharaoh who reigned at the time of Moses, whether he was a Copt, a Syrian, or an Amalekite.[21]

We shall disregard these attempts of some Arabian authors to insert stories culled from the biblical narrative into stories indigenous to the Arabian peninsula, and we shall devote our attention only to the narratives which did not have their source in the Bible or the Haggada. They must have been autochthonous and transmitted from generation to generation on the Arabian peninsula.

There it was told that Syria and Egypt came simultaneously under the domination of the Amalekites, who escaped from Arabia when it was visited by plagues of insects, drought and famine, an earthquake, and a flood at Safa and Tehama. An important historical moment is revealed in the sequence: a natural catastrophe composed of many significant phenomena was the cause of a hurried migration of Amalekites toward Syria and Egypt.

The Arabian traditions that have survived to our time were written down by authors of the ninth to the fourteenth centuries; they refer to these ancient traditions and also to older authors, sometimes naming them.

After the Amalekites invaded Syria and Egypt they established a dynasty of their pharaohs. Al-Samhudi (844–911) wrote:

> The Amalekites reached Syria and Egypt and took possession of these lands, and the tyrants of Syria and the Pharaohs of Egypt were of their origin.[22]

Masudi, who wrote about the plagues that befell Arabia, and the flight of the Amalekites from Mecca, and the flood, recounted also the conquest of Egypt by the Amalekites.

> An Amalekite king, el-Welid, son of Douma, arrived from Syria, invaded Egypt, conquered it, seized the throne and occupied it without opposition, his life long.[23]

We are reminded of the words of Manetho previously quoted: "A people of ignoble origin from the east, whose coming was unforeseen, had the audacity to invade the country, which they mastered by main force without difficulty or even a battle."

In another work of his Masudi[24] gives a more detailed account of the conquest of el-Welid.

> El-Welid, son of Douma, advanced at the head of a numerous army, with the intention to overrun diverse countries and to overthrow their sovereigns.

The end of this passage recalls the sentence in the Haggada: "Amalek ... in his wantonness undertook to destroy the whole world."[25]

Masudi continues:

> When this conqueror came to Syria, he heard rumors about Egypt. He sent there one of his servants named Ouna, with a great host of warriors. El-Welid oppressed the inhabitants, seized their possessions and drew forth all the treasures he could find.

61

Masudi tells of strife among the Amalekites and of the invasion of Egypt by a second wave of this people led by Alkan, surnamed Abou-Kabous.

> The Amalekites entered Egypt, destroyed many monuments and objects of art. . . . The Amalekites invaded Egypt, the frontier of which they had already crossed, and started to ravage the country . . . to smash the objects of art, to ruin the monuments.[26]

These words recall those of Manetho as cited by Josephus in *Against Apion* and quoted above:

> [The Hyksos] savagely burned the cities, razed the temples of the gods to the ground, and treated the whole native population with the utmost cruelty.[27]

The words of Masudi accord with the mention of the destruction of monuments in the inscription of Queen Hatshepsut, a ruler of the Eighteenth Dynasty. In this inscription, referred to on a previous page, it is said:

> There had been Amu in the midst of the Delta and in Hauar [Auaris], and the foreign hordes of their number had destroyed the ancient works; they reigned ignorant of the god Ra.[28]

Tabari (838–923) related stories and legends of Amalekite pharaohs and gave their genealogies. The following sentence is characteristic:

> Then the king of Egypt died and another king, his relative, ascended the throne. He was also of Amalekite race and was named Kabous, son of Mosab, son of Maouya, son of Nemir, son of Salwas, son of Amrou, son of Amalek.[29]

Abulfeda (1273–1331), in his history of pre-Islamic Arabia, wrote:

> There were Egyptian Pharaohs of Amalekite descent.[30]

He also mentioned a most violent tempest that had swept Egypt in remote days.[31] He gave the names of a succession of Amalekite pharaohs and told of the domination of Syria by the Amalekites.[32]

Ibn Abd-Alhakam,[33] cited by Yaqut (1179–1229), and other names could be added to the authors quoted here, but those given suffice to demonstrate that the tradition of Amalekite dynasties of pharaohs was spread among the Arabian scholars.

The historical background of their stories about Amalekite pharaohs in Egypt was regarded with distrust.[34] There were scholars who took an even more radical viewpoint and asserted that the Amalekites had never existed.[35] They based their conclusion on the assumption that the name of the Amalek tribe was never mentioned in Egyptian inscriptions.

On the other hand, an equally extreme viewpoint doubted the very fact of the invasion of Egypt by the Hyksos and interpreted it as a story of legendary origin; the Hyksos were supposed to be just another dynasty of native rulers.[36]

The Amalekites were, presumably, not known to the Egyptian people. The Hyksos (Amu) were equally unknown to other peoples. Therefore the historical existence of the one or the other was sometimes doubted.

Hyksos in Egypt

The Hyksos rule in Egypt endured through all the time that elapsed between the Middle Kingdom and the New Kingdom. The text of the Papyrus Ipuwer was composed at the time of the invasion of the Hyksos and it refers to this invasion. The expulsion of the Hyksos and the period immediately preceding it are also described in some contemporaneous documents. But the period between the invasion and the expulsion is very poor in records; it is a dark age in two senses.

Manetho is a late source on the dominance and expulsion of the Hyksos; about one thousand years separated the historian from his subject. He provides the information that after the Hyksos invaded the country, destroying, burning, raping, and ravaging, they established a dynasty of Hyksos pharaohs; that the first of these kings, named Salitis or Salatis, resided in Memphis and "exacted tribute from Upper and Lower Egypt, and left garrisons in the places most suited for defence. In particular he secured his eastern flank" to protect the realm from the north, as "he foresaw that the Assyrians, as their power increased in the future, would covet and attack his realm".[37]

There, to the east of the Delta, King Salitis discovered a favorably situated place called Auaris, a strategic point from which to control both Egypt and Syria.

> He rebuilt and strongly fortified it with walls, and established a garrison there numbering as many as two hundred and forty thousand armed men to protect his frontier. This place he used to visit every summer, partly to serve out rations and pay to his troops, partly to give them a careful training in manoeuvres, in order to intimidate foreigners.[38]

The fourth king is called Apophis by Manetho, and he is said to have ruled for sixty-one years. The first six king-shepherds are considered the first Hyksos Dynasty of pharaohs. In Manetho-Josephus it is said of them:

> The continually growing ambition of these six, their first rulers, was to extirpate the Egyptian people.[39]

The rule of the Hyksos was cruel. They knew no mercy. Substantiation of this may be found even in graves. The excavator of one of the smaller garrison-fortresses of the Hyksos thus described the contents of a grave: "A heap of bones stacked closely together, most of them were of animals, but among them I found a piece of human jaw and patella."[40] In another grave he found an "apparently separated arm, superfluous loose hand".

When we remember what Manetho said about the extreme cruelty of the invaders, and compare it with the Hebrew narratives about Amalekites mutilating their prisoners by cutting off members of the body,[41] the finding of an odd hand or jaw does not seem an accidental occurrence. The garrison-fortresses were places of torture.

The dominion of the Amu-Hyksos was not confined to Egypt. Scarabs, or official seals, have been found in various countries, with the names of King Apop and King Khian. The name of Khian is engraved on a sphinx discovered in Baghdad and on a jar lid found at Knossos in Crete. An inscription of Apop says that "his father Seth, lord of Auaris, had set all foreign countries under his feet". In Auaris was the sacrarium of Seth, whom the Hyksos worshiped, and who, until the time of the Ramessides, was regarded by the Egyptians as the personification of the dark power (the contestor of Isis and Horus, or equivalent of the Greek Typhon). The finding in distant countries of objects bearing the names of Apop and Khian seems to prove that Apop's words were no vain boast. Some historians have found themselves compelled to believe that the Hyksos, if only for a transient period, commanded a very great empire,[42] and that at least the area of political influence of the Amu-Hyksos was very extensive.

The last Hyksos Dynasty, the Seventeenth Egyptian Dynasty, according to Manetho, was a dynasty of "shepherds and Theban kings", meaning that in Thebes there were princes of Egyptian nationality subordinate to the Hyksos pharaohs. The last of these Hyksos pharaohs was Apop II, also a prominent king.

Malakhei-Roim—King-Shepherds

The Israelites left Egypt a few weeks, or perhaps only days, before the invasion of the Hyksos; they could not avoid meeting these Hyksos coming from Asia, and actually did meet them before they reached Mount Sinai.

Did the Israelites know that Egypt had undergone an "eleventh" and very severe plague—one which endured for centuries—the invasion of the king-shepherds? When they sighted the Amalekites in the whirling and trembling desert they might have been unaware of the new ordeal that these despoilers would bring to Egypt.

But while in Canaan, during the whole time of the Judges, the Israelites, who suffered from the onslaughts of the Amalekites, must have known that Egypt, too, was afflicted with the same plague and to an even greater degree. Is there any reference preserved in the old Jewish sources that would hint at the Hyksos invasion of Egypt immediately after the departure of the Israelites?

It is said in the enumeration of the plagues in Egypt:

> PSALMS 78:49 He [the Lord] cast [sent forth] upon them fierceness of his anger, wrath, and indignation, and trouble, by sending evil angels among them.

What may that mean, evil or bad angels? There is no plague known as the "visit of evil angels". There is no expression like "evil angels" to be found elsewhere in the Scriptures. There is an "angel of death" or "Satan", but no "evil angels". It would appear that the text is corrupted.

"Sending of evil angels" is (presumably) *mishlakhat malakhei-roim*.

"Invasion of king-shepherds" is *mishlakhat malkhei-roim*.

The only difference in spelling is one silent letter, *aleph*, in the first case. It would thus seem that the second reading is the original.

The first reading is not only unusual Hebrew, but it is also contrary to the grammatical structure of the language. If *roim* ("evil", plural) were used as an adjective here, the preceding word could not take a shortened form; *roim* must therefore be a noun. But if *roim* were a noun, it would be in the singular and not in the plural; and finally, the correct plural of "evil" is not *roim* but *raoth*. "Evil angels" in correct Hebrew would be *malakhim roim*; "angels of evils" *malakhei raoth*. Not only the sense but the grammatical form as well speaks for the reading, "invasion of king-shepherds".

When the editor or copyist of the sentence could not find sense in "king-shepherds", he changed the words to "evil angels" without sufficient grammatical change.

Verse 49 of Psalm 78 must therefore be read:

> He [the Lord] cast [sent forth] upon them the fierceness of his anger, wrath, and indignation, and trouble, invasion of king-shepherds.

An old Hebrew legend throws a side light on the same problem:

> Amalek fetched from Egypt the table of descent of the Jews [Israelites] . . . these lists lay in the Egyptian archives. Amalek appeared before the Jewish [Israelite] camp, and calling the people by name, he invited them to leave the camp and come out to him.[43]

This legend implies knowledge on the part of the Israelites of the fact that the Amalekites came to Egypt and became the rulers of the land. In what other way could they have come into possession of the census lists in the Egyptian archives?

In Papyrus Ipuwer it is said:

> PAPYRUS 6:7 Forsooth, public offices are opened and the census-lists are taken away. Serfs become lords of serfs [?].

When the Amalekites vanquished Egypt, they may have looked on themselves as the legatees of the former Egyptian Empire with its colonies; in their wars with the Israelites in succeeding centuries they might have argued that the Israelites had deserted their bondage in Egypt.

Palestine at the Time of the Hyksos Domination

The problem of why, in the Books of Joshua and Judges, which cover more than four hundred years, there is no mention of Egyptian domination over Canaan or any allusion to military expeditions headed by pharaohs has remained unsolved. Yet during this long period of time, according to the conventional chronology, Palestine was dominated by Egypt.

The revision presented here places the time of wandering, of Joshua, and of Judges in the period of the Hyksos-Amalekite rule over Egypt. In harmony with this revised scheme the Amalekites must have been regarded at that time as the mightiest among the nations.

Balaam, the sorcerer, was called upon to curse the Israelites approaching Moab on their way from the desert. He set his face toward the wilderness, but instead of cursing, he blessed Israel with these words:

> NUMBERS 24:7 . . . his seed shall be in many waters, and his king shall be higher than Agag, and his kingdom shall be exalted.

Agag [Agog][44] was the name of the Amalekite king.

Standing on the edge of the mountain, Balaam turned his face in another direction:

> NUMBERS 24:20 And when he looked on Amalek, he took up his parable, and said, Amalek [is] the first of [among] the nations; but his latter end shall be that he perish for ever.

These verses did not seem clear. The Amalekites are supposed to have been an unimportant band of robbers; why were they called "the first among the nations" and what could the blessing "higher than Agag" mean? No satisfactory explanation was presented.

The Amalekites were at that time the first among the nations. The highest degree of power was expressed by comparison with the power of the Amalekite king Agog. He was the ruler over Arabia and Egypt.

The name of the king Agog is the only Amalekite name that the Scriptures have preserved.[45] Besides the king Agog mentioned in the Book of Numbers, there was another Amalekite king Agog, their last king, who reigned some four hundred years later and was a contemporary of Saul.[46]

In the history of Egypt the most frequently mentioned name of the Hyksos kings is Apop. One of the first and most prominent of the Hyksos rulers was Apop; the last king of the Hyksos was also Apop.

The early Hebrew written signs as they are preserved on the stele of Mesha show a striking resemblance between the letters g (gimel) and p (pei). No other two letters are so much alike in shape as these: each is an oblique line connected to a shorter, more oblique line, and is similar to the written number 7; the size of the angle between the two oblique lines constitutes the only difference.

Nevertheless, it seems that not the Hebrew reading but rather the Egyptian must be corrected; I have set forth some reasons in another place.[47] Almost every hieroglyphic consonant stands for more than one sound, and only empirically are all the sounds symbolized by a consonant found.

Agog I appears to be Apop I, and Agog II, Apop II. King Agog reigned at the beginning of the period; according to Manetho, Apop was the fourth king of the Hyksos Dynasty and ruled for sixty-one years. Agog II reigned at the very end of the period, some four hundred years later.

The Amu-Hyksos held Egypt in submission from their fortress Auaris, which they built near the border of the country. Throughout the land they maintained garrisons (Manetho).

In Palestine there was likewise a fort which the Amalekites built for a garrison; it was strategically situated in the heart of the country, in the land of Ephraim.

The Song of Deborah, like the blessing of Balaam, is an old fragment. An obscure verse reads: "Out of Ephraim their root is in Amalek" (Judges 5:14).[48]

"Their" obviously refers to the Canaanites, and to Jabin, king of

Canaan, who reigned in Hazor, and to his captain Sisera, who commanded nine hundred chariots of iron. They oppressed Israel. The Israelites under the guidance of Deborah and Barak succeeded temporarily in breaking the yoke. The verse cited seems to mean that the strength of the Canaanites was based upon the support they received from the Amalekite citadel in the land of Ephraim.

This citadel is also mentioned in another verse of the Book of Judges: "Pirathon in the land of Ephraim, in the mount of the Amalekites" (Judges 12:15).

The Amalekites supported the Canaanites; this explains the reversal in the progress of the Israelite penetration into Canaan and their occasional status as vassals. The Amalekites ruled over vast territories and in their colonial politics allied themselves with kindred nations. This is the ground for the Hebrew tradition that the Amalekites posed as Moabites, Canaanites, and other peoples, and in these disguises carried on war against Israel, or that they supported the Canaanites in their war against the Israelites.[49] The Midianites were close kin of the Amalekites, related since the days when the one people occupied Mecca and the other lived in Medina;[50] together they often invaded the land of Israel just before the harvest.

> JUDGES 6:3–6 And so it was, when Israel had sown, that the Midianites came up, and the Amalekites, and the children of the east. . . . And they encamped against them, and destroyed the increase of the earth, till thou come unto Gaza, and left no sustenance for Israel, neither sheep, nor ox, nor ass.
>
> For they came up with their cattle and their tents, and they came as grasshoppers for multitude; for both they and their camels were without number: and they entered into the land to destroy it.
>
> And Israel was greatly impoverished. . . .

Their cattle and their camels without number are responsible for their Egyptian name of king-shepherds. Marching to devastate a country, they drove their cattle before them. Also in the next chapter of the Book of Judges (7:12) they and their cattle are compared to "grasshoppers" and to the "sand by the sea side".

They employed the same system of exploitation and plunder as in Egypt. They waited until the people of the land had sown; then, shortly before the time for gathering, they appeared in a multitude with their herds to devour the harvest and to take away the oxen used for plowing and every other animal of the household.

> PAPYRUS ERMITAGE 1116b, recto: The Amu approach in their might and their hearts rage against those who are gathering in the harvest, and they take away [their] kine from the ploughing. ... The land is utterly perished, and naught remains.[51]

The time of wandering in the desert is reckoned as forty years; the time of Judges is estimated variously, generally as four hundred years. The dark age in the Near East continued as long as the supremacy of the Amalekites endured. The Israelites seem to have been the only people who incessantly struggled for their independence against the Amalekite and allied tribes, and by their resistance they also safeguarded the maritime cities of Tyre and Sidon.

When under valiant leadership, the Israelites dared to take the offensive. Under Gideon they reached even the cities of Midian. It was a heroic time. Nothing is known of uprisings in Egypt or in other places of the Amalekite Empire during these centuries. But every effort on the part of the Israelites to achieve and retain real independence was doomed to failure as long as the Amalekites ruled northern Africa and Arabia up to the land of the Euphrates, as long as garrisons were stationed at fortified points scattered throughout many countries, and the military wedge they drove toward the coast, between the lands of Africa and Asia, remained unbroken.

It was during this time that the saying was coined (Exodus 17:16): ". . . the Lord will have war with Amalek from generation to generation."

The Length of the Hyksos Period

According to Manetho as cited by Josephus (*Against Apion*, I, 84), the Hyksos period lasted five hundred and eleven years. But in modern books on Egyptian history this period is drastically reduced. The reduction was not based on any consideration of cultural changes or archaeological finds, ancient charts or dates, but mainly on the fact that the end of the Twelfth Dynasty of the Middle Kingdom is ascribed, on the ground of the astronomical computations of the Sothis period, to −1780. It was followed by the Thirteenth Dynasty (the last of the Middle Kingdom) and by the Hyksos period ere the New Kingdom was inaugurated with the advent of the Eighteenth Dynasty in −1580, again according to the calculations based by modern scholars on the Sothis period calendar. If the dates are right, about two hundred years are left for the Thirteenth Dynasty and the Hyksos period, and since some of the kings of the Thirteenth Dynasty had long reigns, the most that can be left for the Hyksos domination in Egypt is one hundred years. This view was offered and defended by Eduard Meyer.

In the view of Flinders Petrie and a few of his followers, this span of time is entirely insufficient for the interval between the Middle and the New Kingdoms. The cultural changes were enormous: as if a curtain fell at the end of the Middle Kingdom and was raised again over the entire different scene of the New Kingdom. Because of the immense changes, Petrie advanced the idea that between the end of the Twelfth Dynasty and the beginning of the Eighteenth were not 200 years but 1660: in other words, an additional Sothis period of 1460 years was interposed by Petrie; the time of the Hyksos domination, instead of being reduced from Josephus' figure, was lengthened.

These two schemes are called the "long" and the "short chronology". They have in common the date −1580 for the beginning of the New Kingdom. Neither the "long chronology" nor the "short chronology" proposed to reduce this date. Both are built on the Sothis period for the computation of Egyptian chronology. At the end of this work we shall examine the validity of this notion that references to the star Sothis or Sirius may provide a basis for a chronological scheme. The great divergence between the schools of historians, as much as 200 or 1660 years for the same period, immediately preceding the New Kingdom, is amazing, especially if we keep in mind that Egyptian chronology served as a basis for the chronology of the entire complex called the ancient East.

Some scholars tried to take the middle road, and, disregarding the involved computations on which the Sothis reckoning is made, suggested a period of four or five hundred years for the Hyksos period. "Were the Sothic date unknown, our evidence would not require more than 400 or at most 500 years between the two—the twelfth and the eighteenth—dynasties."[52]

The conciliatory view did not take root among the scholars; the long chronology after the death of Petrie had very few supporters; and the short chronology, also called the chronology of the "Berlin School", became supreme.

In this book we occupy ourselves with Egyptian history from the moment when the Middle Kingdom came to its end with the Hyksos conquest of Egypt. If the Hyksos period is measured by the time the Amalekites dominated the Near East, or by the sum of years allotted by the Scriptures to the wandering in the desert and to the leadership of the Judges, a time span of over four hundred and forty years should be assigned to the period in question.

The Expulsion of the Hyksos in the Egyptian and Hebrew Records

All the time of subjugation to the Hyksos, Egypt was ruled from Auaris, where a strong garrison was kept by the king-shepherds. Here these kings received tribute from Egypt and gave instructions to their regional governors. The princes of the nomes were dependent vassals and were treated in a disdainful manner, described in the Sallier Papyrus I.[53]

King Apop II (Agog II) sent a messenger from Auaris with a humiliating demand to the Egyptian prince Seknenre.

> SALLIER PAPYRUS I The prince of the southern city [Thebes] remained silent and wept a long time, and he did not know how to return answer to the messenger of king Apophis [Apop].

The Egyptian prince was arrested by the messenger of King Apop II and brought to Auaris. The end of the papyrus is missing.

This roll of papyrus tells the story of the abuse and derision to which the dependent princes of the nomes were subjected.

But it was the darkness before dawn. The last plague, the domination of the shepherds, which had endured since the Exodus, was approaching its end.

The Carnarvon tablet records the participation of the vassal pharaoh Kamose, son of Seknenre, in action against the Hyksos.[54] He was assisted by some foreign troops. An Egyptian monument has also preserved a description of the final act: the story of the expulsion

72

of the Hyksos is engraved on the wall of the tomb of an officer of Ahmose, a vassal pharaoh of one of the nomes and probably a brother of Kamose; the name of the officer was also Ahmose. The story is in the form of a narrative about the sieges and battles in which the officer took part.

In the Ahmose inscription, which is the best available Egyptian source on the war of deliverance, an enigma is inserted concerning the most important circumstance. Obviously, not rebellious Egyptian princes but some warriors coming from abroad were the real deliverers of Egypt. The inscription reads:

> I followed the king on foot when he rode abroad in his chariot. One besieged the city of Avaris. I showed valor on foot before his majesty. . . . One fought on the water in the canal [riverbed] of Avaris. . . . Then there was again fighting in this place; I again fought. . . . One fought in this Egypt, south of this city; then I brought away a living captive. . . . One captured Avaris. . . . One besieged Sharuhen [s'-r'-h'-n] for six years [[55]] [and] his majesty took it. . . .[56]

The indefinite pronoun would not have been used if the Egyptian king had been at the head of the besieging army. Had the Egyptian prince been the main figure in this war for freedom his triumph would not have been attributed to the indefinite "one". The writer would have said: "His Majesty besieged . . ." or "Our troops fought . . ." The Egyptian document says in fact that in the war against the Hyksos a foreign army was active.[57] However, Egyptian inscriptions did not memorialize the deeds of foreign kings, and hence the name of the king who destroyed the Hyksos is missing. The war was fought by a foreign "one", and the history written on that tomb did not ascribe the sieges and the expulsion of the Hyksos to the dead man's own chief, who only aided the foreign liberator.

Samuel, the priest and prophet, said to Saul, whom he had anointed to be king over Israel:

> I SAMUEL 15:2–3 Thus saith the Lord of hosts, I remember that which Amaleck did to Israel, how he laid wait for him in the way when he came up from Egypt. Now go and smite Amalek, and utterly destroy all that they have.

73

Saul gathered "two hundred thousand footmen, and ten thousand men of Judah".

> I SAMUEL 15:5 And Saul came to [the]city of Amalek, and laid wait in [the bed of] the stream (*nakhal*).[58]

These words, "city of Amalek", have always been a stumbling block for commentators and Bible students. The Amalekites are supposed to have been a small tribe of unsettled Bedouins; therefore, what does the "city of Amalek" mean?[59]

It is said that since days of old the Amalekites dwelt in the south. One indication as to the location of the place is its topography: the city was besieged from the bed of a stream (*nakhal*). The city must have been situated near a river. In southern Palestine, the Sinai desert, northern Arabia, and up to the boundaries of Egypt there are no rivers save the "river of Egypt"—the wadi of el-Arish, the only river to which the Scriptures persistently apply the name *nakhal*. In winter it is torrential, in summer its bed is dry.

A geographical indication may also be tentatively drawn from the reference in the verse that follows: ". . . until thou comest to Shur, that is over against Egypt." This was the southernmost point of the victorious campaign of Saul, as a result of the capture of the city of the Amalekites.

> I SAMUEL 15:7–8 And Saul smote the Amalekites from Havilah until thou comest to Shur, that is over against Egypt. And he took Agag the king of the Amalekites alive. . . .

The identity of the foreign liberator of Egypt is thus revealed by the Scriptures. The "one" was King Saul. Apop II was Agog II. The Amalekite city was Auaris. In Egyptian and Hebrew sources alike the strategic use of the bed of the stream in the siege of this city is stressed. Rich spoils of the shepherds' city are mentioned in both sources; it consisted of oxen, sheep, and lambs (I Samuel 15:9). In both sources it is said that during this campaign the "one" (in the inscription of the officer Ahmose) or Saul (in the Book of Samuel) fought the Amu-Amalekites and destroyed them "to the south of Auaris" or "until thou comest to Shur, that is over against Egypt".

Contraposing the Hebrew and Egyptian sources will help to identify the location of Auaris. The material for exact identification is brought together later in this book.

74

The reference "from Havilah" was also a difficult problem for exegesis.[60] How could a skirmish with insignificant Amalekites or the siege of some settlement of nomads result in a victory sweeping across Havila in the land of the Euphrates and up to the border of Egypt? It has been supposed that the text is corrupt and that instead of Havila another name must be read,[61] or that another Havila besides the one in the land of the Euphrates was somewhere near Egypt.[62]

If the true role of the Amalekites during the long period of Judges is recognized, no difficulty will be encountered in accepting the text as correct. The capture of the Amalekite stronghold with its king was the signal for the collapse of the Amu-Amalekite Empire, with the immediate result that all of Syria to the land of the Euphrates—and Egypt, too—regained liberty.

The Hyksos Retreat to Idumaea

A detail of the siege of Auaris is preserved in another document written at a much later date. Manetho narrates in his history of Egypt, as quoted by Josephus, that the Hyksos, after having been beleaguered for a long time in Auaris, were allowed under an agreement to leave the place.

> They [the besieged in Auaris] were all to evacuate Egypt and go whither they would unmolested. Upon these terms no fewer than 240,000, entire households with their possessions, left Egypt and traversed the desert to Syria.
>
> Then, terrified by the might of the Assyrians, who at that time were masters of Asia, they built a city in the country now called Judaea, capable of accommodating their vast company, and gave it the name of Jerusalem.[63]

This confused statement of Manetho is in obvious conflict with the inscription of the officer Ahmose, who recorded the capture of the city Auaris by siege, but did not mention any agreement with its defenders.

The biblical record may reconcile the conflicting statements of the contemporary warrior and of the later historian as far as they concern the immediate fate of the besieged people.

75

Before storming the Amalekite city Saul made an agreement with the tribe of the Kenites, affiliated with the Amalekes, regarding their departure from the city under siege.

> I SAMUEL 15:6 And Saul said unto the Kenites, Go, depart, get you down from among the Amalekites, lest I destroy you with them ... So the Kenites departed from among the Amalekites.

According to Ahmose, after the capture of Auaris the Hyksos-Amu who saved themselves from death escaped to Sharuhen in southern Palestine. In Manetho's story it is stated that the Hyksos, retreating from Auaris, escaped into Judea to a place which they built and named Jerusalem. There is not the shadow of a doubt that the contemporaneous inscription on the grave of Ahmose contains the correct rendition of the name of the place to which the Hyksos retired, and that the later rendering of Manetho is wrong. Either Manetho's source or his text was corrupted, and the lesser-known Sharuhen was replaced by the better-known Jerusalem (Jerushalaim).

This mistake, accidental or deliberate, has played a harsh role in the fate of the Jewish people beginning with the Ptolemaic age; it has left deep marks on the behavior and spiritual evolution of other peoples; only seldom has a mistake of a writing hand had so many tragic consequences as this corrupt text, which is exposed here by comparing the two Egyptian sources on the escape of the Hyksos—to Sharuhen in one source, to Jerusalem in the other. I shall add a few more words on this subject at the end of this chapter.

The defeat of the Hyksos-Amu was brought about in two successive sieges. After the capture of Auaris, the stronghold and residence of Apop, the Hyksos-Amu retreated to Sharuhen[64] in southern Palestine, and here the last siege took place. Ahmose's version of the retreat of the Amu into southern Palestine corresponds with the scriptural narrative. After the capture of the Amalekite city and Saul's overwhelming victory, the Amalekites were not wholly destroyed. Those who escaped with their lives fell back to the hill country of southern Palestine.

From there they undertook a raid on the neighboring cities. This was still during the reign of Saul, and David was one of his officers.

I SAMUEL 30:1–3 ... the Amalekites had invaded the south, and Ziklag, and smitten Ziklag, and burned it with fire; ... and went on their way.

So David and his men came to the city, and behold, it was burned with fire; and their wives, and their sons, and their daughters, were taken captives.

The practice of burning the cities, carrying off the women and children, and withdrawing was the same as the Hyksos had employed in Egypt when they invaded it, four or five hundred years before: "They savagely burned the cities ... carrying off the wives and children into slavery" (Manetho, quoted by Josephus[65]).

David with four hundred men pursued the Amalekite band that had carried off his wives. In the desert they found an unconscious man who "had eaten no bread nor drank any water three days and three nights".

I SAMUEL 30:11–13 And they found an Egyptian in the field, and brought him to David ... and when he had eaten, his spirit came again to him. ...

And David said unto him, To whom belongest thou? ... And he said, I am a young man of Egypt, servant to an Amalekite; and my master left me, because three days agone I fell sick.

David followed the Egyptian slave and found the Amalekite band and rescued the captive women and children.

This episode is most instructive. It shows that the Amalekites invaded the south of Palestine after they had lost their stronghold on the border of Egypt. It reveals also a very striking detail: the Egyptian young man said that he was a slave to an Amalekite master.

Let us return the two histories—that of the Hebrews and that of the Egyptians—to their conventional places in chronology. What does it mean that an Egyptian, a son of the ruling and proud nation, is a servant of an Amalekite, a poor nomad? This man, in identifying himself, spoke of his being a "servant" and the Amalekite being the "master" as of something that was the order of the day.

But they were already the last in their respective roles, the Egyptian slave and his Amalekite master. The Amalekites were in retreat and flight; the few raids, such as that from Sharuhen to Ziklag, both towns in southern Palestine, were the last offensives. Their warriors were dispersed; some of them came into the Philistine region on the coast. Ziklag was on the outskirts of the land of the Philistines, and Sharuhen was a point between Philistia and Seir, the Amalekites' old homeland.

Shortly after the great natural catastrophe the Philistines arrived from the island of Caphtor and occupied the coast of Canaan.[66] They intermarried with the Amalekites, sought their favor, accepted their political leadership, provided them with metalwork and pottery, and, during the centuries that followed, lost more and more of their own spiritual heritage and became a hybrid nation.

This merging of the Philistines and the Amalekites provided, I think, the basis for the Egyptian tradition (Manetho) that the Hyksos Dynasties in the later period of their rule in Egypt were of "Phoenician origin",[67] and also for the assertion that the Philistines were of Amalekite blood.[68]

Saul, after his great victory over the Amalekites, engaged himself in a war with the Philistines. He went into this war with a heavy heart. Samuel, the prophet, had spoken stern words to him: his kingdom would be taken from him for the kindness he had shown by sparing the life of Agog, king of the Amalekites, the eternal enemies of the Jewish people. Samuel killed Agog, never met Saul again, and died. Saul, with the help of necromancy, tried to get in touch with the deceased Samuel. The day after his visit to the witch at Endor, Philistine archers struck him down in the middle of battle. He and his three sons fell on the field. It was an Amalekite, out of the host of the Philistines, who, according to one scriptural version, killed the wounded Saul at his request and brought the evil tidings to David (II Samuel 1:8).

Historical credit for freeing the Near East from the yoke of the Hyksos belongs to Saul, but his great deed was not esteemed, not even recognized. The capture of Auaris and the destruction of the Amalekite host changed the course of history. Once more Egypt rose to power and splendor after being freed from hundreds of years of abject slavery by a descendant of the Hebrews who had been slaves there.

Posterity did not learn of the deeds of Saul; even his contemporaries did not reward him with gratitude. Cursed by Samuel for his soft heart when he spared the life of Agog, driven to depression, and having a premonition of his fate, he went to his last battle. His head and the head of his son Jonathan were cut off by the Philistines and carried through the villages; the headless bodies were hung on the walls of Beth-Shan, in the valley of the Jordan.

This was the sad end of the man anointed to be the first king of Judah and Israel. In a little-known page of some Haggada he is said to have been more pious than David, gentle and generous, the real "elect of God".

Only after overpowering the Philistines could the Israelites consider themselves a free people. The double task fell to David: to destroy the Amalekites in their last strongholds in southern Palestine, and to drive the Philistines from the hilly land.

The officer Ahmose, after the fall of Auaris, followed his prince to Sharuhen in southern Palestine to participate in the siege of this stronghold. The Amu-Hyksos defended it for three long years. This last stronghold of the Amu had to be taken by storm.

The Scriptures have preserved the story of the Amalekites "invading the south" after their disaster on the border of Egypt. Other Hebrew sources have retained the story of the siege of the "Amalekite capital" in the south of Palestine.[69]

"Among all the heroic achievements of Joab, the most remarkable is the taking of the Amalekite capital."[70] For a long time the chosen troops, twelve thousand strong, besieged the stronghold without result. A legend about the adventures of Joab, a captain of David, who penetrated alone into this "very large" city, became a favorite motif for storytellers.

The Israelites stormed the Amalekite city, destroyed the heathen temples, and killed the inhabitants. King David was not present at this prolonged siege; King Ahmose was probably with Joab's army as an ally. The officer Ahmose wrote: "One besieged Sharuhen for three years and his majesty took it." The Egyptian king got his portion of the spoils and the officer Ahmose his share, and he described it.

The Egyptian king returned to Egypt and undertook a campaign against Ethiopia; Joab turned his army to the east and after a time was able to put the crown of the Ammonite king at the feet of David.[71]

On the ruins of the great Amalekite Empire two kingdoms rose simultaneously to freedom and power: Judah and Egypt. The inheritance was divided between them.

Judah absorbed the Asiatic provinces of the Amalekites from the Euphrates in the north to the border of Egypt in the south. Expansion was also eastward—David and Joab led the army against Moab, Amon, Edom, and Aram. Aram meant Syria, and the area stretching toward Mesopotamia; Edom's land was along the entire shore of the Red Sea, the greater part of Arabia.[72]

> II SAMUEL 8:14 And he [David] put garrisons in Edom; throughout all Edom put he garrisons. . . .

Joab remained six months in Edom (I Kings 11:16) and smote "every male in Edom". Hadad, a child of royal blood, was among those who escaped from Midian and came to Paran; and "out of Paran they came to Egypt, unto Pharaoh, king of Egypt".

> I KINGS 11:19 And Hadad found great favour in the sight of Pharaoh, so that he gave him to wife the sister of his own wife, the sister of Tah-pe-nes the queen.

This was in the days of David. The pharaoh must have been Ahmose.[73] Among his queens must have been one by the name of Tahpenes. We open the register of the Egyptian queens to see whether Pharaoh Ahmose had a queen by this name. Her name is actually preserved and read Tanethap, Tenthape, or, possibly, Tahpenes.[74]

The Location of Auaris

Where was Auaris, the stronghold of the Amu-Hyksos, the city which served to keep Egypt in bondage? It was large enough to contain within its walls tens and even hundreds of thousands of warriors, besides the population of women and children and slaves, and to provide for large herds and flocks.

The description Manetho gives in his history points to a site on the eastern frontier of Egypt.

[Salitis, king of the Hyksos] left garrisons in the places most suited for defence. In particular he secured his eastern flank, as he foresaw that the Assyrians, as their power increased in future, would covet and attack his realm. Having discovered in the Sethroite nome a city very favourably situated on the east of the Bubastis arm of the river, called after some ancient theological tradition Auaris, he rebuilt and strongly fortified it with walls, and established a garrison there numbering as many as two hundred and forty thousand armed men to protect his frontier.[75]

According to Manetho, Auaris had a great stone wall "in order to secure all their possession and spoil".

Several scholars have made various conjectures: it was located at Pelusium, at Tanis, and at Tell-el-Yehudiyeh. In the last-named place Hyksos tombs were unearthed,[76] but archaeological study did not confirm the opinion of the excavator that it was Auaris. It was a small stronghold and not the huge fortress of the Hyksos rulers, pharaohs from the Fourteenth up to the Seventeenth Dynasty.

Auaris is futilely sought on the eastern branch of the Delta.[77] However, the precise translation of Manetho-Josephus is: "to the east from the Bubastis arm of the river."

Auaris can be located in the following way:

Saul conquered "the Amalekite city", the residence of the king Agog (I Samuel 15). The capture of this city terminated the domination of Amalekites over "the land from Havilah until thou comest to Shur, that is over against Egypt".

By comparing the last sentence with one in I Samuel (27:8)—". . . and the Amalekites . . . of old the inhabitants of the land, as thou goest to Shur, even unto the land of Egypt"—we find a clue to the location of the Amalekite city on the border of Egypt, but not in Egypt proper.

This also corresponds with the statement in the inscription of the officer Ahmose: "I followed the king on foot when he rode abroad in his chariot. One besieged the city of Auaris."

The meaning of the name "Auaris" is "the town of the desert strip".[78]

Auaris on the northeastern frontier of Egypt was designed by its builder, the Hyksos king Salitis, to secure the eastern flank against any assault from the north.[79] The fortress dominated Egypt and Syria. It was situated on a river—this is told in the Egyptian record of Ahmose and in the story of the siege by Saul. "One fought on the water in the river" (Ahmose). "And Saul came to [the] city of Amalek, and laid wait in the stream" (I Samuel 15:5).

The only river in the whole area is the seasonal stream of el-Arish. Because of the mention of the river battle in the inscription of Ahmose, Auaris was sought on the eastern branch of the Nile, though Manetho-Josephus wrote that it was to the east from the eastern branch of the Delta.

Moreover, the name of the river "nakhal" on which the Amalekite city was built is repeatedly applied in the Scriptures to the stream of el-Arish, which is at the border of Egypt (Nakhal Mizraim).[80]

From the later history of Egypt we also may derive clues for the location of Auaris.

Haremhab, the king who ruled before the Nineteenth Dynasty in Egypt, used to cut off the noses of offenders against the law and banish them to the region of Tharu. The station of Tharu was the northeastern outpost of Egypt, as may be seen from the description of various campaigns into Syria undertaken by kings of the Nineteenth Dynasty. It must have been near Auaris, if it was not another name for the same place.[81]

The noseless offenders were sent to a place for the "unclean"; one deprived of his nose was in appearance similar to a leper, unclean for religious service, and kept apart from society. The unclean were sent into exile to the remotest part of the country. Manetho, writing of the rebellion of these unclean at a later date, said:

> The king . . . assigned them for habitation and protection the abandoned city of the shepherds, called Auaris.[82]

The place where the noseless exiles lived was called by Greek and Roman authors Rhinocolura (cut-off nose) or Rhinocorura. It was identified as el-Arish: Septuagint translated "Nakhal Mizraim" (the stream of el-Arish) by this name, Rhinocorura.[83] Hence Auaris of the ancients is el-Arish of today.

When the archaeologists excavate on the banks of el-Arish they will find the remnants of Auaris, one of the largest fortresses of antiquity.[84]

As a supplementary bit of evidence I cite the following passage from Masudi about a sovereign of the First Dynasty of Amalekite pharaohs.

In the neighborhood of el-Arish he constructed a fortress.[85]

Hyksos and Amalekite Parallels

Is the identity of the Amalekites and Hyksos established, or were they perhaps two different peoples? The historical evidence of the preceding pages will be juxtaposed.

A people called the Amu or Hyksos invaded Egypt after a great natural catastrophe,[86] when the river "turned into blood" and the earth shook.[87] They overran Egypt without encountering any resistance.[88]

The invaders were utterly cruel; they mutilated the wounded and cut off the limbs of their captives;[89] they burned cities, savagely destroyed monuments and objects of art, and razed the temples to the ground;[90] they held the religious feelings of the Egyptians in contempt.[91]

They enslaved the Egyptians and imposed tribute on them.[92] These invaders came from Asia[93] and were called Arabians,[94] but had some Hamitic features too.[95] They were herdsmen,[96] and they were skilled with the bow:[97] their kings ruled as pharaohs of Egypt.[98] They also ruled over Syria and Canaan, the islands of the Mediterranean, and other countries, and had no peer for a long period.[99]

The Amu built a great city fortress east from the Delta of the Nile.[100] They impoverished the native population of Egypt, usually invading the fields with their cattle before harvest time.[101]

Among their kings were at least two rulers by the name (tentatively read) of Apop; both were outstanding, one at the beginning, the other at the very close, of the period.[102]

The domination of this people extended over many countries of the Near and Middle East. Their dynasties ruled some five hundred years,[103] and their rule ended when their fortress-residence on the river was put under siege by some foreign host.[104] A part of the beleaguered population of the fortress was allowed to depart;[105] the bed of the river was the main scene of the siege and of the final storming of the fortress.[106]

With the crushing of the Amu Empire after this siege, Egypt became free, and the expelled invaders moved to southern Canaan to the stronghold of Sharuhen, where they maintained their hold a few years longer.[107] This fortress in the south of Canaan was beleaguered. The siege was protracted. Finally the city was stormed, its defenders killed, and the few survivors were dispersed and lost their importance.[108]

They left a deep feeling of hatred in the people of Egypt.[109]

The other people were called Amalekites. They left Arabia after a series of plagues[110] and immediately after a violent earthquake.[111] Many of them perished during the migration in a sudden flood that swept the land of Arabia.[112] They sighted the Israelites coming out of Egypt, which was laid in ruins by a great catastrophe.[113] In this catastrophe the water in the river turned red as blood, the earth shook, the sea rose in a sudden tidal wave.[114]

The invaders from Arabia occupied the south of Palestine and simultaneously moved toward Egypt.[115] They conquered Egypt without meeting resistance.[116]

The Amalekite conquerors came from Arabia, but apparently they had Hamitic blood in their veins.[117] They were a nation of herdsmen and roamed with their large herds from field to field.[118]

They mutilated the wounded and the prisoners, cut off limbs, and were unspeakably cruel in many other ways.[119] They stole children and carried off women;[120] they burned cities;[121] they destroyed monuments and objects of art that had survived the catastrophe, and despoiled Egypt of her wealth.[122] They were contemptuous of the religious feelings of the Egyptians.[123]

The Amalekites built a city-fortress on the northeastern border of Egypt.[124] Their chieftains were pharaohs and ruled from their fortress.[125]

They held sway over western Asia and northern Africa and had no peer in their time.[126] They kept the Egyptian population in bondage, and their tribesmen used the Egyptians as slaves.[127] They also built smaller strongholds in Syria-Palestine,[128] and by periodically invading the country with their herds before harvest time, they impoverished the people of Israel.[129] Their domination over many countries of the Near and Middle East endured, according to various reckonings, for almost five hundred years.[130]

Among the kings of the Amalekites there were at least two by the name of Agog, both of them outstanding: one a few decades after the Exodus of the Israelites from Egypt, the other at the very end of the period of Amalekite domination.[131]

This people intermingled with the Philistines.[132]

Their supremacy came to an end when their fortress-residence on the border of Egypt was besieged by Saul, king of Israel.[133] The bed of the river (*nakhal*) was the main scene of the siege.[134] A large part of the besieged garrison was allowed to depart.[135] After this siege and the fall of the fortress the Amalekite Empire from Havila in the land of the Euphrates to "over against Egypt" collapsed.[136] Their remnants moved into mountainous southern Palestine,[137] where they gathered their strength in a stronghold-city. But that stronghold, too, was surrounded and, after a protracted siege, was captured by storm.[138] After that they lost their significance.[139]

They left in the people of Israel an intense feeling of hatred.[140]

On the basis of the foregoing, the conclusion is inescapable that the Amu of the Egyptian sources and the Amalekites of the Hebrew and Arab sources were not two different peoples, but one and the same nation. Even the name is the same: Amu, also Omaya, a frequent name among the Amalekites, was a synonym for Amalekite. Dshauhari (Djauhari), an Arabian lexicographer of the tenth century of the present era, wrote: "It is handed down that *this name* [Amu, or Omaya] *was a designation* for an Amalekite man."[141]

The Amu, or the Hyksos, were the Amalekites.

This identity, established on a large number of correlations and parallels, is the answer to the two-thousand-two-hundred-year-old riddle: Who were the Hyksos? As early as the days of Josephus Flavius, in the first century, it was already a debated question of long standing. The arguments of the present chapter for the identity of the Amu-Hyksos and the Amalekites have been stated and repeated here point by point because of the momentous consequences this identification carries. The succeeding parts of this book will show how important the consequences are.

The Israelites could never forget the time of their suffering in Egypt, but they did not bear any hatred against the Egyptians[142] or other peoples of antiquity; the Amalekites alone became the symbol of evil and the object of their hate.

> DEUTERONOMY 25:17–19 Remember what Amalek did unto thee by the way, when ye were come forth out of Egypt;
>
> How he met thee by the way, and smote the hindmost of thee, even all that were feeble behind thee, when thou wast faint and weary; and he feared not God.
>
> Therefore it shall be, when the Lord thy God hath given thee rest from all thine enemies round about . . . that thou shalt blot out the remembrance of Amalek from under heaven; thou shalt not forget it.

The utter wickedness of this people was iterated and reiterated throughout the old literature:[143] how they came to "suck the blood" of the weary in the desert; how cowardly they acted in attacking from ambush; how vile and ignoble and merciless they were.

They mutilated the wounded and blasphemed by throwing the mutilated parts of the bodies toward the heavens and jeered at the Lord.[144]

A legend presents the feeling of the Israelite nation in these symbolic words:

> So long as the seed of Amalek exists, the face of God is, as it were, covered, and will only then come to view when the seed of Amalek shall have been exterminated.[145]

There was also a tradition that "God bade Moses impress upon the Jews to repulse no heathen should he desire conversion, but never accept an Amalekite as a proselyte". For his sins Amalek shall "the first descend to hell". "God Himself took up the war against Amalek."

The Hyksos engendered the same hatred in the Egyptians; their extreme cruelty and their wantonness were cut deep into the memory of the people. They befouled and burned rolls of papyri and objects of art; in their camps they tortured their captives. Heads were cracked open, teeth smashed, eyes gouged out, and limbs hacked off. They professed faith only in their superior strength and employed it in their camps on their defenseless victims.

Even the Arabian authors exposed the evil and the recklessness of the Amalekites in dealing with the holy and the profane in Mecca and in Egypt. They, too, proclaimed that the Lord had sent them away from Mecca for their iniquity.

It was the lot of Saul to carry on the war of liberation of Israel and Egypt. That the Israelites were not remembered with praise for what they had done for Egypt and were referred to as "one" and "they" in Egyptian history was the least of the injustices; their real reward at the hands of the Egyptian historians was to be identified with the ravagers whom the Israelites had driven out of Auaris.

Ahmose wrote that when Auaris was taken, the Hyksos retreated to Sharuhen, in southern Palestine. But Manetho, many centuries later, wrote that the Hyksos retreated into Palestine and built Jerusalem; also that at a later date, when a leper colony in Auaris revolted, these rebels summoned the Solymites (the people of Jerusalem) and together conquered Egypt; that these Solymites were extremely cruel to the population, and that one of the lepers, Osarsiph, had changed his name to Moses.

This confused story reflects the Assyrian conquest of Egypt, when Sennacherib and Esarhaddon invaded Palestine and Egypt "with a great host of Assyrians and Arabs". The people of Jerusalem never conquered Egypt.

The first wave of anti-Semitism in the east was spread much later in the Persian Empire by the vizier, Haman, "the Agagite, the Jews' enemy".[146] Haman, who was of the seed of Agog, the Amalekite,[147] conspired to destroy the Jewish population in Persia and Media.

We may imagine that the traditions of Haman's house were likely to inspire this hatred. These traditions told how Haman's royal forefather Agog had been dispossessed by a Jewish king and killed by a Jewish prophet.

In the Greek world no signs of racial antipathy toward Jews can be traced until the stories of Manetho began to circulate. The Jews were occasionally looked upon as a mysterious people, but no expression of animosity or contempt is preserved in the writings of the old authors. The earliest references are by Theophrastus, Clearchus of Soli, and Megasthenes,[148] philosophers who flourished at the end of the fourth and the beginning of the third century before this era.

Theophrastus wrote:

> They are a race of philosophers; they do not cease to occupy themselves with the divinity.

And Clearchus of Soli wrote:

> The Jews descended from the philosophers of India. The philosophers are called in India Calanians and in Syria Jews. . . . The name of their capital is very difficult to pronounce: it is called Jerusalem.

Clearchus told the story of a Jew from whom Aristotle learned some of his wisdom during his journey in Asia. Clearchus quoted the words of Aristotle on "the great and admirable soberness of this Jew, and his self-restraint".

Also, of Pythagoras and Plato it is recorded that they were in close contact with wise Jews and were anxious to learn from them.[149]

Megasthenes,[150] who lived in India between 302 and 291 before the present era, wrote in his book *Indica:*

> All the opinions expressed by the ancients about nature are found with the philosophers foreign to Greece, with the Brahmans of India, and in Syria with those who are called Jews.

These references to Jews in the Greek language are the only ones in the extant Greek literature from before the time Manetho's history appeared.

The hatred, never extinguished in the memory of posterity, for the inhuman shepherd-conquerors was revived: the Jews were identified with the descendants of the Hyksos. Inaugurated by Manetho, an extensive Jew-baiting literature followed, and the stories of Manetho were told and retold and adorned by many writers.[151] Among them was Apion, against whom the apology (*Against Apion*) of Josephus Flavius was directed. Josephus did not try to cast doubt on the identification of the Jews with the Hyksos; on the contrary, he approved it and even defended it in a most categorical manner; his only reason for doing so was his wish to establish the antiquity of the Jewish people with the help of Manetho's stories.

Josephus played a tragic role in the days of the war in Galilee and Judea and the destruction of Jerusalem by Titus in the first century of the present era. Starting as a soldier at the head of the Galilean army, he ended as a renegade. His apology for the Jewish case was considered a masterpiece and was often translated and quoted by defenders of the Jewish people; but the defense of his pen was worth as much as the defense of his sword.

Hatred of the Hyksos, burning in the generations of the ancient Orient, found a target in the Jews.

An equally glowing hatred of the Amalekites persisted in the memory of the Jews. A Jewish mother even today frightens her child with an "Amalekite".

Hatred can last a long time even when its object is not alive. How much stronger is that hatred if the hated ones did not dissolve their national existence a thousand years before on the Arabian peninsula, but are still supposed to exist? The Egyptian author saw in the Jew the wretched seed of cruel tyrants. Subsequently Greek and Roman authors established in their writings the object for the never dying necessity to hate. Insinuations were heaped one on the other, and monstrous stories were invented about the head of an ass which the Jews kept in their temple and worshiped, and about the human blood they sucked.

The curse on the Amalekites became a curse on the Israelites. "Thou shalt blot out the remembrance of Amalek from under the heaven." It was blotted out. No one knew any longer that the Amalekites were the Hyksos.

The Israelites endured much suffering from this distortion of history. They bore their pain for being identified with the Hyksos. The persecution started with the misstatements of Manetho, the Egyptian, whose nation was freed from the Hyksos by the Jews. In later years anti-Semitism has been fed from many other sources.

World History in the Balance

The proofs of the identity of the Hyksos and the Amalekites have been summarized to make the case as strong as possible. It is not merely the solution of the enigma of the identity of the Hyksos that is at stake. The entire structure of ancient history hangs in the balance. If the catastrophes of the Papyrus Ipuwer and of the Book of Exodus are identical; if, further, the Hyksos and the Amalekites are one, then world history, as it really occurred, is entirely different from what we have been taught.

Thus, establishing the period in which the Exodus took place becomes of paramount importance: Israel did not leave Egypt during the New Kingdom, as all scholars maintain, but at the close of the Middle Kingdom. The entire period of the Hyksos lies in between; the expulsion of the Hyksos was neither identical with the Exodus nor did it take place before the Exodus. The Hyksos were expelled by Saul; their later destruction was the work of Joab, soldier of David. David lived in the tenth century, and Saul was his predecessor on the throne. The expulsion of the Hyksos is put at —1580, which leaves almost six centuries unaccounted for.

Whose history is to be moved by these centuries? Is it possible to place David in the sixteenth century before this era? No student of ancient history will see the slightest possibility of altering the history of the Kings of Jerusalem by a single century, much less by six, without disrupting all established data and concepts. The biblical annals record the succession of the kings of Judah and of Israel, king after king, and give the years of their reigns. If there are, here and there, some discrepancies or difficulties in the double account of the kings of Judah and of Israel, they are of entirely different dimensions, and may amount at most to one or two decades, but not hundreds of years.

Hebrew history is closely related to Assyrian history, and with the help of common data the chronological tables have been made so accurate that while there may be room to question whether Sennacherib in his third campaign invaded Palestine in —702 or in —700, there can be no question whether this Assyrian king came to the Jerusalem of Hezekiah in —1280 or thereabouts. While there may be a difference of opinion whether to lengthen or shorten the reign of one or another of the kings by comparing the accounts of the Book of Kings and the Book of Chronicles, the dates of events in which Assyria or Babylonia participated are, in a number of instances, established with precision, to within a year.

The period of the kings of Jerusalem ended with the exile to Babylon in the days of Nebuchadnezzar, who destroyed Jerusalem in —587 or —586. Cyrus, the Persian, in the second half of the same century, conquered the Chaldean-Babylonian Empire. The Persian rule, king after king, the years of each of them known from many contemporary Greek authors, lasted until Alexander the Great. Where are the six hundred years to be inserted?

Is it conceivable that about six centuries have disappeared from Jewish history, and that because of this disappearance the history contracted so greatly? Where is the historical place of this chasm?

There is not a trace of a historical chasm. Even with the greatest effort of the imagination the succession of centuries cannot be torn apart to make room for additional ones.

On the other hand, how can a history be shortened? Egyptian history is firmly settled too. Dynasty after dynasty ruled in Egpyt; from the beginning of the New Kingdom in about—1580 until the time of the Persian rule in —525, when Cambyses overran Egypt, and down to the Greek conquest in —332 by Alexander the Great the time is filled with successions of dynasties and kings. Not only is the Egyptian past fixed; Egyptian chronology is the rule and the standard for the entire world history.

The ages of the Minoan and Mycenaean cultures of Crete and the Greek mainland are placed in time in accordance with the chronological scale of Egypt. The Assyrian, Babylonian, and Hittite histories, too, are distributed and divided on the world timetable according to their contacts with Egyptian history. Some events of the Assyrian and Babylonian past concerned the Jewish people, and the history of the Double-Stream countries is synchronized with Jewish history; on the other hand, some events of the Assyrian and Babylonian past involved Egypt, and the history of the countries of the Double Stream is synchronized with Egyptian history, which is about six hundred years behind when compared with the history of Judah and Israel. By what prodigious, or rather illogical, procedure could this have come about?

If the fault lies in Egyptian history, the only possibility is that events of that history are described twice, and six hundred years are repeated. It would follow, then, that the events of many other peoples are described in wrong succession. But this seems to be a preposterous statement, which insults the sound judgment of many generations of the entire scholarly world, whoever learned, investigated, wrote, and taught history.

Both these alternatives appear to be chimerical: that six hundred years disappeared from the history of the Jewish people, or that six hundred years were doubled in the history of Egypt and in the history of many other peoples as well.

But that in Jerusalem it was the tenth century while in Thebes it was the sixteenth is an absolute impossibility. We shall do better to admit that the mistake lies, not with history, but with the historians, and that by juxtaposing the two histories, century by century, either six hundred missing years will be found in Palestine or six hundred "ghost" years will be discovered in Egypt.

I shall set down the events of the time following the expulsion of the Hyksos-Amalekites, reign by reign and age by age, in Egypt and in Palestine, and we shall see whether they, too, coincide and for how long. And if the tenth and the ninth centuries in Palestine are found to be coeval with the sixteenth and fifteenth centuries in Egypt we shall have additional proof that the identity of the Hyksos and the Amalekites is not at all an arbitrary assumption. By going through the ages we shall soon be able to establish where the fault lies. Even before we know which of the schemes is wrong we may conclude that the histories of peoples which are in harmony with the two schemes are in a most chaotic state.

NOTES

1. *Manetho* (trans. Waddell).
2. Josephus, *Against Apion* (trans. St. Thackeray), 1, 74–75.
3. Meyer, *Geschichte des Altertums*, Vol. II, Pt. 1 (2nd ed.), p. 42.
4. A hypothesis put forward by I. Rosellini, *I monumenti Storici* (Pisa, 1832), p. 176.
5. R. Weill, "Les Hyksos et la Restauration nationale", *Journal asiatique*, 1910–13, and his *La Fin du Moyen Empire égyptien* (Paris, 1918), pp. 1–262. The same author more recently published on this theme: "Remise en position chronologique et conditions historiques de la XIIᵉ Dynastie," *Journal asiatique* CCXXXIV (1943–45), 131–49; and "Le Synchronisme égypto-babylonien", *Chronique d'Egypte*, XXI (1946), 34–43.
6. Josephus, *Against Apion*, I, 82.
7. Ibid. At present the preferred etymology sees in the name Hyk-sos the Egyptian equivalent for "the rulers of foreign countries".

8. *Against Apion*, I, 76.

9. Midrash Aba Gorion, III (Vilna, 1886), 27. See Ginzberg, *Legends*, VI, 23.

10. Ibid., III, 62.

11. Ibid., 272.

12. Genesis 36:12.

13. Abulfeda, *Historia anteislamica*, ed. H. O. Fleischer (Leipzig, 1831), p. 17.

14. See article, "Amalik", by M. Seligsohn in *The Encyclopaedia of Islam* (Leiden and London, 1908–38).

15. In the Arab text the word used is *ghayth*. Fresnel translates it as *pâturage*, but writes: "*Le mot ghayth, que j'ai rendu par celui de pâturage signifie aussi la pluie et le nuage qui l'apporte.*"

16. Trans. F. Fresnel, *Journal asiatique*, 3rd Series, Vol. VI (1838), 207.

17. Maçoudi (Masudi), *Les Prairies d'or* (Paris, 1861–77), III, Chap. XXXIX.

18. Ibid., p. 101.

19. Ibid., p. 101–102. In these lines Masudi quotes el-Harit, an ancient poet.

20. I intend to bring together more Arabian recollections of the tidal flood in an essay on the Desert of Wandering.

21. Maçoudi, *Les Prairies d'or*, II, Chap. XXXI.

22. Yaqut, quoted by al-Samhudi, *Geschichte der Stadt Medina*, ed. F. Wüstenfeld in *Abhandlungen der Gesellschaft der Wissenschaften zu Göttingen*, Historisch-philologische Klasse, Vol. IX (1860), 1861, p. 26.

23. Maçoudi, *Les Prairies d'or*, II, Chap. XXXI.

24. Maçoudi, *L'Abrégé des merveilles* (French translation by Carra de Vaux; Paris, 1898), p. 342.

25. Ginzberg, *Legends*, III, 62.

26. Maçoudi, *L'Abrégé des merveilles*, p. 361.

27. Josephus *Against Apion*, I, 76.

28. Petrie, *History of Egypt*, II, 19.

29. Tabari, *Chronique* (French trans. L. Dubeux; Paris, 1836), I, 261.

30. *Historia anteislamica*, ed. Fleischer, pp. 17, 179.

31. Ibid., p. 101 (*ventus vehementissimus*).

32. Ibid., p. 179.

33. Commentary to Sura II, 46.

34. T. Noeldeke, *Ueber die Amalekiter* (Gottingen, 1864): "*Wer nun etwas auf das Amalekitertum der Pharaonen geben wollte, der wäre nicht viel kritischer, als wet sie . . . für Römer oder Perser hielte.*" His argument was: The Arab reports are of no value. Only that is true which was appropriated by the Arab writers from the Old Testament.

35. H. Winckler, *Geschichte Israels* (Leipzig, 1895), I, 212. "The nation of Amalek probably rests on a mythological idea."

36. B. Gunn and A. H. Gardiner, "The Expulsion of the Hyksos", *Journal of Egyptian Archaeology*, V (1918), 36 note 1: "R. Weill holds the entire story of the Hyksos to be a legendary construction." See note 5, above.

93

37. *Manetho*, in Josephus, *Against Apion*, I, 77. On the confusion of Assyrians with Syrians (Palestinians) by writers in Greek, see Herodotus (trans. A. D. Godley; 1921–24), VII, 63.

38. Josephus, *Against Apion*, I, 78–79.

39. Ibid., I, 81.

40. W. M. Flinders Petrie, *Hyksos and Israelite Cities* (London, 1906), pp. 12f.

41. Cutting off the hands of the fallen or captured enemy soldiers became a practice in a later period of Egyptian history and Assyrian as well. This practice probably goes back to the time of the Hyksos.

42. Gunn and Gardiner, *Journal of Egyptian Archaeology*, V (1918), 39.

43. Ginzberg, *Legends*, III, 56.

44. Cf. the vowels in the Massorete Bible, Numbers 24:7, and I Samuel 15, and Esther 3.

45. Numbers 24:7. The name of the Hyksos king, Khian, was like that of a planet: "Khiun . . . star of your god" (Amos 5:26). However, the spelling of the king's name has the sound expressed by the letter *khet* and the name of the star has *khaf*.

46. I Samuel 15.

47. Cf. *Worlds in Collision*, p. 151.

48. The King James translation of this verse is cumbersome: "Out of Ephraim was there a root of them against Amalek."

49. Targum Yerushalmi, Numbers 21:1 and 33:4. Ginzberg, *Legends*, VI, 114.

50. The region of the Midianites is incorrectly located on the desert strips on both sides of the Aqaba Gulf. The traditions of the Arabs connecting the Amalekites with Mecca relate the Midianites to the region of Medina. Compare also the name of the high priest of the Midianites in the days of Moses—Jethro, called also Reuel, Raguel, and Hobab—with Jathrib, another ancient Arab name for Medina.

51. Trans. Gardiner, *Journal of Egyptian Archaeology*, I (1914), 103.

52. H. R. Hall, "Egyptian Chronology", *Cambridge Ancient History*, I ,169.

53. Gunn and Gardiner, *Journal of Egyptian Archaeology*, V (1918), 40–42.

54. A. H. Gardiner, "The Defeat of the Hyksos by Kamose", *Journal of Egyptian Archaeology*, III (1916), 95–110.

55. Gardiner reads "three years". See Kurt Sethe, "Die Dauer der Belagerung von Sharuhen", *Zeitschrift für ägyptische Sprache und Altertumskunde*, XLVII (1905), 136.

56. J. H. Breasted, *Ancient Records of Egypt* (Chicago, 1906), Vol. II, Secs. 7–13.

57. Gunn and Gardiner, *Journal of Egyptian Archaeology*, V (1918), 47.

58. The King James translation, "in the valley", is incorrect. *Nakhal* is "a bed of a river", "a river", and more especially the "river of Egypt" or the wadi of el-Arish, as distinguished from Yeor, or the Nile. Levy, *Wörterbuch über die Talmudin und Midrashim*, translates *nakhal* as "*Fluss, Bach, Flussbett*".

59. "One would not expect that the settlement of such a wandering nation would deserve the name of a city." W. Max Müller in the Jewish Encyclopedia, "Amalek, Amalekites", I, 428.

60. "The territory ascribed to Amalek in I Samuel 15:7, 'from Havilah until thou comest to Shur', is perplexing." W. Max Müller, "Amalek, Amalekites", The Jewish Encyclopedia, I, 483.

61. J. Wellhausen changed "from Havilah" to "from Telem", a city in Judah. (*Text der Bücher Samuels* [Göttingen, 1871], p. 97.)

62. A. S. Yahuda, "The Two Hawilas", *The Language of the Pentateuch in Its Relation to Egyptian* (London, 1933), I, 190: "The mention of Hawila . . . has always presented Biblical scholars with great difficulties. . . . Our own very exhaustive investigation and close scrutiny of all the suggested possibilities . . . has in every case yielded unsatisfactory results."

63. Josephus, *Against Apion*, I, 88–90, quoting Manetho.

64. The town is mentioned in Joshua 19:6.

65. *Against Apion*, I, 76.

66. Jeremiah 47:4; Amos 9:7.

67. Manetho (trans. Waddell), pp. 91, 95–99.

68. Abu-el-Saud, Commentary to Sura II, Abulfeda, *Historia anteislamica*, ed. Fleischer, p. 17.

69. The natural retreat for an army pressed at once from Egypt and from the shore of Palestine would be in the direction of Edom, more particularly toward Petra. In the days of Strabo, trade caravans coming from Arabia "come to Petra, and then to Rhinocolura [el-Arish], which is in Phoenicia near Egypt, and thence to the other peoples", and according to this author, this route was preferred in earlier times (Strabo, *The Geography*, 16, 4, 24). Sharuhen was probably situated close to Petra. Petra is "fortified all round by a rock" (Ibid., 16, 4, 21). The early builders of Petra are not known. See Sir H. B. W. Kennedy, *Petra, Its History and Monuments* (London, 1925), p. 81, and G. Dalman, *Petra und seine Felsheiligtümer* (Leipzig, 1908), p. 33; also M. Rostovtzeff, *Caravan Cities* (Oxford, 1932), pp. 37–53.

70. Ginzberg, *Legends*, IV, 98.

71. II Samuel 12:30.

72. According to the Arabian tradition, Medina was conquered by David; see al-Samhudi, *Geschichte der Stadt Medina*, ed. Wüstenfeld, pp. 26ff.

73. Hadad left Egypt after the death of David (I Kings 11:21). Ahmose reigned more than twenty years, according to Manetho twenty-five years.

74. Gauthier, *Le Livre des rois d'Egypte* (Cairo, 1902), II, 187, note 3. But see Stricker, *Acta Orientalia*, XV (1937), 11–12.

75. Josephus, *Against Apion*, I, 77–78.

76. Petrie, *Hyksos and Israelite Cities*, pp. 10–16.

77. Montet, *Le Drame d'Avaris*, p. 47: "*Le lecteur s'étonnera d'apprendre qu'une ville historique ait été promenée par les égyptologues tout le long du Delta*

oriental, de Péluse à Heliopolis, en passant par Tell el Her, El Kantarah, San el Hagar (Tanis), Tell el Yahoudieh."

78. K. Sethe, Urkunden (Leipzig, 1906–1909), IV, 390; Gardiner, Journal of Egyptian Archaeology, III (1916), 100.

79. Josephus, Against Apion, 1,78f.

80. Compare Numbers 34:5; II Kings 24:7; II Chronicles 7:8: "Nakhal" was the border of Egypt.

81. The symbol of Auaris follows immediately that of Sekhet-za; the latter site is closely associated with the site of Tharu on several steles of the Ramesside period. See Gardiner, Journal of Egyptian Archaeology, III (1916), 101.

82. Josephus, Against Apion, I, 237.

83. Epiphanius said: "Rhinocorura" means "Nakhal" (bed of a river); Saadia translated "Nakhal Mizraim" as "Wadi el-Arish", and similarly Abu-faid. See F. Hitzig, Urgeschichte und Mythologie der Philistäer (Leipzig, 1845), pp. 112ff. Hitzig recognized that el-Arish must have been an old city (Laris of the Crusaders), but was unable to identify the ancient city that had been situated on the site of el-Arish or Rhinocolura.

84. Il est facile d'identifier les noms géographiques de l'antiquité, quand ils se sont conservés en arabe. Le tell Basta recouvre les ruines de Bubaste.... Mais le nom d'Avaris était tombé en désuétude bien avant la fin des temps pharaoniques." Montet, Le Drame d'Avaris, pp. 47–48.

85. Maçoudi, L'Abrégé des merveilles, p. 388. The Amalekite pharaoh is called by Masudi Talma successor to Latis. The Hyksos king who built Auaris is called Saltis or Salatis (Cambridge Ancient History, I, 233) by Manetho. The two forms, Latis and Salatis, handed down through two such different channels, are nevertheless noticeably similar.

86. Papyrus Ipuwer (Leiden 344 recto) 3:1; 15:1.

87. Papyrus Ipuwer 2:10; 4:2; 6:1; Papyrus Ermitage 1116b recto.

88. Manetho-Josephus, Against Apion, I, 73ff.

89. Compare the findings of Petrie in Hyksos graves (Huyksos and Israelite Cities, p. 12).

90. Manetho-Josephus; inscription at Speos Artemidos.

91. Sallier Papyrus I; Papyrus Ipuwer 17:2; Manetho-Josephus.

92. Papyrus Ipuwer; Papyrus Ermitage; Manetho-Josephus.

93. Manetho-Josephus ("from the East"); Papyrus Ipuwer 14:10; 15:13.

94. Manetho-Josephus.

95. Papyrus Ipuwer 14:10; 15:3.

96. Manetho-Josephus.

97. Papyrus Ipuwer 14:10; 15:3; Carnarvon Tablet.

98. Scarabs of Hyksos kings; Sallier Papyrus I; Manetho-Josephus.

99. Inscriptions of Apop; see J. H. Breasted, A History of Egypt, p. 218; Eduard Meyer, Geschichte des Altertums, Vol. I, Pt. 2, p. 319.

100. Manetho-Josephus; Tomb of Ahmose; Sallier Papyrus I; inscription of Hatshepsut at Speos Artemidos.

101. Papyrus Ermitage.

102. Sallier Papyrus I; compare Petrie, *A History of Egypt*, I, 243.

103. Manetho-Josephus.

104. Tomb of Ahmose; compare *Against Apion*, I, 88.

105. Manetho-Josephus.

106. Tomb of Ahmose.

107. Tomb of Ahmose; compare Manetho-Josephus.

108. Tomb of Ahmose.

109. Manetho-Josephus.

110. Maçoudi, *Les Prairies d'or*, III, 101; *Kitab-Alaghaniy* (trans. Fresnel), pp. 206ff.

111. El-Harit, cited by Maçoudi, *Les Prairies d'or*, III, 101; compare Exodus 12:29.

112. *Kitab-Alaghaniy* (trans. Fresnel), p. 207.

113. Exodus 15:7–12; 17:8–16; Numbers 14:43–45.

114. Exodus 7:20; 12:29; 14:27.

115. Numbers 13:29; 14:43; Tabari, *Chronique* (trans. Dubeux), p. 261; Abulfeda, *Historia anteislamica*, ed. Fleischer, p. 179; Mekhilta Beshalla, I, 27.

116. Maçoudi, *Les Prairies d'or*, II, 397.

117. See "Amalik", The Encyclopedia of Islam.

118. Judges 6:3, 33; 7:12; I Samuel 15:9, 14.

119. Deuteronomy 25:15f.; Numbers 11:1; Targum Yerushalmi of Exodus 17:8; Midrash Tannaim, 170; Pirkei Rabbi Elieser 44; and many other sources.

120. Numbers 14:3; I Samuel 30:15.

121. I Samuel 30:1.

122. Maçoudi, *L'Abrégé*, pp. 342, 361.

123. *Kitab-Alaghaniy* (trans. Fresnel), p. 206.

124. I Samuel 15:5 and 7; cf. Maçoudi, *L'Abrégé*, I, 331.

125. Maçoudi, *L'Abrégé*, I, 331f., 338; Abulfeda, *Historia anteislamica*, ed. Fleischer, pp. 101ff. and 179; Tabari, *Chronique* (trans. Dubex), p. 209; Ibn Abd-Alhakam, Yaqut, Koran Commentary to Sura II, 46; Alkurtubi, Koran Commentary to Sura II, 46 (Leiden Ms.).

126. Literature in Ginzberg, *Legends*, III, 63; Numbers 24:20; 24:7; I Samuel 15:7.

127. The above Arabian sources of the ninth to the thirteenth centuries; I Samuel 30:13.

128. Judges 5:14; 12:15.

129. Judges 6 and 7; I Samuel 14:48.

130. Compare: Exodus 17:8ff.; I Samuel 27:8; I Samuel 30; I Kings 6:1.

131. Numbers 24:7; I Samuel 15:8.

132. II Samuel 1:13; Abu-el-Saud, Commentary to Sura II, 247; compare Abulfeda, *Historia anteislamica*, ed. Fleischer, p. 17; compare also "Amalik", The Encyclopedia of Islam.

133. I Samuel 15:5.

134. I Samuel 15:5.

135. I Samuel 15:6.

136. I Samuel 15:7.

137. I Samuel 27:8; see also Ginzberg, *Legends*, IV, 99; Compare al-Samhudi, *Geschichte der Stadt Medina*, ed. Wüstenfeld, p. 26.

138. II Samuel 11; Ginzberg, *Legends*, IV, 98f.

139. I Chronicles 4:42f.

140. Deuteronomy 25:17–19; I Samuel 15:2; I Samuel 28:18; for Talmud and Midrash sources see Ginzberg, *Legends*, III, 61f., 333; IV, 230; VI, 480.

141. Italics supplied. See D. F. Tuch, "Ein und zwanzig Sinaitische Inschriften", *Zeitschrift der Deutschen Morgenländischen Gesellschaft*, III (1849), 151. Tuch took the quotation from Djauhari without any thought of the question debated here as to the identity of the Amalekites and the Hyksos.

142. ". . . thou shalt not abhor an Egyptian; because thou wast a stranger in his land" (Deuteronomy 23:7).

143. See the Register to Ginzberg, *Legends*, under "Amalek, Amalekites".

144. Ginzberg, *Legends*, III, 57.

145. Ibid., p. 62.

146. Esther 3:10.

147. Ginzberg, *Legends*, IV, 68, 397, 398, 422.

148. See Theodore Reinach, *Textes d'auteurs grecs et romains relatifs au Judaïsme* (Paris, 1895).

149. "Plato derived his idea of God from the Pentateuch. Plato is Moses translated into the language of the Athenians." Numenius in Eusebius, *Preparation for the Gospel* (trans. Gifford), XIII, 12.

150. Historian and politician in the service of Seleucus Nicator.

151. Quoted by Th. Reinach, *Textes*.

CHAPTER THREE

The Queen of Sheba

Two Suzerains

The beginning of the famous Eighteenth Dynasty, whose kings were
native Egyptians and who freed Egypt from the Hyksos, coincided
with the beginning of the line of Kings in Judea. Saul delivered the
fatal blow to the hegemony of the Amalekites-Hyksos; David established Jerusalem as his capital; in the time of Solomon the realm
achieved its splendor.

According to the scriptural narrative, Solomon had one thousand
and four hundred chariots and twelve thousand horsemen; he
reigned over all the land from the river Euphrates to the land of the
Philistines and the border of Egypt. The kings of Arabia paid him
tribute, and presents were brought from near and far, vessels of silver
and of gold, garments and spices, armor and horses. He made cedars
in Jerusalem "to be as sycamore trees in the vale of abundance". He
built a palace with a great throne of ivory, and a house of worship.
Vessels therein were of gold; of gold were the drinking cups of his
palace. Six hundred threescore and six talents of gold came yearly to
his treasury as tribute, besides the income from the traffic of the
merchants (I Kings 10:14–15).

The kingdom of Egypt, after regaining independence under
Ahmose, a contemporary of Saul, also achieved grandeur and glory
under Amenhotep I, Thutmose I, Hatshepsut, and Thutmose III.
Egypt, devastated and destitute in the centuries under the rule of the
Hyksos, rapidly grew in riches.

The two realms, freed from the same oppressor, entered into trade
relations and into relations of kinship.

King Solomon took an Egyptian princess to be one of his wives,
probably his chief wife. The Scriptures do not preserve her name. It
is known only that her father, the pharaoh, made an expedition

against southern Palestine, the home of the Philistines and the Canaanites, burned Gezer, and gave it as dowry to his daughter (I Kings 9:16).

The pharaoh, whose name is also omitted in the Scriptures—talmudic tradition calls him Shishak—was, according to the scheme presented here, Thutmose I, the third king of the New Kingdom. Only a few lines of his annals are extant. Besides a military expedition into Nubia, which he subdued, he undertook an Asiatic campaign "and overthrew the Asiatics". "After these things" the pharaoh "journeyed to Retenu to wash his heart among the foreign countries".[1]

Nothing positive can be established with these meager remnants except that the pharaoh actually crossed the Sinai Peninsula, waged a campaign of conquest in Philistia, and then proceeded peacefully to Palestine (Retenu) where he had some cause for rejoicing.

Although very sparse records are preserved from the reign of Thutmose I,[2] he was often mentioned as the father of Queen Hatshepsut. He shared the throne with her and made her his successor. Hatshepsut, the great and celebrated queen, left inscriptions, bas-reliefs, and figures of herself in abundance.

If the Exodus took place at the end of the Middle Kingdom; if, furthermore, the Hyksos rule was that of Amalekite invaders, then Queen Hatshepsut, whose huge statues confront us in the high-ceilinged halls of museums, must have been a contemporary of Solomon. Can it actually be that no memory of her was preserved in the annals of Jerusalem? Two lands in the process of developing their foreign relations and trade could hardly have been out of touch during the reigns of Solomon and Hatshepsut, neither of whom broke the peace of their countries.

Both of them built palaces and magnificent temples; both enriched their countries, not by war, but by peaceful enterprises; each possessed a fleet on the Red Sea and sent it on adventurous expeditions;[3] the reigns of both were the glorious periods of these two countries.

If Solomon was really a renowned king, as the Hebrew sources describe him, then the absence of any contact between this queen and this king is difficult to explain. It would, indeed, be very singular, for these two rulers were no ordinary occupants of throne halls, but very excellent suzerains.

Nor would it fit our notion of the adventure-loving character of Queen Hatshepsut, or the words of praise: "Thy name reaches as far as the circuit of heaven, the fame of Makere (Hatshepsut) encircles the sea,"[4] and "her fame has encompassed the Great Circle" (ocean).[5] Neither would it accord with our idea of King Solomon, whose capital was visited by ambassadors from many countries[6] and who had personal contact with many sovereigns: "And all the kings of the earth sought the presence of Solomon" (II Chronicles 9:23), and "all the earth sought to Solomon. . . ." (I Kings 10:24.)

Was the queen of Egypt excluded from "all the kings"?

From Where Did the Queen of Sheba Come?

The visit of the most illustrious of Solomon's guests is recorded twice in the Scriptures. Chapter 9 of the Second Book of Chronicles repeats almost verbatim the story of Chapter 10 of the First Book of Kings.

> And when the queen of Sheba heard of the fame of Solomon, she came to prove Solomon with hard questions at Jerusalem, with a very great company, and camels that bare spices, and gold in abundance, and precious stones: and when she was come to Solomon, she communed with him of all that was in her heart.

Is this story a fairy tale? Did a fairy queen come from a mysterious land with jewels and other marvelous things? If she really came, did she find a splendid and generous king in Jerusalem and not a poor vassal prince who was never mentioned in the history of other countries?

Many are inclined to believe that there is no historical basis for this legendary romance; others, who saw not fiction but real adventure in the story, unsuccessfully sought some historical remnants of the life and reign of the Queen of Sheba.[7]

The land over which the Queen of Sheba ruled is identified by most scholars as the district of Saba (Sheba) in southern Arabia, the land of the "Sabeans from the wilderness, which put bracelets upon their hands, and beautiful crowns upon their heads" (Ezekiel 23:42).

101

The Yemenites, the inhabitants of the land of Saba (Sheba) in Arabia Felix, believe that the Queen of Sheba was their queen, and with rich oriental imagery they ornament the story of her life and of her visit to Jerusalem. This view is supported by the Koran (Sura XXVII).

But Ethiopia vies with Arabia for the fame of the Queen of the South. The kings of Ethiopia claim descent from Menelik, a son of Solomon and the Queen of Sheba, who, they insist, was *their* queen. They possess medieval manuscripts, with texts composed in the early Christian centuries, which carry this tradition.[8]

Neither of the two Talmuds contains any clear historical reference to the mysterious adventurous queen.[9] However the opinion is expressed in the Talmud that "Sheba" in the name Queen of Sheba is not a geographical designation but a personal name.[10] The very numerous inscriptions of southern Arabia are silent on the Queen of Sheba. Travelers have been unsuccessful though they have turned every stone in southern Arabia which promised an answer to the riddle.[11] The genealogy of the Abyssinian emperors, who wished to be recognized as the seed of Solomon and the Queen of the South, is accepted with the same lack of belief as all similar genealogies of kings and demigods.

Research, which produced many treatises and commentaries, after exhausting all material, arrived at the conclusion expressed by one writer: "We shall never know whether the queen who visited Solomon was a pure-blooded Abyssinian or an Arab queen from Yaman [Yemen] or Hadramaut or some other part of the great Arabian peninsula. But the tradition that some 'Queen of the South' *did* visit Solomon is so old and so widespread, that a kernel of historical fact, however small, must be hidden somewhere in it."[12]

Was the Queen of the South a queen of Saba (Sheba) in Arabia, or of Ethiopia, or was she a legendary character from fairyland?

In the *Jewish Antiquities* of Josephus we find the story of the Queen of Sheba[13] introduced by these words:

> Now the woman who at that time ruled as queen of Egypt and Ethiopia was thoroughly trained in wisdom and remarkable in other ways, and, when she heard of Solomon's virtue and understanding, was led to him by a strong desire to see him which arose from the things told daily about his country.[14]

102

Here we have a clear indication: "queen of Egypt and Ethiopia". But Egyptian history, being removed some six hundred years from the synchronical point with Israelite history, could not present a woman ruling in Egypt and Ethiopia.[15]

The history of Egypt, moved along six hundred years and put at the right place, would make it difficult to explain the absence of any reference to Queen Hatshepsut in the Hebrew annals. Is the story of the Queen of Sheba the story of Queen Hatshepsut?

The most convincing proof, of course, would be presented if a record of the voyage to Jerusalem by Queen Hatshepsut were found and if it conformed with the account in the annals of the kings of Jerusalem.

The record was written. It is preserved. And it conforms.

Where Did Queen Hatshepsut Go?

A magnificent temple called "The Most Splendid of Splendors" at Deir el Bahari near Thebes in Egypt was built against a semicircular wall of cliffs. "These cliffs of white limestone, which time and sun have coloured rosy yellow, form an absolutely vertical barrier. They are accessible only from the north by a steep and difficult path leading to the summit of the ridge that divides Deir el Bahari from the wild and desolate valley of the Tombs of the Kings."[16]

On the walls of this temple are engraved bas-reliefs describing the life and the most important events of the reign of Queen Hatshepsut. One series tells the story of her divine birth, her father being the god Ra. Another series, called the Punt reliefs, which are opposite to and symmetrical with the first series, tells the story of a journey to the land of Punt or the Holy Land (Divine, God's land).

"These are undoubtedly the most interesting series of reliefs in Egypt. . . . They are as beautiful in execution as they are important in content."[17]

It must have been a blessed land, the goal of the expedition in ships propelled by sail and oar. It had all the features of a mythical land of happy fields and hills. White men of a north-Semitic[18] or Caucasian race[19] lived there, proud and handsome; they composed the bulk of the population. Then, the pictures show, there were a few entirely different, dark-skinned men in Punt. Animals were led on ropes; they are identified as having belonged to African fauna—a

103

number of monkeys and a panther.[20] The plants were indigenous to the southern Arabian coast.[21] Plants about which the inscription says, "Since the beginning of the world never were seen trees like these," were brought from Punt by the Egyptian expedition. Besides these trees, there were gifts of ivory and ebony, gold and silver and precious stones. These Queen Hatshepsut brought home, a heavy load, in her ships.

Where was the land of Punt?

Many theories concerning the location of the land of Punt have been advanced, but it has remained uncertain. Were it not for the bas-reliefs with the exotic animals and plants, the whereabouts of Punt probably would be no problem.

The name of Punt or Divine (God's) Land is not accompanied by the sign designating a foreign country, showing that the Egyptians regarded Punt as a land affiliated in some way with Egypt. In a number of Egyptian inscriptions Punt is mentioned as situated to the east of Egypt. In an utterance by Amon, dating from the time of Amenhotep III, of the later part of the Eighteenth Dynasty, there are the following words: "When I turn my face to the sunrise . . . I cause to come to thee the countries of Punt."[22]

An official of the Sixth Dynasty left a laconic record stating that he visited Byblos and Punt eleven times.[23] Byblos was the old capital of Phoenicia; its ruins are eighteen miles north of Beirut. Having been visited eleven times by the officer coming from Egypt, and being mentioned together with Byblos, Punt must have been associated with Byblos.

The name Punt or Pont can be traced to "Pontus, father of Poseidon and Sidon", as narrated by Sanchoniaton, the early Phoenician writer.[24] Sidon was a Phoenician metropolis.

There are also numerous inscriptions that mention products of Punt and God's Land as being obtained in Palestine (Retenu), and we shall read some of them in the following chapter. In an inscription in the tomb of a general of Thutmose IV, Palestine (Retenu), is called God's Land; a building inscription of Amenhotep III one generation later also refers to Palestine as God's Land.

But all these points were not even considered by scholars who tried to locate Punt or God's Land. As the bas-reliefs of Deir el Bahari showed exotic animals and plants, zoologists and botanists were called by the historians to determine the place to which they were indigenous.[25] Palestine and Syria were never included among the possibilities.

The elaborate pageant, the fauna and flora, created a division of opinion among the scholars. The majority of the population is north-Semitic or Caucasian. The animals and the few negroids are African. The plants are south-Arabian. God's (Divine) Land is often mentioned in the Egyptian inscriptions as the land which produced frankincense, and if the rare trees were these frankincense plants, they must have been of the Arabian variety, which is rich in foliage; the Somalian frankincense does not leaf as depicted in the bas-relief.[26]

To account for the presence of the white people was easiest: they might have migrated to that country.[27] The people of Punt were Phoenicians, as the name implies, and their presence in Somaliland could be inferred, as some authorities thought, from a passage in Herodotus, who wrote that the Phoenicians at an early date came from the Eritrean Sea to the Mediterranean,[28] and this seemed to sustain the theory that Punt was in eastern Africa. The Puntites would have been the ancestors of the Phoenicians.[29]

The statement of an official of the Old Kingdom that he visited Byblos and Punt eleven times was understood to mean that the official visited Byblos in Syria eleven times and Punt in eastern Africa or southern Arabia eleven times.[30]

The indication that Punt was to the east of Egypt complicated still further the obscure problem of African fauna and Asiatic flora.

A review of opinions, taking a position chiefly for or against southern Arabia and Somaliland, shows how hopelessly confused the problem is.[31] Did the expedition of Queen Hatshepsut undertake the voyage in the faraway kingdom of Saba, whence six hundred years later a fairy queen came to Solomon in Jerusalem?[32] Or did the expedition of Hatshepsut arrive at the Somali shore, the other land which claimed to be the birthplace of the Queen of Sheba?

With regard to the origin of the Queen of Sheba, the same two lands—southern Arabia and the highland of Somali with Ethiopia—entered into the dispute.

Like the biblical scholars who exhausted all means of finding out whether the Queen of the South came from Saba or from Ethiopia,

105

the Egyptologists exhausted every source without arriving at a final decision as to whether Queen Hatshepsut sent her expedition to southern Arabia or to the African coast. It was certain only that somewhere, outside of Egypt but in touch with it, was a blessed land of riches called Punt, also the Divine, Holy, or God's Land,[33] with a population of handsome men, with exotic animals, with marvelous plants.

Finally, a Solomon's judgment was urged, and the land of Punt was divided between the Arabian Land of Saba and African Somaliland.[34]

In the previous chapters a synchronism was presented between Egyptian and Hebrew histories. Queen Hatshepsut was found to be a contemporary of Solomon.

If Queen Hatshepsut were herself the Queen of the South, she, of course, would have come neither from Saba in Arabia nor from the Somali coast of Ethiopia, but from Thebes in Egypt; her destination most probably would have been neither southern Arabia nor Somaliland, but Jerusalem.

Was the Divine Land the region of Jerusalem? Palestine is east of lower Egypt; it had a white north-Semitic population. But the flora and fauna of the bas-reliefs, both foreign to Palestine, would appear to be a double anomaly.

I shall now pass on to the description of the expedition. The rare plants, the animals, and the men (negroids) will not only cease to be a hindrance to the identification here proposed, but will strongly support it. After that the identification will be tested against other references to Punt and Divine Land in Egyptian inscriptions.

The Way from Thebes to Jerusalem

The shortest route from Thebes to Jerusalem is not along the Nile and the coast of the Mediterranean; the Red Sea route is only a little more than half its distance: from Thebes to Coptos, a short distance up the Nile, and then to el-Qoseir,[35] a harbor on the Red Sea, then by ship across the Red Sea and along the Aqaba (Aelana) Gulf, and from Aqaba overland to Jerusalem.

Besides being much shorter, this route was preferred for various other reasons. The way across the Sinai Peninsula was not safe. It led through the last settlements of the Amalekites and Philistines, through Auaris and Gezer. Thutmose I, the father of Hatshepsut, carried on a campaign in this region, clearing it of rebellious bands of Amalekites and Philistines, and burning Gezer. The seaway was safer. Also a voyage by ship was more comfortable for the queen than a long journey in a chariot or in a palanquin, stopping at night to rest in tents pitched in the middle of the desert.

Still another reason for traveling by sea was probably Hatshepsut's desire to display the splendor of her new fleet.

Solomon built a harbor in the Aqaba Gulf of the Red Sea.

> I KINGS 9:26 And king Solomon made a navy of ships in Ezion-Geber, which is beside Eloth, on the shore of the Red Sea, in the land of Edom.

Since the great cataclysm, when the maritime nations were lost together with their ships, daring ventures had been few and far between. For a long time the Atlantic Ocean was agitated by tectonic ruptures in a subsiding bottom.[36]

When the Amalekites were destroyed and their allies, the Philistines, subjugated, the Phoenicians of Tyre and Sidon resumed their full maritime activity and sent their ships to "the Great Circle" (the ocean). Soon they recognized the advantages of the southern route, by the Red Sea, to the overseas countries.

The Phoenician king of Tyre, Hiram, sought an alliance with King Solomon and his friendship, possibly because of the naval base of Ezion-Geber in the land of Edom, vassal to Solomon. In disregard of their practice of allowing no one to know the secrets of their voyages, the Phoenicians took the sailors of Solomon with them.

> I KINGS 10:22 For the king had at sea a navy of Tharshish with the navy of Hiram. . . .
> 9:27 And Hiram sent in the navy his servants, shipmen that had knowledge of the sea, with the servants of Solomon.

To all the other reasons for choosing the maritime route from Thebes to Palestine, this, too, may be added: the queen was interested in visiting the new harbor from which the Phoenician and Hebrew fleets were starting their three-year sail to Ophir.

Solomon inspected Ezion-Geber before the queen arrived. "Then went Solomon to Ezion-Geber, and to Eloth ..." (II Chronicles 8:17). And it is here that one would expect the expedition to land. A few lines after the above passage in Chronicles the story of the visit of the Queen of Sheba begins.

In the Scriptures it is not said that the Queen of Sheba made the first part of her journey by ship. But in rabbinical writings it is recounted of the queen that she arrived on a ship after a long voyage by sea. The Queen of Sheba "assembled all the ships of the sea, and loaded them with the finest kinds of wood, and with pearls and precious stones."[37]

Before the queen undertook the voyage a preliminary exploring mission was sent out. The record of the visit of this mission is preserved in the Haggada.[38] The Koran[39] also tells of the preliminary expedition: "Said she [the Queen of Sheba]: 'O ye chiefs ... I am going to send to them a gift, and will wait to see with what the messengers return.' "[40]

In the lower corner of the bas-relief in Deir el Bahari is depicted a landing place. From the right a "king's messenger" advances at the head of his soldiers; from the left a chief approaches. A line of water with fish swimming about serves to indicate that the place is on the coast. The chief is called "a chief of Punt P'-r'-hw" (Perehu or Paruah). On a tent is written: "Pitching the tent of the king's messenger and his army on the myrrh-terraces of Punt on the side of the sea."[41] Since it is placed on the extreme lower part of the mural, a position of minor importance, this picture probably shows the preliminary expedition or the arrival of the herald of the queen.

Paruah must have been Solomon's representative in the land of Edom, possibly an Edomite vassal of his.

Among the twelve governors of King Solomon – at a later period in his reign (when some of these officials were his sons-in-law)—one was a son of Paruah (I Kings 4:17). Jehoshaphat, the son of Paruah, was governor of Ezion-Geber and Eloth; his father, apparently, administered the same region.[42]

On the picture Paruah is an elderly man with no insignia of sovereign power. He is accompanied on this mission by his wife, disfigured by elephantiasis, and a few other members of his family, which indicates that he and his family were residents and probably natives of the place; he appears to have been the chief of Ezion-Geber.

The landscape is consequently that of Edom. We see the small buildings with penthouses, sometimes constructed on piles. The penthouse, a room reached by a ladder from the outside, may be the "alia" which we find in the Scriptures.

Hatshepsut Led the Expedition to the Divine Land

The next picture shows the departure of a fleet of five vessels; three are under sail, while two are still moored. The inscription reads: "Sailing in the sea, beginning the goodly way toward God's Land, journeying in peace to the land of Punt. . . ."[43]

This is an exceedingly beautiful bas-relief of the fair fleet of the queen, with prows curved like lotuses, with tall masts and flying sails. Each ship had a crew of about fifty men to a shift, half of them on either side of the ship. It is probable that the fleet consisted of more than five ships as the pictures are partially destroyed. This was "the very great company" (II Chronicles 9:1).

The queen herself is not shown on a ship. This is in accord with the rules of Egyptian art: the Egyptian artist did not picture a royal person among the common people; the august likeness must not humble herself by being in the company of common sailors. Kings and queens were drawn oversize like giants among dwarfs. Queen Hatshepsut was portrayed beside the ships, and her figure was taller than the masts. Of course a figure as tall as that could not have been placed on the deck.

This manner of representation, normal from the Egyptian point of view, was the reason for a curious oversight. The expedition to the Divine Land is described in modern histories of Egypt as an expedition sent by the queen. But it is clearly apparent that she herself participated in this voyage.

The very importance ascribed to this expedition, which makes it the most prominent event of her reign,[44] leads one to infer that it was her own experience. Of all the events of her reign it was chosen to be chiseled on the wall of the temple of the Most Splendid of Splendors as a counterpiece to the depiction of her divine birth. If it were an ordinary commercial expedition to Punt, why should the records have perpetuated it so elaborately and with such pomp?

In previous times persons and missions had already been sent to Punt, but only meager and prosaic data are preserved.[45] Punt must have been close to Egypt; this is intimated in various references on Egyptian monuments, such as the one, already cited, in which an official, who lived at the time of the Sixth Dynasty, mentioned that he "had gone eleven times to Byblos and Punt".

The fact that Punt was "written without the sign indicating a foreign country" (together with its frequent mention in inscriptions) "seems to show that the Egyptians considered they were in some way connected with that country".[46] Why, then, would the queen create such excitement over the visit there and acclaim it in great festivals, unless she herself were the visitor? Would the meeting of some royal messenger with a chief Paruah have been an event that the queen would have wanted to immortalize as a thing "that had never happened"?

Queen Hatshepsut undertook the journey like a devout pilgrim who, hearing an inner bidding, takes staff in hand:

> . . . a command was heard from the great throne, an oracle of the god himself, that the ways to Punt should be searched out, that the highways to the myrrh-terraces should be penetrated:
>
> "I will lead the army on water and on land, to bring marvels from God's land for this god, for the fashioner of her beauty . . ."[47]

It was an oracle or a mysterious voice that Queen Hatshepsut heard within her, and she thought it was her god.

Like the Punt inscription, the Haggada, and Josephus, too, describe this strong, imperative desire that inspired the queen and that was considered a divine command.[48]

There is no known precedent of a woman on the throne of Egypt;[49] there is also no known previous instance of a ruler of Egypt paying a visit of homage to a foreign sovereign.

The ships of Hatshepsut rode the waves under full sail; the crew toiled at the oars. The route led the royal fleet into the narrow channel of the Aelana (Aqaba) Gulf. In the days of Amalekite sovereignty this region was visited by neither Israelites nor Egyptians. This journey of the queen, therefore, was an exploring expedition.

It is a perilous gulf. If a north wind enters this channel it may capsize the careless sailor, breaking his mast in a whirlwind. In the time of Solomon this danger was probably not apprehended.

The steep, snow-capped massif of Mount Sinai, towering at the entrance to the gulf, the shallow water at the mouth lying among reefs and coral isles, the intense blue of the water as the gulf suddenly deepens, and the red slopes of the Edom highland, crimson at dawn, scarlet in the twilight, give this silent chasm an eerie character.

> I have led them [the company of the expedition] on water and on land, to explore the waters of inaccessible channels, and I have reached the myrrh-terraces.[50]

The words, "I have led them on water and on land", indicate that the voyage did not end at the coast (Ezion-Geber); from there the queen with her royal train proceeded on land. The caravan moved toward Jerusalem "with great splendor and show of wealth" (Josephus); "the camels were laden with gold and various spices and precious stones"; the resplendent pilgrims to the city of wisdom were escorted by the royal guard.

The queen and her train, including her royal artist, were obviously impressed by the handsome appearance of the Israelite warriors. What did the ancient Israelites look like?

We have been accustomed to seeing the Syrian prisoners sculptured by the Egyptian craftsmen of later kings of the Eighteenth Dynasty, and also of the Ninteenth and Twentieth Dynasties. The beards of the captives are round and untrimmed, "socratic", the figures helpless; for the most part they are portrayed at the moment of their execution. But here on the Punt reliefs the appearance of the dwellers in God's Land is dignified; their noses are aquiline, their eyes set deep, their chins protruding, their beards resembling the beard of the god Ra.[51] It is unprecedented, and it never happened again, that the appearance of foreign soldiers, pictured by the Egyptians, should express more nobility and grace than the figures of the Egyptians themselves.

The impression the Judean guard made on the guests is also told in a legend about the visit of the Queen of Sheba. King Solomon, who remained in Jerusalem to await his august guest, sent a procession of handsome youths to meet her. They were "like the sunrise, the evening star and the lily",[52] and the queen was amazed.

The Glorious Region of God's Land

The route goes up the valley of the Araba, leaving stony Petra on the right, and runs along the Dead Sea, a scene of desolation, where lava gushed from the interior of the earth and congealed in fantastic forms. In the valley of ruined Jericho abundant springs watered gardens in the desert, and they budded and blossomed the greater part of the year.

The road rises to Jerusalem. Here the slopes of the hills were cut in the form of terraces, which are seen even today. Myrrh, cassia, nard, saffron, and cinnamon, sweet spices, fragrant herbs, fruits, and roots were used in the Temple of Jerusalem, and a great part of them was cultivated on these terraces.[53] Almonds, produced for trade, grew there,[54] and aloes used by the maidens of Judea to perfume their beds.[55]

The queen who came from the plains of Egypt later wrote on stone:

> PUNT RELIEFS: I have reached the myrrh-terraces. It is a glorious region of God's land.[56]

She wondered at the pageant of the blooming hills; but the most splendid groves of trees planted upon terraces were in the middle of Jerusalem.

> II CHRONICLES 9:11 And the king made of the algum trees terraces to the house of the Lord, and to the king's palace . . . and there were none such seen before in the land of Judah.

"And the king received her gladly on her arrival and was studious to please her in all ways, in particular by mentally grasping with ease the ingenious problems she set him."[57]

Whether there was a likeness of Solomon on the bas-reliefs of the temple of the Most Splendid is impossible to say. The Egyptians would have considered it a dishonor to see a picture of their queen in society and as a guest in the house of a foreign ruler. Would she picture herself with her host?

On the bas-reliefs she communes only with the god Amon. But several of the pictures on that part of the mural which portrayed the sojourn of God's Land have been erased: "Two thirds of the short wall on which was sculptured the description of the Land of Punt is destroyed."[58] In the lower row there appear to have been huge figures, one of which was the queen, as part of her cartouche between two defaced fields is still recognizable. But if it was a text that was erased and not a likeness, what was peculiar in this text that it had to be erased by the order of her jealous successor, Thutmose III?

Whether details of the queen's visit in the palace of Jerusalem were originally pictured on the bas-relief or not, the strong impression which the queen received there and which she expressed in Jerusalem, she repeated in Thebes.

When the queen saw the palace of the king, "and the sitting of his servants, and the attendance of his ministers, and their apparel, and his cupbearers, and his ascent by which he went up into the house of the Lord, there was no more spirit in her" (I Kings 10:4–5).

She praised "the greatness of the marvels, which happened to her" and wrote: "Never did the like happen under any gods who were before, since the beginning" (Punt reliefs).[59] In Thebes she had heard about the land of the terraces, but what she saw exceeded her expectations.

> PUNT RELIEFS: It was heard of from mouth to mouth by hearsay of the ancestors. . . .

The queen wished to see with her own eyes the land of which she had heard marvelous reports. She decided to tread and to explore that land ("I have led them on water and land"); she reached that country ("I have reached the myrrh-terraces"); and she found it glorious.

The stories of the Divine Land, which she had heard before, she compared with what she herself witnessed.

In the Scriptures the words of the queen are not dissimilar:

> I KINGS 10:6–7 And she said to the king, It was a true report that I heard in mine own land of thy acts and of thy wisdom.

Howbeit I believed not the words, until I came, and mine eyes had seen it; and behold, the half was not told me; thy wisdom and prosperity exceedeth the fame which I heard.

Josephus wrote:

[She] was not able to contain her amazement at what she saw, but showed clearly how much admiration she felt, for she was moved to address the king. . . . "All things indeed, O, King," she said, "that come to our knowledge through hearsay are received with mistrust . . . it was by no means a false report that reached us."[60]

The emphasis in these quotations and in the Punt inscription alike is laid on the comparison of that which came "to our knowledge through hearsay" with that which was witnessed. To know for herself, not through hearsay, she had to visit the Divine Land, and so she undertook the pilgrimage to the myrrh-terraces.

If the sovereign of Egypt, the land of plenty, was astonished at the sight of Jerusalem's splendor, there might well be truth in the words of the First Book of Kings: "King Solomon exceeded all the kings of the earth for riches." The land seemed to the queen an abode of happy men: "Happy are thy men, happy are these thy servants. . . . Blessed be the Lord thy God" (I Kings 10:8–9).

The queen described her impressions in the following sentences addressed to the god Amon: "It is a glorious region of God's Land. It is indeed my place of delight. . . . I conciliated them by love that they might give to thee praise."

The manner of expression ascribed to the Queen of Sheba in speaking about King Solomon is not unlike that which Queen Hatshepsut used in respect to herself: "because the Lord loved Israel for ever, therefore made he thee king. . . ." (I Kings 10:9); "because he [Amon] so much loves the King of Upper and Lower Egypt, Hatshepsut . . ."

The queen, the guest, and the king, the host, exchanged precious gifts. "And she gave the king a hundred and twenty talents of gold, and of spices great abundance, and precious stones" (II Chronicles 9:9).

After the interpolation concerning the almug trees brought by Solomon's navy from Ophir, the story proceeds: "And King Solomon

114

gave to the queen of Sheba all her desire, whatsoever she asked . . ."

What was the desire of the Queen of Sheba?

"The Desire of the Queen of Sheba"

The "desire of the Queen of Sheba" is pictured on the walls of the temple of the Most Splendid. The gifts are shown in the presentation scene, the scene of the loading for the return voyage, the scene of the counting and weighing after the return, and the scene of dedication to Amon.

The gifts were exchanged with mutual generosity; but just as King Solomon knew the weight of the gold he received, so Queen Hatshepsut knew, after measuring and weighing, the exact weight of the precious metals she received in the Divine Land. A scene of the bas-relief shows the queen weighing the objects herself. The amount and size of the gifts were exceedingly great.

> PUNT RELIEFS: Reckoning the numbers, summing up in millions, hundreds of thousands, tens of thousands, thousands and hundreds: reception of the marvels of Punt.[61]

She gave gold and received "green gold of the land of Amu" (landing scene), and "silver and gold" (weighing scene). Silver, rare in antiquity, was abundant in the Jerusalem of Solomon. "And the king made silver to be in Jerusalem as stones" (I Kings 10:27); "silver . . . was nothing accounted of in the days of Solomon" (I Kings 10:21). It sounds like an exaggeration that silver was used for construction, but the inscriptions of Hatshepsut's officials after the return of the expedition tell about a "house of silver" and a "double house of silver" (inscriptions of Senmut[62] and Thutiy [63]), and a "floor wrought with gold and silver" (Thutiy).[64]

The queen gave "precious stones" (I Kings 10:10), and she received "lapis lazuli, malachite, and every costly stone" (Punt reliefs).

Solomon and Hatshepsut vied with each other not only in gifts but also in appreciation of the generosity of the other.

The queen gave spices in abundance and of the choicest varieties:

II CHRONICLES 9:9 . . . neither was there any such spice as the queen of Sheba gave king Solomon.

I KINGS 10:10 . . . there came no more such abundance of spices as these which the queen of Sheba gave to king Solomon.

She desired myrrh and she received it, also in great quantities and of the choicest kinds.

PUNT RELIEFS: Fresh myrrh (anti) in great quantities, marvels of the countries of Punt. Never did the like happen under any gods who were before since the beginning.

The "best of myrrh" was counted in "millions" and found "without number".

But more than all "marvels" the queen valued some precious trees.

PUNT RELIEFS: Thirty-one anti trees.[65] brought as marvels of Punt. Never was seen the like since the world was.

I KINGS 10:11–12 And the navy also of Hiram, that brought gold from Ophir, brought in from Ophir great plenty of almug trees, and precious stones. . . . There came no such almug trees, nor were seen unto this day.

In both scriptural narratives the story of Hiram's navy that brought back the exotic almug trees (algum trees in Chronicles) from Ophir is placed just between the verses about the presents which the Queen of Sheba gave to Solomon and the words ". . . and king Solomon gave unto the queen of Sheba all her desire". Now we see that this interpolation was made not without reason. The trees were brought from a distant island or continent and a part of them was given to the queen guest. A picture shows how they were handed over.

The words, "never was seen the like since the world was", or "unto this day" are alike in the Hebrew and Egyptian sources; the trees were a genuine sensation and were admired like the plants and other "marvels" brought quite two and a half millennia later by sailors returning from the Western Hemisphere.

The queen received not only metals, minerals, and plants; living animals were also among the royal gifts. Apes were brought to her by

116

servants of the king, and the accurate drawings show that they were of the variety known as cynocephali.

In the same tenth chapter of the First Book of Kings we read that apes were brought to Solomon by the navy of Tharshish.

> I KINGS 10:22 For the King had at sea a navy of Tharshish with the navy of Hiram: Once in three years came the navy of Tharshish, bringing gold, and silver, ivory, and apes, and peacocks.

The apes given to the queen were brought from afar. Ivory, too, was brought by the ships of Tharshish, and it was not lacking in the bounty bestowed on the queen.

On the mural, over the vessels laden with jars of myrrh, ivory tusks, wood, trees, and apes, is inscribed:

> PUNT RELIEFS: The loading of the cargo-boats with great quantities of marvels of the land of Punt, with all the good woods of the Divine Land, heaps of gum of anti, and trees of green anti, with ebony, with pure ivory, with green gold of the land of Amu, with cinnamon wood, Khesit wood, with balsam, resin, antimony, with cynocephali, monkeys, greyhounds, with skins of panthers of the south, with inhabitants of the [south] country and their children.
>
> Never were brought such things to any king since the world was.

The rare trees, the myrrh for incense, the ivory, the apes, the silver and gold and precious stones were enumerated in both records, the hieroglyphic and the scriptural. But Queen Hatshepsut mentioned also "inhabitants of the [south] country and their children".

In the picture showing the presentation of gifts there are four rows of kneeling men. The officials of the Divine Land are in the two lower rows, and behind them is a line of men approaching with gifts; the kneeling men of the upper middle row are called "chiefs of Irem", and are not unlike the Egyptians; the uppermost row represents the men of Nm'yw or Khenthenofer who look entirely different—they are dark-skinned and have round heads and thick lips, and seem to be themselves gifts like the animals and plants.[66]

The capital of the Divine Land being Jerusalem, who could be the representatives of two other ethnic groups?

Two foreign lands and peoples are referred to in the chapter on the Queen of Sheba. One is the neighboring people of Hiram, the king of Tyre: ". . . . navy of Hiram brought gold from Ophir." Ophir is the other land, a faraway place mentioned in the account of the gifts. It is quite conceivable that the officers of Solomon's ally, who brought the precious things from afar, would participate in the ceremony of presentation. The chiefs of Irem were therefore the messengers of Hiram.[67] The men of Nm'yw or Khenthenofer were probably the men of Ophir.[68]

Were the men from Ophir brought to Palestine? It is not mentioned in the Scriptures that the navy of Hiram and Solomon brought natives from Ophir. But Josephus Flavius wrote:

> The king [Solomon] had many ships stationed in the Sea of Tarsus, as it was called, which he ordered to carry all sorts of merchandise to the inland nations and from the sale of these there was brought to the king silver and gold and much ivory and *kussiim* [Negroes] and apes.[69]

It has been supposed that Josephus mistook some other Hebrew word of an older source and read it *kussiim*.[70] The picture of the expedition to Punt proves, however, that Josephus was not wrong: *kussiim*, the dark men of Ophir, were apparently brought by the sailors of Hiram and Solomon.

The murals therefore show us the ancient Hebrews, the ancient Phoenicians, and probably some of the people of Ophir together.

The gifts were exchanged, and finally the days and weeks in Jerusalem, an uninterrupted feast, came to their end.

JOSEPHUS: And the queen of Egypt and Ethiopia . . . returned to her own country.

The trees were transported in pots, each carried by four men and along the gangway by six; the apes clung to the rigging; tusks of ivory and jars filled the decks.

PUNT RELIEFS: Ye people, behold! The load is very heavy.

118

The inscription to the next bas-relief states simply and clearly: "The ships arrived at Thebes." Thebes is situated on the banks of the Nile. To reach it by water, the ships must have sailed along the Nile, entering from the Mediterranean.

A voyage from Punt in southern Arabia or in Somaliland to Thebes by the sea route would have meant disembarking at el-Qoseir and journeying overland from there to Thebes. As it is written and shown on the relief that the fleet of ships landed at Thebes, a difficult dilemma faced the commentators: either this part of the story was invented for some obscure reason, or in the days of Hatshepsut there must have existed a canal connecting the Nile with the Red Sea.[71] But there is no mention of such a canal in the days of Hatshepsut. It is known that a waterway connecting the Nile and the Mediterranean with the Red Sea was dug by Pharaoh Necho II, many hundred years after Hatshepsut, and was finished much later, in the time of the Persian conquest.[72]

The Divine Land being recognized as the region of Jerusalem, no problem is encountered in the arrival of ships at Thebes on the Nile. Hatshepsut was obviously eager to see the two sea routes to Palestine, and to display both her fleets, that of the Red Sea and that of the Mediterranean. From Jerusalem she probably traveled to one of the Phoenician seaports; from the Syrian shore the water route to Thebes needs no artificial canal.

Terraces of Almug Trees

The completion of the journey was celebrated in Thebes by two festivals, in the temple and in the palace. These celebrations were memorialized in two great murals: the first contains a formal announcement to Amon of the success of the expedition, the second to the royal court.

> Makere [Hatshepsut] ... the best of myrrh is upon all her limbs, her fragrance is divine dew, her odor is mingled with Punt, her skin is gilded with electrum, shining as do the stars, in the midst of the festival hall, before the whole land.[73]

119

The success of the expedition to the Divine Land was a personal triumph for the queen, and she emphasized it. She decided to thank her "heavenly father", the god Amon-Ra, for the success of the expedition by erecting a new temple and by constructing terraces and planting in these gardens the costly trees which she had brought from Punt.

The temple of the Most Splendid of Splendors, the ruins of which bear these bas-reliefs, was built, terraces were laid out, and trees were planted.

> I have hearkened to my father . . . commanding me to establish for him a Punt in his house, to plant the trees of God's land beside his temple, in his garden.[74]

The terraces in the Divine (God's) Land impressed the queen.

> I KINGS 10:4–5 And when the queen of Sheba had seen . . . the house that he had built . . . and his ascent by which he went up unto the house of the Lord, there was no more spirit in her.

This path ascended from the lower to the upper terraces planted with algum trees (II Chronicles 9:11).

Queen Hatshepsut wrote on the wall of her temple that "the highways of the myrrh-terraces" of the Divine Land were "penetrated", and " have reached the myrrh-terraces".

Similar terraces were built and planted facing the temple of the Most Splendid of Splendors. They were planted with the trees about which it is said in Kings: ". . . there came no such almug trees, nor were seen until this day"; and Queen Hatshepsut wrote: "Never was seen the like since the world was."

On the wall of the Deir el Bahari temple these trees are shown planted, and the inscription reads:

> Trees were taken up in God's Land and set in the ground [in Egypt].[75]

The ruins of this temple disclose where and how the terraces were situated, forming garden plateaux at rising levels.

Not only were the gardens imitated; the plan of the Temple in Jerusalem, even the service, were followed as models.

The design of the Most Splendid of Splendors in Deir el Bahari did not follow the contemporary Egyptian style. The earlier Egyptologists recognized the striking foreign elements in that building, and the opinion was expressed that the original of this imitation had been seen in Punt;[76] during the expedition there a different style of art had been observed, and after the return from Punt the temple in Deir el Bahari had been erected. The queen even emphasized that she built a "Punt". The walls of the temple were adorned with murals of the expedition to the Divine Land, and the style of the temple itself was a memorial to the foreign influence in architecture. "It is an exception and an accident in the architectural life of Egypt."[77]

The temple of Deir el Bahari is regarded by many as the most beautiful building in Egypt; it has the nobility of simplicity and it is free from the heavy ornamentation of the temples of the Ramessides.[78]

The Divine Land being the area of Jerusalem, the temple in Deir el Bahari must have had features in common with the Temple of King Solomon. Although most of the temple of Queen Hatshepsut lies in ruin, proportions of it which are still standing give a fair idea of the structure as it was before it was deserted and fell into decay.

In every generation attempts have been made to reconstruct the Temple of Solomon in drawings or models, but the data in the First Book of Kings did not provide sufficient details, and the reconstructors have had to draw upon their own imagination.

The Temple of Jerusalem was built upon terraces planted with trees. These terraces were cut by an ascending path. The processions of the Levites started on the lowest terrace, and as they sang they mounted the path. This explains the fact that some of their Psalms are called *Shir ha-maaloth*, "song of the ascent".[79]

The Temple in Jerusalem contained a hall which was three times as long as it was wide; in front of the hall was a vestibule; behind the hall was a sanctuary; the large "sea" was most probably placed in the inner court.

The temple at Deir el Bahari was built against a mighty cliff; the Temple of Jerusalem stood upon an elevation with a distant chain of hills running northeast, east, and south of it. This difference in location must have influenced the architects to alter their plans.[80] Slavish imitation would have dictated a location similar to that of the original. But adoption of the style and general features of the plan is probable, and this is implied in the words of the queen that she "built a Punt". A comparison of the data in Kings with the remnants of the temple of the Most Splendid may help to a better understanding of the form of both buildings.

The temple of the Most Splendid of Splendors was a famous sanctuary. Several scholars have tried to reconstruct its plan.[81] This temple was built upon terraces planted with the trees brought from the Divine Land; the terraces were placed at progressively higher levels, and a path leading to the temple mounted from one level to another. Rows of pillars standing on a lower terrace supported the wall of the terrace above. The court of the temple was surrounded by a colonnade; the temple was divided into a vestibule, a hall, and a sanctuary. The ratio of the width to the length of the hall was almost one to three.

The pillars supporting the terraces and surrounding the inner court were rectangular in form; with their shadows, which changed with the movement of the sun, they imparted a harmonious and majestic appearance, which only rectangular stone rhythmically arranged can give.

It is erroneous to assert that the Temple of Solomon was a poor building of an obscure Asiatic chieftain who tried to construct a copy of some Egyptian temple.[82]

Not only the temple architecture but also the temple service in Egypt were given many new features. It was not until the temple of the Most Splendid at Deir el Bahari was constructed that twelve priests, with a high priest heading them, officiated before the altar. A relief on a fragment, now in the Louvre Museum, shows twelve priests divided into four orders, three in each order, and a damaged inscription over their heads reads: ". . . in the temple of Amon, in Most Splendid of Splendors, by the high priest of Amon in Most Splendid of Splendors . . ."[83]

The office of the high priest was established in the Egyptian service only at the time of Queen Hatshepsut.[84] This reform in the religious service was introduced after the visit of the queen to the Divine Land, where shortly before the House of the Lord had been completed.

In connection with her announcement that she made "a Punt" in the garden of Amon, the queen issued a new edict: "Ye shall fulfill according to my regulations without transgression of that which my mouth has given," and this "in order to establish the laws of his [Amon's] house. . . ."

The Origin of the Words "Pontifex" and "Punt"

The obscure origin of the word "pontifex", which means high priest, can be traced here. This question had already been debated before the time of Plutarch, and he quoted the opinions of the authorities without finding any of them satisfactory. One ancient authority regarded "pontifex" as composed of the Latin roots *pons*, *pontis*, "a bridge", and *facio*, "I make". The philological conjecture was: The Pontifex was a man who built bridges, or the principal magistrate; the chief of the people, who united in his person civil and religious prerogatives. This explanation is obviously very strained. Another authority thought that the first pontiffs were so called because sacrifices were supposed to have been made on a bridge (*faciebant in ponte*).[85] This explanation is even more strained.

The word "pontiff" is not of Latin origin. It is not derived from *pons*, but probably from Punt. When it is said that Queen Hatshepsut, after visiting Punt, built a "punt" for the god Amon, this means a sacred place of worship. By erecting a "punt" in Egypt, Queen Hatshepsut also introduced the institution of the high priest, copying the service of the Temple in Jerusalem, built on a Phoenician model.

Solomon's alliance with Hiram, the king of the Phoenicians, explains the strong Phoenician influence in the life of the kingdom of Judah and Israel. This influence is stressed in the Scriptures in the story of the erection of the Temple, built with the help of Hiram, who provided Solomon with building material and with the chief craftsman, a man of Hebrew-Phoenician origin (I Kings 7:13–14). Also the common expedition to Ophir and the peaceful transfer of territory from the domain of one king to that of the other (I Kings 9:11) might have brought it about that the whole of Palestine at that time was called Phoenicia.

The name Punt being the origin of the word "pontiff" (pontifex), what is the origin of the name Punt?

In the Scriptures the Phoenicians are called "the men of Sidon and Tyre" or "the men of Hiram"; the name Phoenicia does not appear. "Phoenicians" was the name used by Greek and Latin authors since Homer.[86] Rome waged so-called "Punic Wars" against Carthage, which was built by immigrants from Tyre.

It is believed that the Greek explanation of the word "Phoenicians" as "the red men"[87] is but a successful adjustment in the style of folk-etymology, although the travels of the Phoenicians to the Western Hemisphere and their contacts with the cultures of the Mayas and Incas were taken into consideration by many scholars who studied pre-Columbian America. The other Greek explanation of the word "Phoenicia" as "the land of palms" is generally rejected.[88]

Pontus, father of Sidon, was a legendary ancestor of the Phoenicians,[89] and their name could have been derived from him, or the name of the mythological ancestor could have been derived from the name of the country.

If Punt was originally the word for Phoenician temples, then it could have been derived from the Hebrew word *panot*, and in this case the Phoenicians received their name from the houses of worship they built.[90]

Even before the conquest of Joshua the land of Jerusalem was called in Egyptian inscriptions Divine Land, God's Land (*Toneter*). Was Jerusalem a holy place before David conquered it, and even before the arrival of the Israelites under Joshua?

In the Bible there is an allusion to the holiness of Jerusalem in early times and to a sanctuary in that place. When the patriarch Abraham returned from pursuing the kings of the north, who had captured his kinsman Lot, "Melchizedek king of Salem [Jerusalem] brought forth bread and wine: and he was the priest of the Most High God" (Genesis 14:18).

The name Divine (or Holy) Land, given to the region of Jerusalem in Egyptian inscriptions of the Old and Middle Kingdoms, casts light upon the religious significance of Jerusalem and Palestine generally in the days before David, even as early as the days when the Israelites were still nomads. Since then and up to the present day they have been called "the Holy City" and "the Holy Land".

The rivalry between the Arabian and Ethiopian traditions for the Queen of the South can be decided against the Arabian claim and in favor of the Ethiopian, but only to the extent that she was the "queen of Egypt and Ethiopia", which is the true description by Josephus; this does not necessitate an approval of genealogical claims embodied in the Ethiopian tradition.

Mohammed, who endorsed the Arabian claims, was obviously wrong. He put into the mouth of Solomon the following words: "I have compassed what ye compassed not; for I bring you from Seba [Saba] a sure information: verily, I found a woman ruling over them, and she was given all things, and she had a mighty throne; and I found her and her people adoring the sun instead of God."[91] The land of Saba (in Hebrew Shwa), because of the similarity in name, had confused the writers on the Queen of Sheba even before Mohammed borrowed the last sentence of the quoted Sura from the Hebrew Haggada, which he probably had heard from the Jewish teachers of Medina.

The Ethiopians are not satisfied merely to claim the Queen of the South as their queen; they insist that a child was born of her liaison with Solomon; this son, Menelik, is the direct ancestor of the dynasties of Abyssinian monarchs, the present royal house included. Being of David's seed, the son of Solomon and the Queen of the South, their legendary ancestor, is regarded by them as kindred to Jesus, who through Joseph, the carpenter of Nazareth, also traced his ancestry to David.[92] Venerating the Queen of the South, who returned from her visit pregnant with royal seed, the Ethiopians honor more than any other passage in the Gospels the words: "He answered and said unto them ... An evil and adulterous generation seeketh after a sign. ... The queen of the south shall rise up in the judgment with this generation, and shall condemn it: for she came from the uttermost parts of the earth to hear the wisdom of Solomon; and, behold, a greater than Solomon is here."[93]

The Abyssinian tradition is put down in writing in *Kebra Nagast* or *The Book of the Glory of Kings*.[94] The existing version in Ethiopian is a translation from an Arabic text, which in turn was translated from the Coptic. It contains quotations from the Gospels and is therefore a fruit of the time when Christianity had already found its way to the African continent, in an early century of the present era.

With colorful imagination *Kebra Nagast* recounts the bridal night of Solomon and the Queen of the South; among the presents he gave her there were "a vessel wherein one could travel over the sea, and a vessel wherein one could travel by air".

When the queen returned to her country, "her officials who had remained there brought gifts to their mistress, and made obeisance to her, and did homage to her, and all the borders of the country rejoiced at her coming. . . . And she ordered her kingdom aright, and none disobeyed her command; for she loved wisdom and God strengthened her kingdom." This quotation from *Kebra Nagast* resembles the story of the festival for the officials and for the whole rejoicing land, arranged by Queen Hatshepsut after her return from her journey; so do the words that "she ordered her kingdom aright" and that she "loved wisdom". But there is nothing so extraordinary in these things as to compel the conclusion that Ethiopian tradition about the Queen of the South knows more than the Scriptures narrate. Even the romance might have been borrowed from a Jewish source,[95] which in a single line says that the king responded to the desire of the guest queen. In the Jewish tradition there is nothing about a child having been born of this intimacy.[96]

It would, of course, strengthen the claim to originality of the Ethiopian tradition if it disclosed some fact not contained in the Scriptures, which could be checked with the help of our knowledge about Queen Hatshepsut, and which would be more than an accidental coincidence. Even in this case it would not necessarily mean that—in the words of *Kebra Nagast*—Solomon "worked his will with her", and a child of that union was enthroned in Aksum, "the New Jerusalem"; but it would show that the Ethiopian legend about the Queen of the South going to Jerusalem is not entirely a fanciful addition to the scriptural story, like the legend of Bilkis, the Queen of Saba of the Arabian authors.

There is a detail in the Ethiopian legend which only by a rare chance could have been invented. The Ethiopians call the Queen of the South Makeda. The royal name of Queen Hatshepsut, mentioned throughout the Punt reliefs, is Make-ra. "Ra" is the divine name of a god.[97] The main part of the name of the Egyptian queen is identical

126

with the first two syllables in the name of the Queen of the South. It was preserved in the Ethiopian tradition; it did not come from the Scriptures.

One can imagine that if the name was not handed down by an uninterrupted tradition it could have been disclosed by some Copt, who might have lived in early Christian times in Egypt, seen the Punt texts in Deir el Bahari, and been able to read them, and in this way might have identified Hatshepsut with the Queen of Sheba ahead of the present author. There may have been a chronological reason, too, for such a hypothetical Copt to identify Hatshepsut with the Queen of the South, or he might have heard a legend that the reliefs of Deir el Bahari do represent a voyage to Jerusalem. The same theory could be applied to Josephus, who might have written "queen of Egypt and Ethiopia" on the basis of the scenes of the bas-reliefs at Deir el Bahari; he might have mentioned the *kussiim* (negroids) because they were in the picture. This is a forced construction (Josephus was never in Egypt); on the other hand, the historical facts known to Josephus and not preserved in the Scriptures must have been transmitted by some means during the one thousand years which separated Josephus from Solomon.

Thutmose I, the father of Hatshepsut, conquered the northern part of Ethiopia known as Nubia. It is of interest that in Egyptian documents the viceroy of Ethiopia (Nubia) was called "king's son", which is supposed to be only a title, without implying a blood relationship with the Egyptian king.[98] The name of the "king's son" in the time of Hatshepsut is not preserved; in the days of her successor, Thutmose III, the viceroy of Ethiopia was named Nehi.

Another incident in the Ethiopian legend—the robbing of the Temple in Jerusalem—will be told in the next chapter. The actual successor to Hatshepsut on the Egyptian throne was the one who sacked the Temple, a deed attributed to the putative son of Solomon and the Queen of the South.

We took a short leave of the historical material to investigate the Abyssinian legend, and now we should like also to take a look at one or two Hebrew legends about the Queen of Sheba. Having become acquainted with the historical person, we are interested to know what stimulated the folk fantasy and how it worked

We have already mentioned the divine command heard by Queen

127

Hatshepsut compelling her to undertake her expedition to the Divine Land. On the murals, in the coronation scene and in other scenes, Hatshepsut is portrayed before the god Horus with the head of a hawk; a serpent of Lower Egypt or a vulture of Upper Egypt, as royal emblems, are also often pictured with her.[99]

A curious legend in the Haggada[100] narrates that the Queen of Sheba, while on her way one morning to pay homage to the god of the sun, received a message from a bird summoning her to visit Solomon in Jerusalem.

In the inscriptions Hatshepsut is called king; the pronoun used for her is sometimes "she" and sometimes "he"; on the pictures her raiment is that of a king. She is called the daughter of Amon, but in the picture of her birth a boy is molded by Khnum, the shaper of men. It was unusual, and contrary to the political and religious conceptions of the Egyptians, to have a woman ruling on the throne; therefore she disguised herself and assumed the attributes of a man. On many of her statues and bas-reliefs she is portrayed with a beard. There is a well-known legend that Solomon, at his first meeting with the Queen of Sheba, said to her about the hair on her skin (her legs were reflected in a mirrored floor): "Thy hair is masculine; hair is an ornament to a man, but it disfigures a woman."[101]

Rabbi Jonathan in the third century of the present era maintained that it was a king and not a queen of Sheba who visited Solomon. Egyptologists of the first half of the nineteenth century pictured and described Hatshepsut as a king, being misled by some of her statues and the masculine pronoun she applied to herself.

Could it be that, a few centuries after Hatshepsut, the pictures of Deir el Bahari, seen by visitors to Egypt, gave rise to these two strange legends?

Did Hatshepsut Visit the Land of the Queen of Sheba?

The Queen of Sheba who visited King Solomon lived in the tenth century before this era. It is not known whence she came—from the land of Saba or from Ethiopia. In the sixteenth century before this era Queen Hatshepsut undertook an expedition to the Divine Land and Punt. It is not known where Punt was, but it is thought that it was either in the land of Saba or on the Somali shore of Ethiopia.

128

Accordingly Queen Hatshepsut might have visited the land of the Queen of Sheba six centuries before this queen made her voyage to Jerusalem.

As the "marvels" brought by Hatshepsut from Punt are not unlike the "marvels" brought by Hiram and Solomon from Ophir, a few authors have ventured to identify Punt and Ophir, though the whereabouts of the latter place has not yet been established. According to that theory, the Phoenician king Hiram undertook a voyage of discovery to Punt, visited by Hatshepsut centuries earlier.[102]

The reconstruction of history offered here, reducing the age of the New Kingdom of Egpyt by almost six centuries, places Queen Hatshepsut in the tenth instead of the sixteenth century and makes her a contemporary of Solomon. Consequently, I have argued: It is said about Solomon that his fame spread to faraway countries, and all kings sought his presence; the same is told of Queen Hatshepsut; if they were contemporaries, it is strange that they should not have come in contact with each other.

To prove that the Queen of Sheba and Queen Hatshepsut were one and the same person, I had to show that the Queen of Sheba came from Egypt and that Queen Hatshepsut visited Palestine. Regarding the former, there is a definite statement by Josephus that the queen came from Egypt. This statement has been unduly neglected. Fortunately we also have an illustrated diary of Queen Hatshepsut, which gives an account of an expedition to a foreign country. That she participated in the expedition is not difficult to establish by explicit statements, in which she calls herself the leader of the expedition.

I also had to show that Punt and the Divine Land are Phoenicia and Palestine. Repeated references to Punt as a country east of Egypt excluded Somaliland. The return of the ships of the expedition to the Nile harbor of Thebes excludes southern Arabia and Somaliland and places Punt on the shore of the Mediterranean.

I then compared the scriptural record of the visit of the Queen of Sheba and the Egyptian account of the expedition to the Divine Land and found complete accord.

The journey of the "remarkable woman who ruled Egypt", who came "with a great company" into the Divine Land; "her astonishment at what she saw" and what excelled "the hearsay"; "the terraces of trees" she wondered at, the gifts she exchanged, among them

129

the marvels brought by Hiram from Ophir, the apes and other animals, silver, rare until that time, ebony, ivory and precious stones, incense and myrrh "without number", and trees "never seen before"—all this is found alike in the scriptural version of that voyage, in Josephus' story, and in the inscriptions and pictures of the Egyptian temple built by the queen after her return to Thebes. This temple impressed the scholars by its foreign architecture, and the queen herself emphasized that it was an imitation of what she had seen in Punt. New forms of services were introduced, with twelve priests and a high priest officiating.

The complete agreement in the details of the voyage and in many accompanying data makes it evident that the Queen [of] Sheba and Queen Hatshepsut were one and the same person.[103] Punt was Palestine-Phoenicia, and the Divine Land or Holy Land, the Holy Land of Jerusalem. The people of the "Caucasian" or "north-Semitic race" were the Jews. The chief Paruah who met the expedition of Hatshepsut in the harbor was the governor of Solomon in Ezion-Geber. The apes and other exotic animals, which led the historians to the conclusion that Punt was in Africa, were actually brought to Jerusalem on the ships of Solomon and Hiram. The exotic plants were also brought from afar.

There remains to be explained the provenance of one plant exported from Punt. Long before the expedition of Queen Hatshepsut and for some time thereafter, Punt and Divine Land were repeatedly mentioned in Egyptian documents as places producing frankincense. Because of the frankincense, Punt was thought to be in southern Arabia. I leave it for the next chapter to show that frankincense grew in Palestine.

NOTES

1. Breasted. *Records*, Vol. II, Sec. 81.
2. Thutmose I described the domain of his influence—from Ethiopia to the land of "the inverted water—the river that flows upstream". It is generally supposed that he meant the Euphrates, as the Egyptians had the idea that a river like the Nile must flow from the south to the north.

    The northernmost area of Thutmose's domain was Edom in southern Palestine. By "the river that flows upstream" is meant the Jordan. This will be explained at length in another place. But if the Egyptians were amazed by a river flowing southward, it would be the Jordan, which is closer to Egypt, rather than the Euphrates.

3. Scholars following the established construction of history could not close their eyes to the similarity of these enterprises: "... ambitious and inventive was Solomon's policy of ... developing a maritime route on the Red Sea. The old vigor of Egypt as displayed by Queen Hatshepsut in her navigation of those waters had long since disappeared." J. A. Montgomery, *Arabia and the Bible* (Philadelphia, 1934), p. 176.

4. The Punt reliefs in Breasted, *Records*, Vol. II, Sec. 269. The complete record may be found in Edouard Naville, *The Temple of Deir el Bahari* (Memoirs of the Egyptian Exploration Fund, London, 1894–1908, Vols. 12–14, 16, 19, 27, 29).

5. The Karnak obelisk. Breasted, *Records*, Vol. II, Sec. 325.

6. I Kings 4:34.

7. Cf. J. Halévy, "La Légende de la reine de Saba", *Annuaire, Ecole pratique des Hautes Etudes*, 1905 (Paris, 1904); L. Legrain, "In the Land of the Queen of Sheba", *American Journal of Archaeology*, 38 (1934).

8. *Kebra Nagast*, translated from the Ethiopian by E. A. W. Budge as *The Queen of Sheba and Her Only Son Menyelek, being the Book of the Glory of Kings* (Oxford, 1932).

9. Compare Babylonian Talmud, Tractate Baba Batra 15b.

10. See Halévy, *Annuaire, Ecole pratique des Hautes Etudes*, 1905 (1904).

11. See L. Legrain, *American Journal of Archaeology*, 38 (1934), 329–37. Systematic excavations have been possible in southern Arabia only in the last few years.

12. *Kebra Nagast* (trans. Budge), p. vii. See also Halévy, *Annuaire Ecole pratique des Hautes Etudes*, 1905 (1904), 6.

13. Josephus did not name the queen.

14. Josephus, *Jewish Antiquities*, VIII, 165.

15. The country on the Nile south of the Second Cataract, or the modern Sudan, was then described as Ethiopia.

16. Naville, *Deir el Bahari*, Introductory Memoir, p. 1.

17. Breasted, *Records*, Vol. II, Sec. 246.

18. Suggested by Chabas.

19. Naville, *Deir el Bahari*, Pt. III, p. 12.

20. "The animals represented in the sculptures are exclusively African, as is also a part of the population" (ibid.). See E. Glaser, "Punt und die südarabischen Reiche", *Mitteilungen, Vorderasiatisch-ägyptische Gesellschaft* (Berlin, 1899), Vol. IV, p. 62.

21. A. Lucas, *Ancient Egyptian Materials and Industries* (2nd ed.; London, 1934), p. 93; W. H. Schoff, *The Periplus of the Erythraean Sea* (New York, 1912), p. 218.

22. Breasted, *Records*, Vol. II, Sec. 892.

23. Ibid.; Montgomery, *Arabia and the Bible*, p. 176, n. 28.

24. Philo of Byblos as quoted by Eusebius in *Preparation for the Gospel*, I, 10, 27.

25. *Um seine Lage genauer zu bestimmen, sind wir ausschliesslich auf die Abbildungen angewiesen, die von den Einwohnern und den Produkten des Landes überliefert werden.*" Glaser, *Mitteilungen, Vorderasiatisch-ägyptische Gesellschaft*, IV (1899), 53.

26. Lucas, *Ancient Egyptian Materials* (2nd ed.), p. 93.

27. Glaser, *Mitteilungen, Vorderasiatisch-ägyptische Gesellschaft*, 1899, p. 33f.

28. Herodotus, I, 1 and VII, 89. The designation "Eritrean Sea" covered all of the Indian Ocean as well as the Red Sea.

29. See R. Lepsius, *Nubische Grammatik* (Berlin, 1880). Compare Glaser, *Mitteilungen, Vorderasiatisch-ägyptische Gesellschaft*, 1899, pp. 33f.

30. Cf. the paper of Newberry, "Three Old Kingdom Travellers to Byblos and Pwenet", *Journal of Egyptian Archaeology*, XXIV (1938), 182–84.

31. See the various opinions presented by G. Maspero, *The Struggle of the Nations* (New York, 1897), p. 247.

32. G. Maspero ascribed to Eduard Meyer the belief that the inhabitants of Punt were the ancestors of the Sabeans (*Geschichte des Altertums*, p. 234); however Meyer thought Punt was in Africa.

33. *Neterto* [Toneter] is translated by Naville as Divine Land and by Breasted as God's Land.

34. J. Dümichen, "Geographie des alten Aegypten", in E. Meyer, *Geschichte des alten Aegypten* (Berlin, 1879–87); J. Lieblein, *Handel und Schiffahrt auf dem Roten Meere in alten Zeiten* (Christiania [Oslo], 1886); Glaser, *Mitteilungen, Vorderasiatisch-ägyptische Gesellschaft*, 1899. See also Naville, *Deir el Bahari*, Introductory Memoir, p. 22: "... Hatshepsu's fleet undoubtedly sailed for the coasts of Africa and not for those of Arabia, but we are not justified in limiting the land of Punt to the African coast alone. . . . The land to which their [the Egyptian] religious texts ascribe an almost legendary character, lay upon both shores of the southern end of the Red Sea."

35. Since olden times the harbor of el-Qoseir on the Red Sea has been mentioned as the starting point of travel to the Divine Land.

36. Plato, *Timaeus*, 25.

37. Ginzberg, *Legends*, IV, 144.

38. Ibid.

39. The Koran, Sura XXVII (trans. E. H. Palmer).

40. Also in *Kebra Nagast* (trans. Budge), the Ethiopian legend about "the Queen of the South", an Ethiopian emissary by the name of Tamrin, possessor of three and seventy ships and leader of a merchant caravan, visited Jerusalem and on returning to his queen "he related unto her how he had arrived in the country of Judah and Jerusalem . . . and all that he had heard and seen".

41. According to the reading "on both sides of the sea", the harbor was situated on opposite shores. Ezion-Geber was built by Solomon on the Gulf of Aqaba, where both shores can be seen.

42. It appears that the last word in I Kings 4:16 belongs to the next verse, and the last word of 4:17 to the following verse. The reading then would be: "... and in Aloth Jehoshaphat the son of Paruah." In this case the son remained governor where his father had served in the same capacity, Aloth and Eloth being the same. In a context having no relation to the question presented here, Albright (*Journal of the Palestine Oriental Society*, V [1925], 35) made the same suggestion that the place Aloth be transferred to the next verse, into the domain of Jehoshaphat, son of Paruah. See also J. W. Jack, *Samaria in Ahab's Time* (Edinburgh, 1929), p. 95.

43. Breasted, *Records*, Vol. II, Sec. 253.

44. "Hatshepsu did not prize her military laurels as highly as her naval expedition to the land of Punt.... The considerable space which these sculptures cover, the fullness of the details, and the exquisiteness of the work, all prove how highly the queen valued the achievements of her ships, and took pride in their results." Naville, *Deir el Bahari*, Pt. III, p. 11.

45. Breasted, in *Records*, Vol. II, Sec. 247, collected the earlier references to voyages to the land of Punt. "None of these sources contains more than the meagerest reference to the fact of the expedition."

46. Naville, *Deir el Bahari*, Pt. III, p. 11.

47. Breasted, *Records*, Vol. II, Sec. 285.

48. Josephus, *Jewish Antiquities*, VIII, 165f.

49. With the possible exception of Sebeknofrure at the end of the Twelfth Dynasty.

50. Breasted, *Records*, Vol. II, Sec. 288.

51. The Egyptians are pictured beardless; only the gods and the pharaohs are shown with beards.

52. M. Grünbaum, *Neue Beiträge zur semitischen Sagenkunde* (Leiden, 1893), p. 213; Ginzberg, *Legends*, IV, 145.

53. Cf. the article "Incense" in Encyclopedia Biblica, Vol. II, col. 2167, concerning the aromatic substances used in the Herodian Temple.

54. Genesis 43:11.

55. Proverbs 7:17.

56. Breasted, *Records*, Vol. II, Sec. 288.

57. Josephus, *Jewish Antiquities*, VIII, 167, which follows I Kings 10:2–3.

58. Naville, *The Temple of Deir el Bahari*, p. 22.

59. Breasted, *Records*, Vol. II, Sec. 274.

60. *Jewish Antiquities*, VIII, 170–171.

61. Breasted, *Records*, Vol. II, Sec. 278.

62. Breasted, *Records*, Vol. II, Sec. 352.

63. Ibid., Sec. 375.

64. A "house of silver" may signify the treasury, but "a floor wrought with gold and silver" must be understood as made of these metals.

65. Anti trees are termed myrrh by Breasted, and frankincense by Naville, identified as Boswellia Carteri by Schoff. Lucas, *Ancient Egyptian Materials*, p. 93.

66. In the opinion of some scholars the tribute-paying Negroes signify that beside the expedition to Punt there was another expedition to the African region of Khenthenofer, the bas-reliefs bringing together what geographically was divided.

67. Hiram was the traditional or often recurring name of the kings of Tyre. See Ginzberg, *Legends*, V, 373.

68. Ophir may signify generally Africa. Different theories place Ophir in Africa on its east coast, in Arabia, in the Persian Gulf, the coast of India, Ceylon, Malaya, China, Spain, the West Indies, and Peru, and also in many other countries. In the West Indies, in Australia, and on Madagascar there are no apes. Peacocks abound in South America and Australia. The presence of silver in Ophir and the three years needed for the voyage and return, starting in the Red Sea, are important indications. Necho II sent a Phoenician expedition around Africa; they sowed and reaped on the way, and it took four years to circumnavigate the continent.

69. *Jewish Antiquities*, VIII, vii, 2.

70. See a note by R. Marcus to his translation of Josephus, *Jewish Antiquities*, VIII, vii, 2 referring to the opinion of Weill.

71. Meyer, *Geschichte des Altertums*, II, i (2nd ed., 1928), 117.

72. Herodotus, II, 158. F. H. Weissbach, *Die Keilinschriften der Achämeniden* (Leipzig, 1911), p. 105.

73. Breasted, *Records*, Vol. II, Sec. 274.

74. Ibid., Sec. 295.

75. Ibid., Sec. 294.

76. "Mariette, struck by the strange appearance of the edifice, thought that it betrayed a foreign influence, and supposed that Queen Hatshopsitu [Hatshepsut] had constructed it in the model of some buildings seen by her officers in the land of Puanit." (*Deir el Bahari* [Leipzig, 1877], pp. 10–11, cited by G. Maspero in *The Struggle of the Nations*, p. 241, note 2.)

77. A. Mariette, *Deir el Bahari*, quoted by Naville in *The Temple of Deir el Bahari*, Introductory Memoir, p. 1. However, a more ancient temple of similar architecture was discovered in the vicinity; it probably represents, too, a Phoenician influence.

78. "Though she [Hatshepsut] was a sovereign fond of building and erecting edifices like that of Deir el Bahari which are accounted the most beautiful left to us by Egyptian antiquity, she did not make a useless display of gigantic buildings in the desire to dazzle the posterity as did Ramses II." Naville in Davis, *The Tomb of Hatshopsitu*, p. 73.

79. Psalms 120–34: "Songs of degrees."

80. See H. E. Winlock, *Excavations at Deir el Bahri, 1911–1931* (New York, 1942), pp. 134ff.

81. Naville, *Deir el Bahari*, Introductory Memoir.

82. "Solomon . . . wanted palaces and gardens and a temple, which might rival, even if only in a small way, the palaces and temples of Egypt and Chaldea, of which he had heard such glowing accounts." Maspero, *The Struggle of the Nations*, p. 741. "Compared with the magnificent monuments of Egypt and Chaldea, the work of Solomon was what the Hebrew kingdom appears to us among the empires of the ancient world—a little temple suited to a little people." Ibid., p. 747.

83. Breasted, *Records*, Vol. II, note to Sec. 679. "The queen was conscious of the resemblance of the temple-gardens in Deir el Bahari and Punt. The service and equipment of the temple receive some light from the mention of its High Priest, with twelve subordinate priests in four orders." Ibid., note to Sec. 291.

84. Ibid., Sec. 388. But according to G. Lefebvre, the office of the high priest was already established by Ahmose (*Histoire des grands prêtres d'Amon de Karnak* [Paris, 1929], p. 69).

85. Plutarch, *Lives*, "Numa", 9. Numa is said to have introduced the institution of the high priest or pontifex in Rome. Compare A. Bouché-Leclercq, *Les Pontifes de l'ancienne Rome* (Paris, 1871).

86. *The Odyssey*, XIII, 272; XIV, 288ff.

87. V. Beard, "Le Nom des Phéniciens", *Revue de l'histoire des religions*, 93 (1926), 187ff.; G. Contenau, *La Civilisation Phénicienne* (Paris, 1926), p. 356; *Syria, Revue d'art oriental et d'archeologie*, VIII (1927), 183; Bonfante, "The Name of the Phoenicians", *Classical Philology*, XXXVI (1941), 1–20.

88. K. Sethe in *Mitteilungen, Vorderasiatisch-ägyptische Gesellschaft*, XXI (1917), 305.

89. Eusebius, *Preparation for the Gospel*, I, 10, 27.

90. *Panot* in Hebrew means to face, incline, address, turn to. It is applied innumerable times in connection with worship. Cf. Maimonides, chapter "On homonyms in the Bible", in *Guide for the Perplexed*. Of the same root is "Presence" (of the Lord)—an idea found with the Phoenicians of Carthage. Cf. W. F. Albright, *From the Stone Age to Christianity* (Baltimore, 1940), p. 228.

91. The Koran, Sura XXVII (trans. Palmer).

92. "They never doubted that Solomon was the father of the son of the Queen of Sheba. It followed as a matter of course that the male descendants of this son were the lawful kings of Abyssinia, and as Solomon was an ancestor of Christ they were kinsmen of our Lord, and they claimed to reign by divine right." Budge, *Kebra Nagast*, p. x.

93. Matthew 12:42; Luke 11:31.

94. "The *Kebra Nagast* is a great storehouse of legends and traditions, some historical and some of a purely folklore character, derived from the Old Testament and the later Rabbinic writings, and from Egyptian (both pagan

and Christian), Arabian and Ethiopian sources. Of the early history of the compilation and its maker, and of its subsequent editors we know nothing, but the principal groundwork of its earliest form was the traditions that were current in Syria, Palestine, Arabia, and Egypt during the first four centuries of the Christian era" *Kebra Nagast* (trans. Budge), pp. XV–XVI.

95. 2 *Alphabet of Ben Sira* 21b. Ginzberg, *Legends*, VI, 289.

96. 2 *Alphabet of Ben Sira* 21b also states that Solomon married the Queen of Sheba.

97. Likewise "Da" could be the divine name Adad or Ada, which is a part of several scriptural names.

98. G. A. Reisner, "The Viceroys of Ethiopia", *Journal of Egyptian Archaeology*, VI (1920), 31.

99. Naville, *Deir el Bahari*, Pt. II, Plates 35, 38, 39; ibid., Pt. III, Plate 58, etc.

100. Ginzberg, *Legends*, IV, 143.

101. Ibid., IV, 145.

102. Maspero, *The Struggle of the Nations*, p. 742. See also Karl Peters, *Das goldene Ophir Salomos* (Munich, 1895); Eng. trans., *King Solomon's Golden Ophir* (New York, 1899).

103. Shwa (the Hebrew for Sheba) might be the last part of the name Hatshepsut. R. Engelbach, *The Problem of the Obelisks* (London, 1923), spells her name Hatshepsowet. The final *t* in her name was not pronounced. Naville (*Deir el Bahari*) spells it Hatshepsu. It was usual to shorten the Egyptian names: so Amenhotep was often shortened to Hui.

CHAPTER FOUR

The Temple in Jerusalem

*Thutmose III Prepares the Disintegration
of the Empires of Solomon*

During the last period of her reign Queen Hatshepsut made Thutmose III co-ruler with her on the throne. At first he played a subordinate role: his name was written after hers; in pictures his figure was placed behind hers. The bas-reliefs of the Punt expedition show the young prince as a small figure in the background, bringing an offering of incense to the bark of Amon-Ra. Later, when he reigned alone, Thutmose III became the greatest of all the conquerors who sat on the throne of the New Kingdom in Egypt.

His military expeditions were directed into Palestine and Syria. He subdued these countries; some of the cities were vanquished by force; others bowed before him, opened their gates, and became tributaries.

The records of his military successes, cut in hieroglyphics on the walls of the great Amon temple in Karnak, recount his campaigns in Palestine and Syria. A list of one hundred and nineteen cities in Palestine is engraved three times on the walls of that temple. Each city is represented by a man with his arms bound behind him, and a shield covering the body of the man bears the name of the city he symbolizes. Another list, imperfect and in one copy, shows almost three hundred Syrian cities as captives, also as men bound and with shields on their breasts.

A bas-relief at Karnak shows the treasures in gold, silver, bronze, and precious stones that Thutmose III brought as booty from one of his campaigns; a series of other murals exhibits the flora and fauna he transported from Palestine to Egypt.

These campaigns are unanimously considered to have been waged in a Canaan which had not yet heard of the Israelite tribes. The places named in the list of cities are regarded as settlements of the Canaanites; the booty taken in the conquest contains examples of the art that flourished in Canaan long before the Israelites, led by Joshua, came to that land. Their exodus and their entry into Canaan were still far in the future when Thutmose I conquered Canaan and Syria. It is also sometimes said that if the Israelites in his day had already become organized into a tribal sept they were still obscure nomads in the highlands between the Euphrates and the Nile.

Thutmose III, the conqueror of Palestine, was the heir of Hatshepsut; according to the chronological scheme advanced here, he reigned in the later years of Solomon and at the time of Solomon's son and successor, Rehoboam. If the reconstruction of history undertaken in this work is correct, Thutmose's victorious march through Palestine must have taken place in the years closely following the reign of Solomon, and a record of it should be preserved in the Scriptures. The conquest by Thutmose III was a triumphant sweep across Palestine, and the chronicles of Judah and Israel could not possibly have omitted it; indeed, a conspicuous record of it must be there.

Once more this reconstruction is put to the test. The absence of a scriptural record of the conquest of Palestine by an invading Egyptian army under the personal leadership of a pharaoh, in the years following the reign of Solomon, would be regarded as irrefutable evidence against it.

But this record is preserved in the Scriptures, and it coincides completely with the Karnak inscriptions.

Two monarchies rose from the ruins of the Amalekite Empire: Israel and Egypt. Israel, the vanquisher of Amalek, got a large share; his legacy ran from the Euphrates to Egypt, embracing the kingdoms of Syria, Canaan, and Edom, deep into the Arabian peninsula; tribute came to Israel from north, east, and south, and his land dominated the trade between Asia and Africa and the terminals of two great sea routes, the Mediterranean and the Red Sea.

Egypt, the ancient house of bondage, was freed from the Hyksos tyranny by the Israelites under Saul. But Egypt rewarded good with evil. The tales of the splendor of Jerusalem, enriched by conquest and trade, inspired Hatshepsut with the desire to behold its wealth herself. Thutmose III, as a young prince, either participated in this

138

expedition into "God's Land" or heard from Hatshepsut about the "luxurious land of delight"; and he saw the gifts which were said to be marvels. The bas-reliefs of the Punt expedition were bound to keep those experiences fresh in his mind. His eyes looked with envy on that land and on the wealth accumulated there.

Solomon, whose name is a synonym for a wise, even the wisest, man, must be charged with most grievous political errors. In the latter part of his reign the builder of the Temple in Jerusalem became a worshiper of strange gods and a builder of high places to them. It is said that he built on the hills around Jerusalem sanctuaries to foreign deities because of the strange women he loved: ". . . his wives turned away his heart" (I Kings 11:4).

Solomon planned, through his marriages with the daughter of Pharaoh and with the "women of the Moabites, Ammonites, Edomites, Zidonians, and Hittites", to build a cosmopolitan center at Jerusalem. In erecting shrines to strange gods, he thought that Jerusalem, as the abode of the deities of foreign nations, would demonstrate his tolerance and that Jerusalem would become the gathering place for various religions and cults.

Since the Egyptian princess is mentioned as the first among the wives of Solomon, the statue of Amon-Ra, the official supreme god of Egypt, should have been among the gods of all of Solomon's "strange wives" (I Kings 11:5–8).

Among the murals of the Punt expedition, one, defaced by a chisel, contained an inscription, and from the few words which remain it can be understood that a statue, obviously of the god Amon-Ra, was erected in the Divine Land.[1] The statues of pagan gods were destroyed by later kings of the House of David[2] and cannot be found in Jerusalem; but the text of the eleventh chapter of the First Book of Kings, just following the story of the Queen of Sheba, helps to cast light on the damaged inscription.

Jerusalem, with its foreign deities and cults, did not become a political center for strange peoples, but it rapidly became the goal of their political aspirations. Already in the oracle scene of the Punt expedition it is said in the name of the god Amon-Ra: "It is a glorious region of God's Land, it is indeed my place of delight. I have made it for myself. . . . I know [them], I am their wise lord, I am the begetter Amon-Ra."

This was written in the days of the peaceful Hatshepsut. The

139

conversion of Jerusalem into a holy city for all nations doomed its national independence.

The dismembering of the state was provoked by Solomon's policy. In the days of Saul and David the state attained independence. The heyday of imperialistic power was very short, political ascendancy reaching its culmination in the reign of Solomon; at the end of his reign the rapid decline started.

Three adversaries rose against Solomon, and they began to rend his realm apart. The first of these was Hadad the Edomite.[3] In the time of David, Hadad, then still a child, fled from Edom into Egypt and later married into the pharaoh's family. On David's death he returned to his country. When Solomon built the harbor of Ezion-Geber the Edomite land was temporarily pacified. Hadad, Solomon's adversary in Edom, succeeded in stirring up this region into a state of unrest. Related to the house of Pharaoh and coming from Egypt, Hadad obviously was supported by the bearer of the double crown of Egypt.

Another adversary, Rezon, who fled from his lord Hadadezer, king of Zobah, became captain of a band and ruled in Damascus.

Thus, these two legacies of the Amalekite Empire, Arabia and Syria, turned away from the House of David. Even the land of the Twelve Tribes was divided when Solomon closed his eyes.

Solomon's third adversary, Jeroboam the Ephraimite, Solomon's subject and servant, whom he made "ruler over all the charge of the house of Joseph", secretly harbored the ambitious intention of making the land of Ephraim independent. Solomon, becoming aware of the germs of conspiracy in time, sought to kill Jeroboam.

> I KINGS 11:40 ... And Jeroboam arose, and fled into Egypt, unto Shishak king of Egypt, and was in Egypt until the death of Solomon.[4]

When Solomon died, leaving the throne to his son Rehoboam, the conspirators, backed by the Egyptian king, were ready to act.

> I KINGS 12:3–4 ... they [his adherents] sent and called him [Jeroboam to return from Egypt]. And Jeroboam and all the congregation of Israel came, and spake unto Rehoboam, saying:
> Thy father made our yoke grievous. . .

The negotiations ended with the proclamation: "What portion

have we in David? ... to your tents, O Israel. ... So Israel departed unto their tents." But the cities of Judah and Benjamin remained faithful to Rehoboam.

Rehoboam sent Adoram, who was in charge of collecting the taxes, and "all Israel stoned him with stones" (I Kings 12:18). "So Israel rebelled against the house of David" (I Kings 12:19). Rehoboam gathered the warriors of Judah and Benjamin to fight against Israel, but the prophet Shemaiah called every man to return to his house and not to fight against his brethren (II Chronicles 11:2–4).

Jeroboam girded Shechem in Mount Ephraim, and Penuel; to keep the people from going to Jerusalem to sacrifice, he set up images of foreign deities in Beth-el and in Dan (I Kings 12:28–29), two old sanctuaries of Israel, holy long before Jerusalem was conquered by David.

In the northern kingdom Jeroboam introduced as the official cult the worship of bullocks, which may have had something in common with the Egyptian cult of the bull Apis. He received political power from the mighty pharaoh, for he had to rend Israel from Judah. From the very outset he was a political and cultural vassal of Egypt. The prophet Shemaiah could hear the steps of Jeroboam's master behind him. Fighting between Judah and Israel accorded completely with the plan of Thutmose III.

The annals of Thutmose III, after giving the date of the first victorious expedition to extend the boundaries of Egypt, record:

ANNALS OF KARNAK: Now at that time the Asiatics had fallen into disagreement each man fighting against his neighbor.[5]

A victory over a foe weakened by internal discord is a diluted triumph. Why, then, do the annals mention this discord in the land of Pharaoh's foes? It was the work of Thutmose III himself to prepare the disunity by setting one part of the population against the other; hence this record does not detract from his right to laurels.[6]

Rehoboam felt the danger. He rushed to wall the cities, and to those fortified in the time of his father Solomon and his grandfather David he added Bethlehem, Etam, Tekoa, Beth-Zur, Socoh, Adullam, Gath, and Mareshah, and other strongholds.[7]

141

Four years had elapsed since Solomon died, and the pharaoh already was on the move northward.

Thutmose III invades Palestine

Thutmose III advanced to the estuary at the eastern mouth of the Nile; in nine days he marched across the triangular peninsula to Gaza, proceeding along the old military road.[8] On the lowland of the coast he held a consultation with his army and decided to go through the narrow pass to Megiddo (Mykty). Megiddo was one of the principal regional cities of Solomon (I Kings 4:12), mentioned together with Taanach as the seat of a governor. It was a stronghold closing the passage from the south of Carmel into the valley of Jezreel. Ordering the consultation with his army, Thutmose III said as follows:

> That wretched enemy [the chief] of Kadesh has come and entered into Megiddo; he is there at this moment.[9]

Who was the chief (king) of Kadesh who came to defend this fortified point? He is not called by name. And where was this Kadesh? The events of the next few days will be followed, and then these questions will again be asked.

The southern wing of the Egyptian army was in Taanach; the army passed Aruna and came to the bank of the brook of Kina.

> "Prepare your weapons! for we shall advance to fight with that wretched foe in the morning!"
> The king rested in the royal tent ... the watch of the army went about saying,
> "Steady of heart! steady of heart! Watchful! Watchful!"
> ... Early in the morning, behold, command was given to the entire army to move.[10]

The king went forth in a chariot of electrum (gold in amalgam with silver) in the center of his army, one wing being at the brook of Kina, the other northwest of Megiddo. The Egyptian army prevailed:

142

... When they saw his majesty prevailing against them they fled headlong to Megiddo, in fear, abandoning their horses and their chariots of gold and silver.

.. Now, if only the army of his majesty had not given their heart to plundering the things of the enemy, they would have [captured] Megiddo [Mykty] at this moment, when the wretched foe of Kadesh [Kds] and the wretched foe of this city were hauled up in haste to bring them into this city. The fear of his majesty had entered their hearts.[11]

The above explanation is given to excuse the escape of the king of Kadesh, who was not taken prisoner.

The victorious army of his majesty went around counting their portions. ... The whole army made jubilee.[12]

The siege of Megiddo began. Although the city was fortified by a heavy wall, the garrison was unable to stand a long siege against the army of the pharaoh, and surrendered. "Behold, the chiefs of this country came to render their portions."

The Egyptian army took three other cities. In the annals the description of the end of the campaign has not been preserved, but it may be reconstructed. The results of this first campaign were the conquest of the walled cities as recorded in the quoted annals; the submission of the entire land of one hundred and nineteen cities with the city of Kadesh first on the list, as appears on the "cities bas-relief" in the Karnak temple; rich spoils of precious vessels shown on a mural also in Karnak.

The most important questions have remained unsolved. Where was the city of Kadesh? Who the king of the city of Kadesh was is not even asked.

How to find the name of a king of an unlocated city, who lived hundreds of years before Joshua, if the Egyptian record does not mention it?

The second question puzzling investigators is how the name of Jerusalem or Salem or Jebus, as it was called before, came to be omitted from a most complete list of the cities of Palestine, a list comprising, it would seem, all the towns of importance in pre-Israelite Palestine.

The third puzzle is the exquisite form of the vessels captured in this campaign, showing a very high degree of craftsmanship; it is

143

surprising to find that the uncultured peoples of the Canaanite era were artisans of such excellent skill.

In view of the documentary inscriptions and carvings of Karnak, it is thought that the Canaanites were superior artists in metalwork, and that Jerusalem escaped the doom of its neighbor cities; but the location of Kadesh remained a matter for scientific debate.

The king of Kadesh came to Megiddo to defend it against the king of Egypt; the stronghold fell; he succeeded in escaping; but in the same campaign his own city was subjugated too.

Kadesh in Judah

The historians claim to know one famous Kadesh, located by them on the Orontes River in northern Syria. But in the list of Thutmose III the city of Kadesh is named as the first among one hundred and nineteen Palestinian (not Syrian) cities; in second place is Megiddo, the scene of the battle; and one hundred and seventeen other cities follow them. This Kadesh could not be a city in Syria, for in the Palestine campaign Thutmose did not reach the Orontes. There was a Kadesh in Galilee, Kadesh Naphtali, mentioned a few times in the Scriptures; but what would be the purpose of placing this unimportant city at the top of the list just before Megiddo, It became a matter for conjecture.

According to one hypothesis, the city referred to was Kadesh Naphtali;[13] according to another, Kadesh on the Orontes;[14] and each theory had to be supported by some explanation as to why a city outside Palestine or an insignificant city in Palestine was placed at the head of a list of Palestinian cities where one would expect to find the capital of the land.

The suggestion was advanced that the first name on the Palestine list did not belong to the register and had been added later.[15] This is highly improbable, especially since the interpolation (if it be such) was made on all three copies. Or, it was said, the Galilean city might have been intended, but the sculptor mistook it for the famous Kadesh on the Orontes and for this erroneous reason put it in first place.[16]

These theories met opposition. The Palestinian lists were executed shortly after the return from the Palestinian campaign and prior to

144

the Syrian campaign; at that time there was no reason to confuse the cities;[17] beyond doubt the list was executed with the personal knowledge of Thutmose III and was checked by his officials.

The coeval history of Judah, short yet clear, records:

> II CHRONICLES 12:2–4 And it came to pass, that in the fifth year of king Rehoboam, Shishak king of Egypt came up against Jerusalem. . . .
>
> With twelve hundred chariots, and threescore thousand horsemen: and the people were without number that came with him out of Egypt; the Lubim [Libyans], the Sukkiim, and the Ethiopians.
>
> And he took the fenced cities which pertained to Judah, and came to Jerusalem.

The conquest of the walled cities is the phase of the war related in the beginning of the annals of Thutmose III which have been preserved. The second phase is the move on the capital.

Jerusalem, against which the pharaoh advanced, must have been the city of Kadesh. This one answer serves two questions: Why was Jerusalem not on the list of Thutmose III, and Where was the king-city of Kadesh?

Is Jerusalem anywhere else called Kadesh? In many places all through the Scriptures. "Solomon brought up the daughter of Pharaoh out of the city of David . . . because the places are *kadesh* [holy]."[18] In the Psalms the Lord says: "Yet have I set my king upon Zion, my mount *kadesh*."[19] Joel called on the people: "Blow ye the trumpet in Zion, and sound an alarm in my mountain *kadesh*."[20] He also said: "So shall ye know that I am the Lord your God dwelling in Zion, my *kadesh* [holy] mountain: then shall Jerusalem be [really] holy [*kadesh*]."[21] Isaiah said of the people of Jerusalem: ". . . for they call themselves of the city Kadesh." He prophesied about the day when the Lord "will gather all nations and tongues. . . . And they shall bring all your brethrens . . . out of all nations . . . to my *kadesh* mountain Jerusalem."[22] Daniel prayed: ". . . let thine anger and thy fury be turned away from thy city Jerusalem, thy *kadesh* mountain,"[23] "thy city *kadesh*."[24] And Nehemiah wrote: ". . . the rest of the people also cast lots, to bring one of ten to dwell in Jerusalem, the city *kadesh*."[25]

145

The "Holy Land" and the "Holy City" are names given to Palestine and Jerusalem in early times, as not only the Scriptures but also the Egyptian inscriptions ("God's Land", "Kadesh") bear witness.

The name Kadesh is used for Jerusalem not in Hebrew texts alone. The names of the most obscure Arab villages in Palestine were scrutinized by the scholars in biblical lore in an endeavor to locate the ancient cities, but the Arab name for Jerusalem was overlooked: it is el-Kuds (the Holy, or the Holiness).

Kadesh, the first among the Palestinian cities, was Jerusalem. The "wretched foe", the king of Kadesh, was Rehoboam. Among the one hundred and nineteen cities were many which the scholars did not dare to recognize: they were built when Israel was already settled in Canaan. Antiquity was ascribed to other cities which were not entitled to it.

The walled cities fortified by Rehoboam (II Chronicles 11:5ff.) may be found in the Egyptian list.[26] It appears that Etam is Itmm; Beth-Zur—Bt sir; Socoh—Sk.[27] Here is a new field for scholarly inquiry: the examination of the list of the Palestinian cities of Thutmose III, comparing their names with the names of the cities in the kingdom of Judah. The work will be fruitful.

At the end of the preceding section we left Thutmose III under the walls of Kadesh-Jerusalem. The strongholds of the land westward from Jerusalem had fallen; the strongest of them, defended by the king of Jerusalem, opened its gates after the king and his escort had escaped and returned to Jerusalem.

II CHRONICLES 12:5 . . . the princes of Judah . . . were gathered together to Jerusalem because of Shishak.

The land only a few years before had been rent in two. Judah, weakened by the separation of the northern kingdom, was unable to defend itself.

The prophet Shemaiah, who, four years earlier, had warned the people against engaging in a fraternal war, came to the king and to the princes of Judah with a wrathful message from the Lord, forsaken by the people of Jerusalem. The king and the princes abased themselves, and the king said: "The Lord is righteous." Then Shemaiah came with another message:

146

II CHRONICLES 12:7f. . . . . I will grant them some deliverance; and my wrath shall not be poured out upon Jerusalem by the hand of Shishak. Nevertheless they shall be his servants.

Because they did not wish to be the servants of the Lord, they would be servants of an earthly kingdom. This was said in the name of the Lord. And the Egyptian king and his army, who at Megiddo "gave praise to Amon for the victory he granted to his son", certainly repeated their praise at the walls of Jerusalem.

After the fall of "the fenced cities which pertained to Judah", Jerusalem opened its gates without offering resistance. "Whereupon the princes of Israel and the king humbled themselves", and they were "not destroyed", and the Lord's wrath was "not poured out upon Jerusalem by the hand of Shishak." The city was not stormed.

There is no record of any storming of Kadesh by Thutmose III. But Kadesh in Palestine fell to Thutmose III and its name heads the list of Palestinian cities captured by him. The humbling of the princes of Judah after the fall of Megiddo is described also by Thutmose III in these words:

> Behold, the chiefs of this country came to render their portions, to do obeisance to the fame of his majesty, to crave breath for their nostrils, because of the greatness of his power, because of the might of the fame of his majesty.[28]

"The greatness of his power" is also mentioned in the Hebrew version—"the twelve hundred chariots, and sixty thousand horsemen, and the people without number that came with him out of Egypt; the Lubim, the Sukkiim, and the Ethiopians."

The indignant Lord humiliated Jerusalem, which erected statues of Egyptian, Sidonian, and other deities on the hills around Jerusalem; soon the god with the head of a ram would receive all the splendor from Solomon's Temple, every removable precious thing.

II CHRONICLES 12:9 So Shishak king of Egypt came up against Jerusalem, and took away the treasures of the house of the Lord, and the treasures of the king's house; he took all: he carried away also the shields of god which Solomon had made.

There remained the old Ark of the Covenant, a worthless piece, a relic of the desert, together with the stones of the Commandments. Was Egypt poor in old steles?

147

The treasures brought by Thutmose III from Palestine are reproduced on a wall of the Karnak temple.

The bas-relief displays in ten rows the legendary wealth of Solomon. There are pictures of various precious objects, furnishings, vessels, and utensils of the Temple, of the palace, probably also of the shrines of foreign deities. Under each object a numerical symbol indicates how many of that kind were brought by the Egyptian king from Palestine: each stroke means one piece, each arch means ten pieces, each spiral one hundred pieces of the same thing. If Thutmose III had wanted to boast and to display all his spoils from the Temple and the Palace of Jerusalem by showing each object separately instead of using this number system, a wall a mile long would have been required and even that would not have sufficed. In the upper five rows the objects of gold are presented; in the next rows silver things are mingled with those of gold and precious stones; objects of bronze and semiprecious stones are in the lower rows.

Wealth accumulated by a nation during hundreds of years of industrious work and settled life in Palestine, spoils gathered by Saul and David in their military expeditions, loot of the Amalekite Auaris, earnings from the trade between Asia and Africa, gold from Ophir, the gifts of the queen Sheba-Hatshepsut, all became the booty of Thutmose III. The work of Huram, of the tribe of Naphtali, is reproduced on the walls of the Karnak temple; Huram and his workmen were skilled artisans, and the hand of their royal master, Solomon, supplied them lavishly with precious metal and stone.[29] Specimens of the skill of David's craftsmen must also be found in this exhibit, for

> I KINGS 7:51 . . . Solomon brought in the things which David his father had dedicated; even the silver, and the gold, and the vessels, did he put among the treasures of the house of the Lord.

The sacred objects wrought by the ancient master Bezaleel, son of Uri,[30] may also have been reproduced here.

An exhaustive indentification of objects pictured in the Karnak

148

temple and of those described in the Books of Kings and Chronicles is a matter for prolonged study and should preferably be done with the help of molds from the bas-reliefs at Karnak. The following short excursus is not intended to be complete and definitive; it is only tentative. Yet it will demonstrate the identity of the booty of Thutmose III with that carried out of Jerusalem by the Egyptian king in the days of Rehoboam, son of Solomon.

A large part of the booty of Thutmose III consisted of religious objects taken from a temple. There were altars for burnt offerings and incense, tables for the sacrifice, lavers for liquid offerings, vessels for sacred oil, tables for showbread, and the like in great quantity. No doubt it was an extremely rich temple that was pillaged by Thutmose III.

The objects taken by Shishak from Jerusalem were the treasures of the Temple of Solomon and of the king's palace (II Chronicles 12:9).

On the Karnak bas-relief Thutmose III is shown presenting certain objects to the god Amon: these objects are the part of the king's booty which he dedicated to the temple of Amon and gave to the Egyptian priests. This picture does not represent the whole booty of Thutmose III. He chose for the Egyptian temples what he took from the foreign Temple, and in this collection of "cunning work" one has to look for the objects enumerated in the sections of the Books of Kings and of Chronicles describing the Temple.

On the walls of the tomb chambers of Thutmose's viziers treasures are shown in the process of transportation from Palestine. Besides the art work familiar from the scene of presentation to Amon, there are also other objects, apparently from the palace. These were delivered to Pharaoh's palace and to the houses of his favorites.

The books of the Scriptures have preserved a detailed record of furniture and vessels of the Temple only. Fortunately the separation of the sacral booty in the scene of dedication to Amon makes the task of recognition easier.

The metals used and the style of the craftsmanship will be compared briefly in the Hebrew and Egyptian sources. The material of which the objects were made is indicated by accompanying inscriptions on the bas-relief; they were made of three different metals, translated as gold, silver, and bronze. The metals used for the sacral furniture and for the vessels in the Temple of Solomon were of gold, silver and bronze ("brass"). The "cunning work" was manufactured of each of these metals.

Often an article is represented on the wall in gold and another of the same shape in brass. The fashioning of identical objects in gold as well as in bronze (brass) for the Temple of Solomon is repeatedly referred to in the Books of Kings and of Chronicles.

When gold was used for the vessels and the furnishings of Solomon's Temple, it was either solid gold[31] or a hammered gold overlay on wood.[32] The pictures of the objects in Karnak are described by the words "gold" and "overlaid with gold".

In the period when Israel had no permanent site for its place of worship, the Ark of the Covenant and other holy objects were moved from one place to another and were sometimes taken into battle. In order to facilitate transportation, the furnishings of the tabernacle were made with rings and bars.[33] The old furniture of the tabernacle was placed in the Temple by Solomon,[34] and was carried off, in the days of his son, by the pharaoh and his army. The Ark of the Covenant, however, was not removed but remained in the Temple until the Babylonian exile.[35] It was probably a model for other transportable sacred paraphernalia used in the holy enclosure in Beth-el and in Shiloh and thereafter in Jerusalem. In the second and seventh rows of the Karnak bas-relief are shown various ark-shaped chests with rings at the corners and bars for transportation.

"A crown of gold round about" was an ancient Judean ornament of sacred tables and altars.[36] Such ornamentation is seen on the golden altar in the second row (9) of the mural, as well as on the bronze (brass) altar in the ninth row (177).[37]

The preferred ornament on the vessels was the *shoshana,* translated as "lily" (lotus).

> I KINGS 7:26 ... the brim thereof [of the molten sea] was wrought like the brim of a cup, with flowers of lilies.

The lotus motif is often repeated on the vessels reproduced on the wall of Karnak. A lotus vial is shown in gold (10), in silver (121), and in colored stone (malachite?) (140). A rim of lily work may be seen on various vessels (35, 75, 175), a very unusual type of rim ornament, found only in the scriptural account and on the bas-reliefs of Thutmose III.

Buds among flowers ("his knops and his flowers"[38]) were also used as ornamentation in the tabernacle. This motif appears on a vase (195) in the lower row of the Karnak mural and also in the fifth row (75).

Of animal figures, lions and oxen are mentioned as decorative motifs of the Temple of Jerusalem (I Kings 7:29 and 36). The Karnak mural shows lion heads (20, 60), and the head of an ox is recognizable as an ornament on a drinking vessel (132).

Gods were often depicted in Egyptian temples in shameless positions. Among the figures of sacred objects on the Karnak bas-relief there are none of phallic form, neither are there any pictures of gods at all. A few animal heads (lions) with the sign of the uraeus on their foreheads and the head of a hawk are wrought on the lids of some cups. These cups might have been brought from the palace Solomon had built for his Egyptian wife.

Idols were and still are used in all pagan worship. The hundreds of sacred objects appearing in the mural were obviously not of an idolatrous cult; they suggest, rather, a cult in which offerings of animals, incense, and showbread were brought, but in which no idols were worshiped. The Temple of Kadesh-Jerusalem, sacked by Thutmose III, was rich in utensils for religious services but devoid of any image of a god.

Piece by piece the altars and vessels of Solomon's Temple can be identified on the wall of Karnak.

In the Temple of Solomon there was an altar of gold for burnt offerings (I Kings 7:48; II Chronicles 4:19). It was the only such altar. In the second row of the bas-reliefs is an altar with a crown around the edge, partly destroyed, but partly plainly discernible (9). The inscription reads: "The [a] great altar." It was made of gold.

Another altar in the Temple of Jerusalem was of "brass" (bronze); it was square and very large.[39] In the ninth row of the Karnak relief an altar of "brass" (bronze) is pictured, the shape of which is similar to that of the gold altar. The inscription says (177): "One great altar of brass [bronze]." Inasmuch as its height is equal to its width, the altar does not fit the description of the altar mentioned in the Second Book of Chronicles, which was half as high as it was wide. However, from the first chapter of the Second Book of Chronicles we know that another brazen altar made by Bezaleel was among the holy objects of the Temple at Jerusalem.

Next to the altar was the table "whereupon the shewbread was" (I Kings 7:48; II Chronicles 4:19). The showbread was obviously not of flour, but of silver or gold; in the Book of Exodus[40] it is said that showbread was made by Bezaleel, who was a goldsmith. Showbread is pictured on the bas-relief of Karnak in the form of a cone. The cone in the seventh row (138) bears the explanation: "White bread." This bread was of silver. The thirty cones of gold (48) and the twenty-four cones of colored stone (malachite) (169), identical in form with the silver cone, also represent showbread.

The "candlestick with the lamps" (II Chronicles 4:20) was an illuminating device with lamps shaped like flowers. Figures 35, 36, 37, and 38 of the mural are candlesticks with lamps. One of them (35) has three lily lamps on the left and three on the right. The other candlesticks (37, 38) have eight lamps to the left and eight to the right. The candlestick with lamps wrought by Bezaleel for the tabernacle had three lamps to the left and three to the right.[41] There were almonds, a knop, and a flower on the arms. A later form showed a preference for seven lamps on both sides of the stem.

Other candlesticks are mentioned in addition to those with lamps. In the Book of Kings they are described as bearing flowers (I Kings 7:49). This form is seen in the third row of the bas-relief (25, 26, 27, and 28). The candlestick is in the shape of a stem with a lotus blossom.

Next to the altar, the tables with the showbread, and the candlesticks were the tables for offerings.

EXODUS 35:13 The table . . . and all his vessels.
37:16 . . . vessels . . . upon the table: his dishes, and his spoons, and his bowls, and his covers to cover withal, of pure gold.

The table, like its vessels, was of gold (I Kings 7:48). "The tables of sacrifice" in the third row (of gold) and in the seventh row (of silver) of the mural have sets of vessels on them: three flat dishes, three large cups, three pots (or bowls), one shovel. Many tables of gold and silver and bronze are reproduced on the bas-relief. The paraphernalia of the Temple contained also "hooks and all instruments" (II Chronicles 4:16). In the third row of the Karnak mural, near the table of offerings, and in the same row at the left end, there are hooks, spoons, and other implements (30, 31, 32, 33, 43, 44); bowls appear in most of the rows, but especially in the second and sixth (of gold).

"The incense altar, and his staves, and the anointing oil" were in the Temple of Jerusalem (Exodus 35:15). As no detailed description of the form of this altar is given in the Scriptures, various objects in the form of altars suitable for incense may be considered. Did the smoke of the burning incense pour through the openings in the ornamental spouts? Was the incense burned in a dish set on a base (41, 181)? Vessels containing anointing oil are shown on pedestal altars (41); over the figures in the lower row (197–99) is written: "Alabaster, filled with holy anointing oil for the sacrifice."

Golden snuffers were used in the Temple of Solomon for spreading the fragrance during the service (II Chronicles 4:22; I Kings 7:50). *Masrek* in Hebrew means a fountain or a vessel that ejects a fluid. Such fountains are mentioned as having been in the Temple of Solomon (I Kings 7:50; II Chronicles 4:22). Among the vessels shown on the wall at Karnak there are one or two whose form is peculiar. The vessel in the fifth row (73) has two side spouts and is adorned with figures of animals. The spouts are connected with the basin by two animals (lions?) stretching toward them; rodents run along the spouts, one pair up and one pair down; amphibians (frogs) sit on top of the vessel. It is not unusual to decorate modern fountains in a like manner. The figures of frogs are especially appropriate for this purpose. The tubes and the mouths of the animals on the vessel could be used to spout perfume or water. The neighboring object seems also to be a fountain.

One hundred basins of gold were made by Solomon for the Temple (II Chronicles 4:8). Ninety-five basins of gold are shown in the sixth row of the mural; six larger basins are shown apart.

The walls and floor of Solomon's Temple were "overlaid with fine gold" and "garnished with precious stones" (II Chronicles 3:5–6; I Kings 6:28ff.). Pharaoh, who "took all", did not leave this gold or these stones on the walls. Some of them were worked into jewels, and the inscription (over 63–65) reads: "Gold and various precious stones his majesty had reworked." Other gold was taken in the form of bricks and links (chains) (23, 24). Chains of gold are also mentioned as having been in the Temple of Solomon (II Chronicles 3:16): "And he made chains."

153

Thirty-three doors are represented in the lower row of the bas-relief and the inscription says they are "of beaten copper" (190).

> II CHRONICLES 4:9 Furthermore he made the court of the priests, and the great court, and doors for the court, and overlaid the doors of them with copper.[42]

Targets or shields of "beaten gold" are named among the booty of the pharaoh (II Chronicles 9:15). These three hundred shields, together with the two hundred targets of gold (II Chronicles 9:15, 16), were not part of the furnishings of the Temple; they adorned "the house of the forest of Lebanon". In the seventh row of the mural there are three disks marked with the number 300, which means that they represent three hundred pieces. The metal of which they are made is not mentioned; some objects in this row are of silver, but the next figure has a legend indicating that it is of gold.

The large "sea of brass" and the brazen bases (I Kings 7:23; II Chronicles 4:2) were not removed by the pharaoh (II Kings 25:16). Among the things which were taken later by Nebuzaradan, the captain of Nebuchadnezzar, were "two pillars, one sea, and the bases which Solomon had made for the House of the Lord".[43]

The ephod of the high priest (a collar with a breastplate) was not mentioned in the Scriptures among the booty of the pharaoh and might not have been taken. But precious garments of the priests were carried off. The fourth row displays rich collars, some with breastplates; they were destined to be gifts for the priests of Amon.

In the bas-reliefs of Karnak we have a very excellent and detailed account of the vessels and furniture of the Temple of Solomon, much more detailed than the single bas-relief of the Titus Arch in Rome, showing the candlestick and a few other vessels of the Second Temple, brought to the Roman capital just one thousand years after the sack of the First Temple by the Egyptians.

Zoological and Botanical Collections from Palestine

Thutmose III succeeded in his plans. The kingdom of David and Solomon was divided; Judah bowed to Egyptian domination. The naval base at Ezion-Geber was no longer controlled by Judah. The maritime expeditions of the Israelites in company with the sailors of Tyre and Sidon were not repeated.

Megiddo, which guarded the road between Jerusalem and Sidon, became the chief Egyptian stronghold in Syria-Palestine. The Sidonians, who helped the garrison of Megiddo, probably as mercenaries, tried after its fall to save their own independence. Jaffa fell to a general of Thutmose III;[44] the fleet of the Phoenicians, or part of it, was seized by Thutmose III a few years later. The northern realm of Israel, ruled by a puppet king, a tool in the hands of the pharaoh, did not need to be conquered. Jeroboam came from Egypt, where he was trained for his task. It is only logical, therefore, to expect that the land of Israel would voluntarily pay tribute to Thutmose III.

In less than five months—in one hundred and forty-eight days, to be exact—the Palestine campaign was over. The next year Thutmose III returned to Palestine on an inspection tour and received tribute. In Palestine the pharaoh took a daughter of the royal family to be one of his wives[45] and brought her home with jewels of gold and lapis lazuli and a retinue of thirty slaves; he brought also horses, chariots wrought with gold or electrum, bulls and small cattle, flat dishes of gold "which could not be weighed", flat dishes of silver, a gold horn inlaid with lapis lazuli, "much silver", 823 jars of incense, 1718 jars of honeyed wine, ivory, precious wood, "all the luxuries of this country".

The next year ("year 25") Pharaoh returned to Palestine for inspection. This time he visited northern Palestine (Upper Retenu). On his way he admired the gardens of Judah, Benjamin, and Ephraim. Many of these gardens, rich in color, form, and fragrance, were transplanted to Egypt:

> All plants that grow, all flowers that are in God's land which were found by his majesty, when his majesty proceeded to Upper Retenu.[46]

As in the Punt bas-reliefs of Hatshepsut, this land is called God's Land (Divine Land). Following the peaceful expedition of Hatshepsut, when only thirty-one almug trees were transferred to the soil of Egypt, the tribute-collecting expedition of Thutmose III transferred entire botanical collections. These collections are reproduced on the walls of the Karnak temple, showing various and peculiar shapes of the flora of Palestine some twenty-eight and a half

155

centuries ago. A zoological collection was also taken along; no inscription mentions it, but the figures of animals appear among the plants on the bas-relief.

> His majesty said, "I swear, as Re loves me, as my father Amon favors me, all these things happened in truth."

Looking at these pictures, we remember what was said about Solomon, whose royal pleasure was to collect and study plants and animals.

> I KINGS 4:33 And he spake of trees, from the cedar tree that is in Lebanon even unto the hyssop that springeth out of the wall: he spake also of beasts, and of fowl, and of creeping things, and of fishes.

Botanists[47] recognize among the plants brought by Thutmose III the blue lotus, the vine date tree, the pomegranate, the dragon plant, the arum, the iris, the chrysanthemum, as well as the cornflower and mandragora, also a variety of pine tree and some sort of "melon tree". Many of the plants, however, cannot be identified at all.[48]

It seems certain that various specimens of flora as pictured on the Karnak bas-relief were not indigenous to Palestine. How, then, explain the presence of these plants among those brought by Thutmose III from Palestine?

"Possibly, the twofold geographical designation, Palestine and God's Land, could be explained by the fact that a number of plants actually came from God's Land. Still another conjecture to explain the presence of these plants may be made, namely that princes of distant countries sent messengers with gifts to the pharaoh while on his war of expedition."[49]

The second surmise is strange; it is unusual for remote countries to send plants and birds to warriors on their march of conquest. The first surmise merely illustrates the type of conjecture necessary in order to evade the identification of Palestine with God's Land.

As has already been said, part of the flowers and other plants appear not to be native to Palestine, and some of them cannot be identified at all. Since the forms were drawn by an accurate hand, this conclusion was reached: "These plants, rare at the time of Thutmose III, do not exist in our day."

Solomon had trees brought on ships from countries lying more than a year's voyage distant. No wonder that in the booty of Thutmose III there were plants strange to the Palestine of today; they were strange to the southeastern corner of the Mediterranean even in the days of Thutmose III. This may be concluded from the circumstance that the Pharaoh brought them home from his military expedition and had them depicted on the walls of the Karnak temple, like the treasures of gold and silver; though cultivated in Palestine, these plants were exotic and rare.

Among the plants are carvings of animals; better preserved are the figures of fowl. A zoologist has recognized many varieties of the birds, but some seemed to him to be fantastic inventions of the sculptor,[50] as they are unknown in the East. We know that the ships of Tharshish brought back peacocks (I Kings 10:22; II Chronicles 9:21); surely Solomon collected not merely one species of birds. And it was his zoological garden that was carried by the pharaoh to Egypt as well as the treasures of his Temple and palace.

Today, in the reliefs of Deir el Bahari and Karnak, we may gaze on the people of Judah of the days of Solomon, the animals and the plants they raised, and the objects they cherished.

Genubath, King of Edom

Edom, like Israel, was ruled by a chief appointed by the Theban king. Hadad the Edomite had a son by the sister of Tahpenes, the queen of Ahmose, and his name was Genubath.

> I KINGS 11:20 And the sister of Tahpenese bore him [Hadad] Genubath his son, whom Tahpenes weaned in Pharaoh's house: and Genubath was in Pharaoh's household among the sons of Pharaoh.

Hadad had returned to Edom in the days of Solomon, after the death of Joab.[51] Since then about forty years had elapsed. Genubath, his son, was now the vassal king of Edom; he dwelt either in Edom or in Egypt.

Tribute from this land, too, must have been sent to the Egyptian crown; there was no need to send an expedition to subdue Edom. When Thutmose III returned from one of his inspection visits to Palestine he found in Egypt tribute brought by couriers from the land "Genubatye", which did not have to be conquered by an expeditionary force.

> When his majesty arrived in Egypt the messengers of the Genubatye came bearing their tribute.[52]

It consisted of myrrh, "negroes for attendants", bulls, calves, besides vessels laden with ivory, ebony, and skins of panther.

Who were the people of Genubatye? Hardly a guess has been made with regard to this peculiar name. The people of Genubatye were the people of Genubath, their king, contemporary of Rehoboam.

The year before, in the seventh year after the campaign of Megiddo and Jerusalem, Thutmose III, using his stronghold of Megiddo as a base and assisted by a fleet captured from the Sidonians, pushed northward to Arvad.

"On his return to Egypt he took with him the children of the native princes to be educated in friendship toward Egypt, that they might be sent back gradually to replace the old hostile generation of Syrian princes."[53]

> Behold, the children of the chiefs and their brothers were brought to be in strongholds in Egypt. Now whosoever died among these chiefs, his majesty would cause his son to stand in his place.[54]

This is a policy similar to that applied in the case of Hadad and Genubath, the Edomites.

Princess Ano

In the Greek version of the Old Testament—the Septuagint—made in Alexandria in Egypt in the third century before the present era, there is found the information that when Jeroboam heard in Egypt of Solomon's death he spoke in the ears of the king of Egypt, saying, "Let me go, and I will depart into my land." And Susakim (Shishak) gave to Jeroboam to be his wife Ano, the eldest sister of Thelkemina, his own wife. She was great among the daughters of the king, and she bore to Jeroboam Abijah.[55] This information is of importance because it gives us the name of the queen's sister; the extant Hebrew

158

A page from the Papyrus Ipuwer containing the story of the plagues

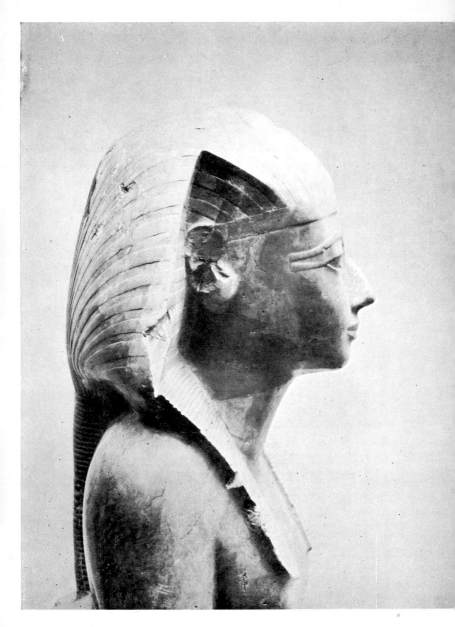

Queen Hatshepsut (*courtesy of the Metropolitan Museum of Art*)

Loading Egyptian boats

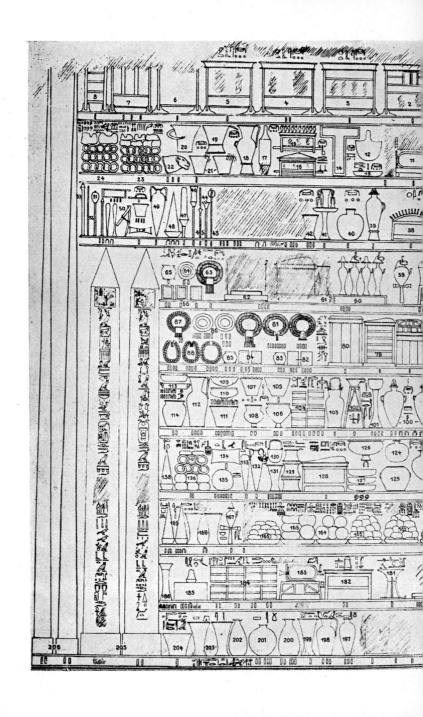

Vessels and furnishings of the
Temple at Jerusalem

(*drawing by* W. *Wreszinski*)

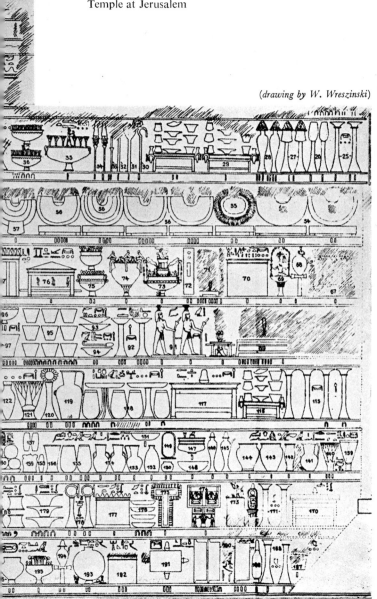

Thutmose III

Vessels and furnishings of the
Temple at Jerusalem

(photograph by W. Wreszinski)

The Mesha stele

Bible told us only that Jeroboam fled abroad "to Shishak, king of Egypt, and was in Egypt until the death of Solomon". According to the Septuagint, Jeroboam received an Egyptian princess for a wife, like Hadad the Edomite only one generation earlier (I Kings 11:19).

In the Metropolitan Museum of Art in New York there is preserved a canopic jar bearing the name of Princess Ano.[56] The time when the jar originated has been established on stylistic grounds as that of Thutmose III. No other references to a princess of such name is found in any Egyptian source or document. The existence of a princess by the name of Ano in the days of Thutmose III lends credence to the information contained in the Septuagint and gives additional support to the identification of Shishak or Susakim of the Septuagint with the pharaoh we know by the name Thutmose III.

"The Astounding Civilization"

In addition to the bas-reliefs on the walls of Karnak, a number of other monuments show the wealth brought from Palestine by Thutmose III. Among these monuments the tombs of Rekhmire,[57] vizier of Thutmose III, and Menkheperre-Seneb, high priest, rank first. Vessels and furniture are shown en route from Palestine to Egypt. Chariots, too, of gold and silver, were among the booty of Thutmose, and were given as presents to his favorites.

The various works of art from Palestine were very much appreciated in the Egyptian capital, and the artisans of Palestine were brought captive to Egypt, there to work at their crafts. On the walls of the sepulchral chambers of the vizier, coppersmiths are shown, and it is written: "Bringing the Asiatic copper[-smiths] which his majesty captured in the victories in Retenu." Over cabinetmakers from Palestine appear the words: "Making chests of ivory, ebony." There are brickmakers, too, working to build the Amon temple,[58] and next to them is the inscription: "Captives which his majesty brought for the works of the temple of Amon." "The taskmaster, he says to the builders: 'The rod is in my hand; be not idle.'"

All this was regarded as proof that the Canaanites, the indigenous population of Palestine, were skilled in the arts, strange as this may seem: "We learn from the booty carried into Egypt—chariots inlaid

159

with silver, gold-plated chariots, etc.—of the astounding civilization of Syria at that period."[59]

We know now that the astounding civilization, the products of which we see on these monuments, was not Canaanite but Israelite. It is therefore of interest to read what role the modern historians ascribe to the Canaanites in the process of the development of Egyptian art and in the refinement of the Egyptian race; it is an opinion as unbiased as criticism of the works of an artist whose name is concealed by a pseudonym.

"At this time (Thutmosis III, 1503–1449) the Syrians stood at a higher stage of civilization than even the wonderfully gifted race of Egypt. The plunder carried back to Egypt of coats of mail, of gold-plated chariots, of chariots inlaid with silver, witnesses to an industrial and artistic development that was able to teach Egypt. With all these precious goods went captives, who fell to working in the Nile valley at the crafts to which they were accustomed at home, and as they worked they taught the Egyptians. . . . The Syrian craftsmen worked so well in Egypt that their wares changed even the taste of the Egyptians, while the language was semitized, and the method of writing gradually developed into a smooth-flowing and graceful style. Under the great influx of foreign blood even the features of the conquering race were changed into a less bold and more delicate form. Egypt had never known such changes since the beginning of the monarchy."[60]

It is also worth while to notice the chariots, gold-plated and inlaid with silver, captured by Thutmose III in Palestine. 'The song of songs, which is Solomon's" is supposed to be a creation of a late period, and the mention of luxury in the time of Solomon is thought to be a product of the poet's fancy. "King Solomon made himself a chariot of the wood of Lebanon. He made the pillars thereof of silver, the bottom thereof of gold, the covering of it of purple." The Egyptian monuments show that in the fifth year after Solomon, in Jerusalem and also in Megiddo, there was not one but many chariots of gold and silver.

We thus discard our supposed knowledge of the Canaanite art of the sixteenth and fifteenth centuries before the present era, and we begin to acquire some knowledge of the Jewish art of the tenth century, about which the history of art was thought to be ignorant.

The present chapter shows the conquest of Palestine by Thutmose III to have taken place, not in the Canaanite period, but in the days of the Jewish kings, and more precisely in the fifth year of Rehoboam, son of Solomon. In the preceding chapter the expedition of Queen Hatshepsut to the land of Punt was shown to have taken place in the days of King Solomon, and the land visited to have been Judea and probably Phoenicia too. In other words, we assume that Queen Hatshepsut on her peaceful journey and Thutmose on his military expeditions visited the same country.

We are now in a position that will either trap us or furnish us with additional proof that Queen Hatshepsut went to Palestine on her famous expedition, and not to eastern Africa. Is not this point essential for the identification of Queen Hatshepsut with the Queen of Sheba?

The assumption that the people of God's Land in the pictures of Hatshepsut were people of Palestine can be easily proved or disproved by comparing these pictures with the figures of men with shields on the Karnak mural symbolizing the conquest of Palestine. In both cases Egyptian artists of practically the same generation did the sculpturing. They were masters in depicting the characteristic features of different races. Drawings of various ages are preserved in which Egyptian artists have made collections of racial types. A glance at the people of God's Land, the "people of the South", and the Egyptians on the bas-reliefs of the expedition to Punt may help us to understand the fine feeling these artists possessed for expressing the types of their own and of foreign races.

The same characteristic profiles, the same hair styles, with a ribbon around the hair tied behind, and the same long beard shaped as a prolongation of a pointed chin make it certain that types of one and the same people were pictured on the bas-reliefs of both Hatshepsut and Thutmose III.

But, one might ask, if Thutmose III went to the same country to which Hatshepsut had gone two or three decades earlier, why did he not call the country of his conquest, Rezenu (Palestine), by the same names that Hatshepsut called it, God's Land and Punt?

Year after year Thutmose III returned to Palestine to collect tribute (II Chronicles 12:8: ". . . they shall be his servants"). Three years after the conquest of Megiddo, Kadesh, and other cities, he had carved on the walls at Karnak pictures of trees and plants that he had brought from Palestine, with this inscription, already quoted: "Plants which his majesty found in the land of Retenu. All plants that grow, all flowers that are in God's Land which were found by his majesty when his majesty proceeded to Upper Retenu."[62]

This sentence induced the translator to conjecture that "God's Land is sometimes applied to Asia".[63]

The sixth campaign of Thutmose III, like the first, was military: he conquered the north of Syria. Three years later he went to Palestine to gather the levy. After describing the tribute obtained from Shinar and Kheta and the land of Naharin, the register reads: "Marvels brought to his majesty in the land of Punt in this year: dried myrrh . . ." The translator was surprised by this phrase.[64] We find that Thutmose III used the same terms as Hatshepsut—Punt and God's Land—for the land they visited—Phoenicia-Palestine.

The question that arises is: Did Palestine produce myrrh, referred to in the account of the tribute, and frankincense, mentioned among the gifts Hatshepsut received in God's Land?

Myrrh and frankincense are repeatedly mentioned in Egyptian inscriptions as products of Punt. Frankincense (olibanum) falls in clear drops which, when gathered and formed into balls or sticks, turn white. Because of its color the precious incense is called "white" in various languages (Greek, Arabic), likewise in Hebrew (*lebana*, white). The less precious incense, ladanum, is yellow or brown in color.[65] Frankincense grows in only a very few places, Somaliland and southern Arabia on opposite shores of the Red Sea being areas which produce it even today. The botanists were guides to the archaeologists in search of the land of Punt.

After his fifth visit of inspection to conquered Syria and Palestine, Thutmose III listed frankincense, oil, honey, and wine as tribute. Following his ninth visit, he stated that he had received as "Retenu tribute in this year" horses, chariots, various silver vessels of the workmanship of the country, and also "dry myrrh, incense 693 jars, sweet oil and green oil 2080 jars, and wine 608 jars".

Of his seventh campaign he wrote: "Tribute of the princes of Retenu, who came to do obeisance. . . . Now, every harbor at which his majesty arrived was supplied with loaves and with assorted loaves, with oil, incense, wine, honey, fruit—abundant were they beyond everything. . . . The harvest of the land of Retenu was reported consisting of much clean grain, grain in the kernel [not ground], barley, incense [frankincense], green oil, wine, fruit, every pleasing thing of the country."[66]

Myrrh and frankincense were products of Palestine. Let us see whether the Scriptures give any indication that they were products of the Holy Land in the days of Solomon. In the "Song of songs, which is Solomon's", the enamored prince says to the little shepherdess (4:6):

"Until the day break, and the shadows flee away, I will get me to the mountain of myrrh, and to the hill of frankincense."

Even if it was written later, the song speaks of the time of Solomon.

Lebanah (frankincense) near Beth-el (Judges 21:19) was probably the place where the incense plant grew. In the time of Thutmose III the rare plants of the Palestinian gardens were transferred to Egypt, as he himself told and pictured. Thereafter, in the days of Isaiah (60:6) and Jeremiah (6:20), incense was imported into Palestine from southern Arabia.

It is appropriate here to explain the name "Retenu" or "Rezenu" often employed in the Egyptian inscriptions of the New Kingdom to designate Palestine. Galilee is called "Upper Rezenu". "Rezenu" is apparently a transcription of the name used by the population of Palestine for their land. The Hebrew language must be questioned on its meaning.

In the Scriptures Palestine is frequently called "Erez" (country), "Erez Israel" (the land of Israel), and "Arzenu" (possessive case, "our country").[67] What the Egyptologists read as Retenu or Rezenu is probably the "Arzenu" of the Bible.

In only one inscription of the Middle Kingdom (Twelfth Dynasty) under Sesostris III is the name Rezenu mentioned—it is a very short account of a raid into that country against M-n-tyw. As we shall find the same name, Mntyw, in Egyptian documents of a much later period, that of King Menashe (Manasseh), the Mntyw of the Middle Kingdom must mean the tribe Menashe. If the inscription is correctly attributed to the time of Sesostris III, the mention of the tribe Menashe would imply that before the Israelites had come to stay in

Egypt they had dwelt in Palestine, not as a single patriarchal family, but as tribes strong enough to be regarded as enemies by the pharaoh. This would accord with the tradition of a defeat inflicted by Abraham and the servants of his household on the kings of Shinar and Elam and their allies (Genesis 14), and with the number of the Israelites (about two million, including women and children) in the days of the Exodus, after some two hundred years of sojourning in Egypt.

Sosenk (Shoshenk)

According to conventional history, Thutmose III and the whole Eighteenth Dynasty ruled over a Canaanite Palestine. Shishak, the pharaoh who carried away the vessels from the Temple at Jerusalem, was sought in the period that followed by some hundred years the epoch of the Ramessides. The Ramessides were the last great pharaohs of the imperial era in Egypt.

Among the names of the kings of the rather obscure period which the historians dealing with Egypt extend over six hundred years (until the conquest of Egypt by Cambyses in −525), a hieroglyphic name is found which reads "Sosenk". This king of the Libyan Dynasty cut the names of cities subject to him on the outside of the southern wall of the Karnak temple. These cities are represented by figures like the city-figures of Thutmose's bas-relief, and it is obvious that Sosenk copied that mural. But whereas Thutmose's list consists of well-known names familiar from the Scriptures, Sosenk's list contains mostly unknown names. The accompanying inscription consists "of stereotyped phrases ... too vague, general and indecisive to furnish any solid basis for a study of Sosenk's campaign. Had we not the brief reference in the Old Testament to his sack of Jerusalem, we should hardly have been able to surmise that the relief was the memorial of a specific campaign."[68]

And yet Sosenk is presented as the scriptural Shishak in all textbooks and manuals. However, it is admitted that "the date of Shishak's accession is dependent on Israelite chronology".[69]

The relief has one hundred and fifty-five names of cities.[70] "Only seventeen can be located with certainty, and two more with probability. Fourteen of these belong to Israel; they are mostly unim-

164

portant towns while the remaining five in Judah are, with one exception, obscure villages."[71]

Among the Palestinian cities the following were identified because of phonetic similarity: Beth-Shan, Hapharaim, Gibeon, Megiddo, "and the most interesting is p'-hw-k-rw'—'b'-r'-m or Hekel Abram, which can be nothing else than 'the Field of Abram'."[72]

There are, as a matter of fact, a number of p'-hw-k-rw', and each of them is identified as *hekel*, "field" (in Aramaic). No locality by the name of Hekel Abram is known to have been in Judah or Israel, nor is any other Hekel.[73] Almost no name could be located in all of Judah, as the few "obscure villages" were mere guesses. This gives the impression that only Israel was subject to Sosenk (Shoshenk), not Judah. Neither Jerusalem, Hebron, Beer-Sheba, Bethlehem, nor any other known place was among the names on the list; nor was Jaffa, Gath, or Askelon.[74]

The inscription refers in general terms to tribute given to Sosenk, but where are the spoils, the furniture and vessels of the Temple of Solomon and of his palace? Was "Shishak" so modest that he did not mention the capital he conquered and the rich booty of the Temple, and at the same time so vainglorious that he piled up a list of names of non-existent cities?

Thutmose III, on the other hand, is said to have invaded pre-Israelite Canaan some five or six hundred years before the time ascribed to Shishak, conquered cities and strongholds built much later in the time of Judges or Kings, and taken from Canaan an immense plunder of sacred vessels and furniture in gold, silver, and brass (bronze), which six hundred years later were copied by Solomon in form and even in numbers and described in the Book of Kings.

Is this not a dubious construction? Is not the attribution of the art products to the Canaanites based on error? But if this is so, then who was this Libyan Sosenk who received tribute from the northern realm (Israel) hundreds of years subsequent to the time of Jeroboam and Rehoboam? In the pages dealing with the period of the Libyan Dynasty he will be identified as Pharaoh So to whom Hoshea, the last king of the northern realm, sent tribute (II Kings 17:4).

The generation that followed Queen Hatshepsut in Egypt was synchronized in this chapter with the generation that followed King Solomon in Palestine. In Egypt it was the time of the pharaoh known to us from modern history books as Thutmose III; in Palestine it was the time of Rehoboam, son of Solomon, and Jeroboam of the northern kingdom. The two countries, Egypt and Palestine, came into close contact. The pharaoh invaded Judea and, according to the Egyptian and scriptural narratives alike, 'took all the cities" and approached the capital, called Kadesh in the annals of the pharaoh and in the Scriptures called both Jerusalem and Kadesh. The conquest of Palestine is described almost identically in the Book of Kings and Chronicles and in the Egyptian annals. The country "fell into disagreement"; after an unsuccessful attempt to defend the land, the fortresses and other towns submitted; princes and their households gathered in the capital. By consent of the king and the princes the capital was thrown open and they "humbled themselves". The palace and the Temple of the capital were sacked and the vessels and furniture carried to Egypt. The detailed description of these furnishings and utensils, as preserved in the Books of Kings and Chronicles, agrees perfectly with the pictures engraved on the walls of the Karnak temple. The objects are of identical form and shape, of the same workmanship, of the same number: altars, lavers, tables, candlesticks, fountains, vases with rims of "buds and flowers", and cups of lotus shape, and vases of semiprecious stone, and priestly ephods, gold shields, and doors overlaid with copper.

The captives on the bas-relief representing captured cities are of the same race and appearance as the people of Punt and the Divine Land visited by Queen Hatshepsut one generation earlier, again proving that Hatshepsut went on her peaceful expedition to Palestine. Among the cities captured by Thutmose III were cities built by Solomon and Rehoboam, which were not found in the complete list of Canaan in the time of Joshua's conquest; however, according to the conventional chronology, Thutmose III preceded Joshua.

The biblical references to golden chariots in the days of Solomon are demonstrated to be true. Such chariots were brought by the pharaoh from Palestine. Also craftsmen from Palestine were employed in Egypt.

Judea became a dependency and its people vassals to the pharaoh. On his repeated expeditions to collect tribute the pharaoh took back with him frankincense, a product of the land. This, by the way, shows that the frankincense brought by Hatshepsut from the Divine Land was a Palestinean product; and the pharaoh actually refers to the products of Punt and of the Divine Land in connection with his expedition to Palestine. He also transferred to Egypt the zoological and botanical collections of King Solomon.

Jeroboam, while in Egypt as a refugee from Solomon, married a queen's sister by the name of Ano. A canopic jar with her name on it, dating from the time of Thutmose III, is preserved in the Metropolitan Museum of Art.

Genubath is referred to in the Book of Kings as the son of the Edomite king Hadad in exile, born in the palace of the pharaohs and reared there in the days of David and Solomon. He is mentioned by name in the annals of Thutmose III as the prince of the vassal land paying tribute to the pharaoh.

The time of Hatshepsut was that of Solomon; and the time of Thutmose III was that of Rehoboam, Solomon's son, and Jeroboam, his rival.

NOTES

1. K. Sethe, who directed attention to this text and to the very surprising fact that a statue for worship was erected in Punt, expressed his hope that the possible future discovery of an Egyptian statue would help to determine the location of Punt. ("Eine bisher unbeachtet gebliebene Episode der Punt Expedition der Königin Hatschepohwet", *Zeitschrift für ägyptische Sprache und Altertumskunde*, XLII [1905], 91–99.)
2. II Chronicles 14:3.
3. I Kings 11:14–25.
4. The Greek version of I Kings 12:24ff. makes Jeroboam a son-in-law of the pharaoh.
5. Breasted, *Records*, Vol. II, Sec. 416.
6. The text of this inscription is mutilated. The translation of Breasted was questioned. See the controversy between Kurt Sethe, *Zeitschrift für ägyptische Sprache und Altertumskunde*, XLVII (1910), 80–82, and Eduard Meyer, *Geschichte des Altertums* (2nd ed.; 1928), II Pt. I, p. 121, note 4. Cf. also the

translation of J. A. Wilson in *Ancient Near Eastern Texts*, ed. Pritchard (Princeton, 1950).

7. II Chronicles 11:6–10.
8. Herodotus (II, 159) described the conquest of Palestine by Thutmose and named him Sesostris.
9. Breasted, *Records*, Vol. II, Sec. 420.
10. Ibid., Secs. 429–30.
11. Ibid., Sec. 430.
12. Ibid., Sec. 431.
13. A. Mariette, *Les Listes géographiques des pylônes de Karnak* (Leipzig, 1875), pp. 12–13.
14. G. Maspero, *Transactions of the Victorian Institute*, XX (London, 1887), 297.
15. W. Max Müller, *Asien und Europa nach altägyptischen Denkmälern* (Leipzig, 1893), p. 145, n. 3.
16. W. Max Müller, "Die Palästinaliste Thutmosis III", *Mitteilungen, Vorderasiatisch-ägyptische Gessellschaft*, Vol. XII, No. 1 (1907), p. 8.
17. J. Simons, *Handbook for the Study of Egyptian Topographical Lists Relating to Western Asia* (Leiden, 1937).
18. II Chronicles 8:11.
19. Psalms 2:6.
20. Joel 2:1.
21. Joel 3:17.
22. Isaiah 66:18ff.
23. Daniel 9:16.
24. Daniel 9:24.
25. Nehemiah 11:1. Like expressions may also be found in Psalms 3:4, 15:1, 43:3 and 99:9; in Isaiah 65:11 and 25; in Ezekiel 20:40; in Zephaniah 3:11; in Zechariah 2:12; and in many other passages of the Bible.
26. A. Jirku, *Die ägyptischen Listen der Palästinensischen und Syrischen Ortsnamen, Klio Beihefte*, XXXVIII (Leipzig, 1937); Simons, *Handbook*.
27. Etam is number 36 on the list, Beth-Zur 110 (it is Beth-Zur, and not Beth-Shan as A. Jirku assumed), Socoh 67.
28. Breasted, *Records*, Vol. II, Sec. 434.
29. I Kings 7:13–45; II Chronicles 4:11–22.
30. II Chronicles 1:5.
31. I Kings 7:48–50; II Chronicles 4:7, 8, 21, 22.
32. I Kings 6:20, 21, 28, 30, 32, 35; II Chronicles 3:7, 9.
33. Exodus 37:3, 13–14.
34. I Kings 8:4.
35. Seder Olam 25. Other sources in Ginzberg, *Legends*, VI, 380.
36. Exodus 37:11, 12, 25.

37. See Plate VIII, "Vessels and Furnishings of the Temple at Jerusalem".
38. Exodus 37:17ff. Rim ornamentation of the vessels is discussed by H. Schaefer, *Die altaegyptischen Prunkgefaesse mit aufgesetzten Randverzierungen* (Leipzig, 1903). No reference to the biblical description of the vessels is suggested in his work.
39. Twenty cubits square, ten cubits in height (II Chronicles 4:1).
40. Cf. Exodus 25:30; 35:13; 39:36, and Numbers 4:7.
41. Exodus 25:35; 37:21.
42. *Nechoshet* is translated both "brass" and copper. However, it was either copper or bronze (alloy of copper with tin); brass (alloy of copper with zinc) was introduced much later.
43. II Kings 25:16. A few gold vessels might have been saved by the priests under Rehoboam, as it is said that Nebuchadnezzar took vessels of gold which Solomon had made for the Temple (11 Kings 24:13). But in Seder Olam it is said that Pharaoh Zerah returned to Asa what Shishak had taken from Rehoboam.
44. See the fantastic story of the capture of Jaffa by a general of Thutmose III in the Harris papyrus, 500, reverse, translation of Goodwin, *Transactions of the Society of Biblical Archaeology*, III, 340–348, and G. Maspero, ibid., I, 53–66; a new translation by T. E. Peet, *Journal of Egyptian Archaeology*, XI (1925), 226f.
45. Breasted, *Records*, Vol. II, Sec. 447.
46. Ibid., Sec. 451.
47. G. Schweinfurth, "Pflanzenbilder im Tempel von Karnak", Engler's *Botanische Jahrbücher*, LV (1919), 464–80. Wreszinski, *Atlas zur altaegyptischen Kulturgeschichte*, Pt. II, text to Plate 26.
48. Wreszinski, *Atlas*, Pt. II, text to Plate 33: "... *entzieht sich die weit überwiegende Zahl der dargestellten Pflanzen der botanischen Bestimmung und damit auch der Bestimmung ihrer Heimat.*"
49. Ibid., Pt. II, text to Plate 33.
50. M. Hilzheimer, quoted by Wreszinski, *Atlas*, Pt. II, text to Plate 33.
51. I Kings 11:21–22.
52. Breasted, *Records*, Vol. II, Sec. 474.
53. Ibid., Sec. 463.
54. Ibid., Sec. 467.
55. Septuagint, Reges III, 12:24e.
56. Metropolitan Museum of Art, No. 10.130.1003.
57. Breasted, *Records*, Vol II, Sec. 760, on the tomb of Rekhmire: "This is one of the most important scenes preserved in ancient Egypt. Similar scenes will be found in other Theban tombs, but none contains so elaborate, detailed, and extensive representation of the wealth of the Asiatic peoples, which was now flowing as tribute into the treasury of the Pharaohs."
58. Ibid., Sec. 756: "... of particular interest are the Semitic foreigners, who appear among the brickmakers, of the captivity which his majesty brought for

the works of the temple of Amon. This is, of course, precisely what was afterwards exacted of the Hebrews."

59. Mercer, *Extra-Biblical Sources*, p. 10. See also P. Montet, *Les Reliques de l'art syrien dans l'Egypte du Nouvel Empire* (Paris, 1937).

60. R. W. Rogers, *Cuneiform Parallels to the Old Testament* (2nd ed., New York and Cincinnati, 1926), p. 255.

61. Eduard Meyer reads "Rezenu". Breasted transliterates "Retenu".

62. Breasted, *Records*, Vol. II, Sec. 451.

63. Ibid., note to Sec. 451.

64. Ibid., Sec. 486.

65. See Lucas, *Ancient Egyptian Materials* (2nd ed.), p. 92.

66. Breasted, *Records*, Vol. II, Secs. 471–73.

67. Joshua 9:11; Judges 16:24; Psalms 85:10, 13; Micah 5:4; The Song of Solomon 2:12; compare also Leviticus 26:5; Numbers 10:9; and Jeremiah 5:19.

68. Breasted, *Records*, Vol. IV, Sec. 709. Wilson, "Egyptian Historical Texts" in *Ancient Near Eastern Texts*, ed. Pritchard: "There is no narrative account of the campaign by the pharaoh. The references in his inscriptions to tribute of the land of Syria or to his victories . . . are vague and generalized."

69. W. F. Albright, *Archaeology and the Religion of Israel* (Baltimore, 1942), p. 211.

70. Jirku, *Die ägyptischen Listen, Klio Beihefte*, XXXVIII (1937).

71. Breasted, *Records*, Vol. IV, Sec. 711.

72. Ibid., Sec. 715.

73. Jirku (*Die ägyptischen Listen, Klio Beihefte*, XXXVIII [1937]) expressed doubt whether an Aramaic word *hekel* would have been used in the tenth century in Palestine.

74. It must be noted that a portion of the bas-relief is destroyed.

CHAPTER FIVE

Ras Shamra

The Timetable of Minoan and Mycenaean Culture

On a spring day in 1928 a peasant plowing his field near the shore of Ras Shamra in northern Syria lifted the stone of a burial vault. In 1929 and in the following years, in twelve seasons of excavation,[1] buildings of a city and its harbor were unearthed together with pottery, utensils, jewelry, and the tablets of a library. This obscure place, not even marked on maps, lies to the north of Latakia, the ancient Laodicea *ad mare*, at a point on the Syrian coast opposite the elongated arm of land stretching out from Cyprus toward the mainland on the east. On a bright afternoon the island can be seen from the hills surrounding Ras Shamra.

The place was tentatively identified as Ugarit of the el-Amarna letters,[2] and written documents found there confirmed this conjecture. In gray antiquity the city had been repeatedly reduced to ruins. The levels at which dwellings were dug up are numbered from I to V starting at the surface. The first or uppermost layer is the most explored, but in the first nine archaeological seasons only about one eighth of this level had been unearthed. Digging in deeper strata has been confined to very small areas. The second layer yielded a few objects of Egyptian origin of the time of the Middle Kingdom; during the Middle Kingdom the north Syrian coast was in the sphere of Egyptian influence. At a depth of more than ten meters still older civilizations have come to light; remnants of the Neolithic (Late Stone) Age were found on the underlying rock.

The age of the remains found in the upper layer, which is only from forty centimeters to two meters under the surface, was established before the inscriptions were read. The material, design, and workmanship of pottery are held to be a reliable calendar in the hands of the archaeologists. The ceramics of the necropolis of Minet el Beida (the harbor of Ras Shamra) and of the acropolis of Ras Shamra were found to be of Cyprian origin and also of Mycenaean manufacture of the fifteenth and fourteenth and part of the thirteenth centuries before the present era.[3]

When a few Egyptian objects were found in this layer, too, the experts' identification of them as belonging to the Eighteenth and Nineteenth Dynasties[4] gave fair support to the time determination made on the basis of the pottery; the period during which Ugarit enjoyed prosperity was placed in the fifteenth century, and the fourteenth century was recognized as the one that saw the sudden decline of the city.

As two different methods had been applied and both led to similar conclusions, there was no further questioning of the age of the site, and all publications dealing with Ras Shamra-Ugarit[5] are based on the premise that the literary and cultural remnants from the excavated layer were products of the fifteenth and fourteenth centuries.

Before going further, we must appraise the real value of ceramics and other objects of art from Mycenae and Crete in dealing with time reckoning. In the course of this discussion I shall also have a few words to say on the age of the Minoan and Mycenaean cultures.

In Knossos on the northern shore of Crete, in Phaestus on the southern shore, and in other places on the island, remnants of a culture were found which is called Minoan, from the name of the semi-legendary king Minos. The remains belong to various epochs. The palace at Knossos and other buildings were suddenly destroyed, giving place to a new palace and buildings, which were again destroyed and again rebuilt. Many reasons led the explorer of these antiquities to the belief that a natural catastrophe was the agent of destruction, which marked the end of one period and the beginning of another.[6] The ages are divided into Early, Middle, and Late Minoan, and each age is divided into three parts, I, II, and III.

Another culture recognizable by its characteristic pottery had its center in Mycenae on the mainland of Greece. It, too, is divided into Early, Middle, and Late Mycenaean or Helladic Ages, which correspond roughly to the Minoan Ages of Crete.

The Minoan and Helladic Ages begin with the end of the Stone Age and are subdivisions of the Bronze Age. There is no internal evidence that would help to fix the dates of the Minoan-Mycenaean

172

Ages. The scripts of Crete have not yet been deciphered, despite some promising efforts, and the contacts with Egypt are regarded as the only source for establishing a timetable in the Minoan-Mycenaean past.[7] With some deviations, the Old, Middle, and New Kingdoms of Egypt are held to be the counterparts of the Early, Middle, and Late Minoan and Helladic Ages.

At Knossos of the Early Minoan period were found vases similar to pottery unearthed at Abydos in Egypt of the First Dynasty. Seals of the type of the Sixth Egyptian Dynasty were found in Crete. During the Middle Minoan period there was active intercourse between Crete and Egypt. At Abydos, in a tomb dating from the Twelfth Dynasty, a polychrome vase of the Middle Minoan II period was found, and at Knossos a statuette dating from the Twelfth or Thirteenth Dynasty was discovered. The dating of the Middle Minoan Age "of course depends upon that assigned to the Twelfth Dynasty".[8]

Crete was ruined by a catastrophe that corresponds in time to the catastrophe of the Exodus (the end of the Middle Kingdom and the end of the Middle Minoan II period). After the Middle Minoan III period, which corresponds to the time of the Hyksos rule in Egypt (the name of the Hyksos pharaoh Khian was found on the lid of a jar at Knossos), Crete freed herself from the influence of Egypt. It had its renaissance in the Late Minoan I period, which corresponds in time to the Egyptian renaissance after the expulsion of the Hyksos.

At Mycenae on the Greek mainland also were unearthed a few Egyptian objects bearing the cartouches of Amenhotep II, Amenhotep III, and his wife Tiy, of the Eighteenth Dynasty (New Kingdom); vases of the Late Mycenaean style were dug up in large numbers in Egypt, in Thebes, and especially from under the ruined walls of Akhnaton's palace at el-Amarna, "which thus gives a fixed date (about 1380 B.C.) for this style of vase-painting".[9]

The present research endeavors to bring to light a mistake of more than half a millennium in the conventional Egyptian chronology of the New Kingdom. If Akhnaton flourished in −840 and not in −1380, the ceramics from Mycenae found in the palace of Akhnaton are younger by five or six hundred years than they are presumed to be, and the Late Mycenaean period would accordingly move forward by half a thousand years on the scale of time.

It is my contention that the glorious Eighteenth Dynasty, the Kingdom of David and Solomon, and the Late Minoan and Late Mycenaean periods started simultaneously, about the year 1000 before the present era.

Returning to the excavations of Ras Shamra, we find that there were not two separate and coinciding time clues in the ceramics and bronze from Crete and Mycenae and in Egyptian pieces of art, but ultimately only a single one: the timetables of Crete and Mycenae are built upon the chronology of Egypt. This will also be shown in more detail in the chapter in which problems of stratigraphic archaeology are explored.

Sepulchral Chambers

The question which now arises is: Are there not other finds besides the ceramics which support or challenge the generally accepted view that the upper layer of the Ras Shamra excavations belongs to the period from the fifteenth to the fourteenth centuries? Does the testimony of architecture and of written documents support the conventional chronology or do they strengthen the view maintained here that this stratum and the past it buried are of the period from the tenth to the ninth or eighth centuries?

The chambers of the necropolis, unlike the buildings of the acropolis, are preserved intact. In a typical tomb well-arranged stone steps lead to a sepulchral chamber with arched ceiling. Similar vaults have been found in Cyprus. The excavator of Ras Shamra wrote of them: "Those in Cyprus are considerably later and continue down to the eighth and seventh centuries, according to the Swedish excavators.[10] One might therefore consider these Cypriote tombs to be late copies of the chamber tombs at Ras Shamra. One fine example is the burial vault of Trachonas on the east coast of the Karpas peninsula, exactly facing Ras Shamra. However, until earlier tombs of this type have been found in Cyprus, direct affiliation cannot be claimed. Some five hundred years lie between the Trachonas tomb and those of Ras Shamra."[11]

The vaults on Cyprus across the strait and in Ras Shamra are of identical construction; they must have been built in the same age. The distortion of chronology makes it necessary to maintain that five

174

hundred years elapsed before the Cypriotes started copying the Ras Shamra vaults, which by then would have been covered with earth and concealed from the human eye. Or it must be said that despite the obvious similarity of the peculiar vaults on both sides of the strait, no affiliation had occurred because of the five hundred years' difference in time.

In addition to the form of the sepulcher and its vault, a characteristic feature of the tombs of Ras Shamra's necropolis is a device which served for the offering of libations. It is an aperture with a pipe in it through which fluid food was delivered to the dead to sustain the soul on its journey to the other world.

It is obviously a very forced explanation to say that the population of Cyprus allowed half a millennium to elapse before they began to copy the tombs of the necropolis of Ras Shamra (Minet el Beida). Indeed, this explanation is utterly untenable, not only because of much archaeological material demonstrating that the influence came from Cyprus to the mainland and not the other way round, but especially because of the pottery found in the graves. The following statement was published after the first year of excavation at Ras Shamra:

"The influence which appears to dominate, if not at Ras Shamra itself, then, at least, at the near-by necropolis of Minet el Beida, is that of the island of Cyprus. The graves of Minet el Beida are of Cyprian shape and construction, and the bases of painted baked clay of which the funeral equipment largely consists, are very clearly and almost all Cyprian."[12]

Greek Elements in the Writings of Ras Shamra

Ras Shamra was not merely a maritime city that traded in arms of Cyprian copper and in wine, oil, and perfume: jars, flagons, and flacons were found there by the hundreds; it was also a city of learning: there was a school for scribes and a library. In the school the future scribes were taught to read and to write at least four languages.

Tablets of clay were found in the dust under the crushed walls of a building, destroyed by human hand or by the unleashed forces of nature. The entire collection is written in cuneiform, in four different languages. Two of the languages were easily read: Sumerian, "the

Latin" or the "dead language" of the scholars, and Akkadian, the tongue of business and politics in the Babylonian world.

Business letters in Akkadian, commercial receipts, and orders were read. Two tablets very similar to those of the el-Amarna collection were also found,[13] and with them the connection of Ras Shamra with Egypt at the end of the Eighteenth Dynasty was firmly established. Some large tablets are lexicons, bilingual and even trilingual. On some of the tablets there is a "copyright" mark: it is a statement that these tablets were made at the order of Nikmed, king of Ugarit.

Nikomedes is an old Greek name.[14] The similarity between the name Nikomedes, regarded as originally an Ionian name, and the name of the Ugaritian King Nikmed, is so obvious that, after deciphering the name of the king, two scholars,[15] working independently, related it to the Greek name. Other scholars, however, rejected this equation of the name of the king Nikmed (who also wrote his name Nikmes and Nikmedes) with Nikomed (Nikomedes) of the Greeks, asking how an Ionian name could have been in use in the fourteenth century before this era. Those who made the identification were unable to defend their position against the mathematics of conventional chronology.[16]

Ugarit was a maritime commercial city; its population was composed of various ethnic groups. One document found there describes the expulsion of King Nikmed and all the foreign groups in the city. Among them were people of Alasia (Cyprus), Khar (explained to be Hurrites), and Jm'an. The last name was identified by the decipherers as Jamanu, which is well known from the Assyrian inscriptions, and means Ionians.[17] This interpretation of Jm'an was disputed for no other reason than that in the fourteenth century a reference to Ionians would have been impossible. In the same inscription, at a point where the names of the expelled are repeated, the name Didyme appears. The decipherers took it to be the name of the city of Didyma in Ionia.[18] This city was renowned for its cult of Apollo Didymeus. Again, the name of the deity Didymeus (Ddms) was inscribed on another Ras Shamra tablet; the decipherers,[19] turning neither left nor right, translated it 'Apollon Didymeus'. Now antiquities have been brought from the site of Didyma, originating from the eighth century.[20] But in the fifteenth or fourteenth century neither Ionians nor the shrine of Apollo Didymeus could have been mentioned. Chronology could not square with the Ionian names of Nikomed, or the name of the Ionian city of Didyma, or the Greek cult of the god of that city, or the very name Ionians in the Ras Shamra texts – but all these were there, and no explanation was put forth in place of the rejected theory about an Ionian colony from the city of Didyma near Milet in Ionia that came to Ugarit and was expelled together with the king of Ionian origin, Nikmed.[21] It could only be stated that there was not a grain of probability in such a reading of texts belonging to the middle of the second millennium.

Among the tablets found in Ras Shamra there is a "catalogue of ships". It is an enumeration, for lexicographic purposes, of the various forms and uses of military and commercial vessels. Cargo ships, passenger ships, racing boats, fishing smacks, ferries, war vessels, and troop transports are listed. In the second book of the *Iliad* there is a similar, famous catalogue of ships. This portion of the *Iliad* was regarded as a later interpolation, but when a scholar[22] pointed out the similarity between the catalogues of Ras Shamra and the *Iliad*, the commentary to this portion of the *Iliad* was revised and one of opposite import substituted: "The catalogue, as certain scholars now agree, would not be a later interpolation, but has a long history behind it; as the Ras Shamra texts show, they were drawing up such catalogues in the port of Ugarit many centuries before the date of the Homeric lists."[23]

The *Iliad* is commonly supposed to have been put into writing in the seventh century. As for the time of its origin, the views of the scholars from antiquity on differ widely, placing it anywhere from the twelfth to the seventh century.

By placing King Nikmed at the end of the fifteenth and the beginning of the fourteenth centuries, the only conclusion possible from comparing the two "naval gazettes" was that hundreds of years before the very earliest date for the *Iliad* there existed a catalogue of ships which served as a model to the epic poet.

Hebrew Elements
Two Cities and Two Epochs Compared

The third language of the Ras Shamra tablets in cuneiform (Sumerian and Akkadian being the first two) did not long retain its secret.

The large tablets were apparently written in an alphabetic script. Their cuneiform could not be an ideographic or syllabic script, for a syllabic script like Akkadian uses hundreds of different signs, but alphabetic script only a few; and in this third script there were only thirty different characters.

An example of the simplification of the cuneiform script was already known to the scholars: the Persians in the sixth century had used cuneiform for an alphabet of thirty-six characters.[24]

The bright idea came simultaneously to more than one scholar[25] that it might be ancient Hebrew written in cuneiform. An attempt to substitute Hebrew letters for cuneiform signs was successful, and before the scholarly world were tablets in a legible language. Some of the texts were even re-edited by modern scholars in Hebrew characters.[26]

Reading was facilitated by strokes placed after each word by the scribes of Ras Shamra-Ugarit. The Cyprian script of the sixth century has the same characteristic stroke after each word, and this similarity was stressed, but it was asserted that, before this peculiarity returned, more than six hundred years had passed.[27] Again six hundred years! As in the case of the sepulchral chambers, it required six hundred years of latency before the Cypriotes started to imitate their neighbors only sixty miles away.

With an eagerness comparable only to the avaricious excitement of discoverers of a hidden treasure, scholars kindled their lamps and read the messages in ancient Hebrew. They thought they knew, even before they began to read, that the tablets were some six hundred years older than the oldest known Hebrew inscription. The discovery was startling: hundreds of years before the Israelites entered Canaan, the Canaanites not only used Hebrew[28] but wrote it in an alphabetic script.[29] Alphabetic writing in the fifteenth century before the present era was a revelation for paleographers and scholars in the history of human culture. "Since these documents date from the fourteenth or fifteenth century, the Ras Shamra alphabet is among the first alphabets to be composed, and actually is the earliest yet known."[30] The Hebrew-cuneiform alphabet of Ras Shamra is not a primitive pioneer effort; it has features that indicate it was already in an advanced stage. "The Ras Shamra alphabet is already so advanced that it implies the existence of a still earlier alphabet yet to be found."[31]

178

What the aborigines of Canaan wrote down was even more unexpected. In the mirror in which, in conformity with biblical references to the Canaanites, it was expected that the face of a wicked generation and of a low spiritual culture would be seen, the face of a dignified people was reflected. In the Book of Leviticus and in other books of the Scriptures iniquity and vice were attributed to the Canaanites: the country "was defiled by them". This appeared to be a "biased attitude of Israelite historians. . . . As it is, the Ras Shamra texts reveal a literature of a high moral tone, tempered with order and justice. By means of these documents we now see that the early Israelites differed in no way from the Canaanites."[32]

The Hebrew texts of Ras Shamra are mostly poems describing the exploits and battles of the gods and the adventures and wars of heroes. The pantheon of Ras Shamra[33] was composed of a number of gods; Baal was one of them, but the supreme deity was El.[34] The land of the Canaanites is sometimes called "the whole land of El", and the supremacy of this deity ("no one can change that which El has fixed"), known by the same name in the Bible as the Lord of the Israelites, is regarded as "a clear indication of a monotheistic tendency in the Canaanite religion".[35] However, besides El being not the sole but the chief god, he is described in the Ras Shamra texts in Homeric terms strange to the Old Testament: "El laughs with his whole heart and snaps his fingers.'

Besides the name El, which is predominant in the poems, especially in the poem of Keret dealing with exploits in Negeb, the name Yahu (Yahwe) is also encountered in the Ras Shamra texts.[36]

A few rare expressions or names found on Ras Shamra tablets are found also on monuments of the seventh century before the present era.[37]

A very unusual expression on one of the Ras Shamra tablets—"Astart, name of Baal"—appears in the epitaph of Eshmunazar, the Phoenician king of Sidon of the fifth century.[38]

The mythological pictures of the Ras Shamra poems often employ the same wording as the so-called mythological images of the Scriptures. Leviathan is "a crooked serpent" (Isaiah 27:1); it has several heads (Psalms 74:14). Lotan of the poems also is "a swift and crooked serpent" and has seven heads. There is, in one of the poems, an expression put into the mouth of El which sounds like a reference

179

to the great feat of tearing asunder the sea of Jam-Suf. And the verb, "to tear asunder", used there and in Psalms (136:13) is the same (*gzr*). The conclusion drawn from the similarity was this: long before the Exodus and the passage through the Red Sea, the Canaanites of Palestine knew this myth.[39]

The language of the poems of Ras Shamra is, in etymology and syntax, "surprisingly akin"[40] to the language, etymology, and syntax of the Scriptures, and the characteristic dual and plural forms, both masculine and feminine, are cited as examples.

The meter of the poems, the division into feet of three syllables or three words, and the balancing of the theme (parallelism) are also found in the Scriptures.[41] "These rules are precisely those of Hebrew poetry, and even the language from some of our Ras Shamra texts is entirely Biblical."[42] It was therefore concluded that Hebrew and Phoenician alike derived from the Canaanite, which could be called an Early Hebrew dialect.[43]

"There are striking similarities in the vocabulary, many words and even locutions being identical"[44] in the Ras Shamra texts and in the Old Testament. Here and there is found a turn of speech known from the Psalms, as, for instance, "I watered my coach with tears."

"The style resembles most the poetic books of the Old Testament, and especially the Book of Isaiah."[45] "We see that the Phoenicians of the fourteenth century before our era used rhythm and poetical forms that have all their development in the Song of Songs. . . . There are even some composite terms which are identical in both languages, such as the expression Beth-Haver, 'house of association', which occurs on one of the tablets and also in the Book of Proverbs."[46] In short, "there are innumerable parallels with the Old Testament in vocabulary and poetic style",[47] and an "intimate relationship existing between the Ras Shamra tablets and the literature of the Old Testament".[48]

The religious cult, as reflected by poems and other texts of Ras Shamra, also bore a certain resemblance to the cult of the Israelites. There was a *Rav Cohanim*, a high priest; adzes with engraved dedications to *Rav Cohanim* were unearthed. The offering called *mattan tam*, known from the service in the Temple of Jerusalem, is mentioned in the Ras Shamra texts. Circumcision was also practiced at Ras Shamra, judging from stone phalli found in this Phoenician city.[49]

The Jewish law forbidding the people to boil a calf in the milk of its mother was directed against a definite custom and a culinary dish. This dish was enjoyed at Ras Shamra, as its writings reveal.

From all this the following conclusion was drawn: "The traditions, culture and religion of the Israelites are bound up inextricably with the early Canaanites. The compilers of the Old Testament were fully aware of this, hence their obsession to break with such a past and to conceal their indebtedness to it."[50]

Even in minute details the life in Ras Shamra of the fifteenth century and the life in Jerusalem some six or seven hundred years later were strikingly similar.

Isaiah, on a visit to the gravely sick king, Hezekiah, ordered a *debelah*, a remedy made of figs, to be applied to the inflamed wound. *Debelah* is registered in the pharmacopoeia of Ras Shamra's medical men and is found mentioned in a veterinary treatise. The deduction was therefore made: "The prophet made use of a very old-fashioned remedy, known previously to the veterinary surgeons at Ugarit in the fifteenth and fourteenth centuries."[51]

This case of correspondence between the medical tablets of Ras Shamra and the Scriptures is not unique: "In the same [veterinary] treatise we also find some technical words corresponding exactly with similar expressions in the Bible, which further emphasize this contact between the Ras Shamra texts and the Old Testament."[52] And the generalization concerning medicine was: "They [the exactly corresponding technical words] establish a very striking similarity in the medical knowledge of the Canaanites or Proto-Phoenicians, and that of the times of the kings of Judah."[53]

The weights and measures of Ras Shamra were also those known from the Scriptures. In the Sumero-Babylonian system a talent was divided into 3600 shekels, but in the Scriptures (Exodus 38:25–27) the talent is composed of only 3000 shekels. Was this an erroneous statement? In the Ras Shamra texts, too, the talent is divided into 3000 shekels.[54]

Jewels of gold to adorn the maidens of Ras Shamra are mentioned in the texts and were unearthed.[55] "Now in the texts three kinds of gold pendants are mentioned by the name of 'Astarte', 'suns', and 'moons'. The word used for a sun pendant is *'shapash'*. 'Shapash' is identified with the word *'shebis'* mentioned in Isaiah 3:18.[56] The same prophet alludes to crescents or pendants in the form of the moon. So at Ras Shamra we find not only mention of these pendants in the Canaanite texts but also the ornaments themselves that Yahwe, in the passage cited in Isaiah, will take away one day from the haughty daughters of Zion."[57]

The ornaments named in the curse of the prophet emerged out of the earth. "Because the daughters of Zion are haughty, and walk with stretched forth necks and wanton eyes, walking and mincing as they go, and making a tinkling with their feet: Therefore the Lord will smite with a scab the crown of the head of the daughters of Zion. ... In that day the Lord will take away the bravery of their *akhasim* [anklets] and *shebisim* [suns] and *shaharonim* [crescents], the chains, and the bracelets ... and the earrings, the rings ... and the mantles, and the wimples, and the crisping pins ... the fine linen, and the hoods, and the veils. ... And her gates shall lament and mourn; and she being desolate shall sit upon the ground" (Isaiah 3:16–26).

In the hour of sorrow and mourning dust was thrown over the head, in ancient Ugarit and in Jerusalem alike, as the texts of the tablets and the scrolls of the Old Testament testify.

All these revivals of style and meter, of religious myths and cult, of old customs, of weights and measures, medieal science, apparel, and jewelry, emphasized and re-emphasized by modern scholars, would definitely point to the coexistence of Ugarit with the Jerusalem of the eighth or ninth century were it not for one obstacle. This was the fact that the Ugarit texts and objects were considered to be contemporaneous with the Egyptian and Mycenaean worlds of the fifteenth and fourteenth centuries.

Bible Criticism and the Documents of Ras Shamra

For the past seventy years the doctrines of Bible criticism have been taught from most cathedrals of modern exegesis and are at last being preached from many pulpits. Two of the fundamental concepts have been: (1) before the time of the Kings (or before −1000) there were no written documents among the Israelites, and (2) most passages of the Scriptures are of a later origin than the Scriptures themselves suggest or rabbinical tradition ascribes to them.

Since 1930 when the tablets of Ras Shamra were deciphered, they have been regarded as proofs (1) that already in the fifteenth century

182

Hebrew was written in a highly perfected alphabetic script that had a long period of development behind it, and (2) that many biblical traditions and legends were alive, and biblical style, poetic form, and ways of expression were in use some six hundred years before the biblical books were composed, even according to rabbinical tradition.

The confusion became great.[58] For three generations famous scholars, to whose lectures students traveled from afar, writers for encyclopedias, and authors of commentaries, all were moved to decrease the age of the Old Testament and even to assume a post-evangelical editing of various parts of the Old Testament. The whole argument was supported by linguistic considerations and by a general theory of the natural development of religious thought. It could be shown expertly that one or another expression in Psalms or in Proverbs could not have been employed in the days of David or Solomon in the tenth century, but was a product of the sixth to third centuries. Now, in the Ras Shamra tablets of the fifteenth or fourteenth century, the same expressions were found. A verdict of later origin had been given on many portions of the prophets: many passages were supposed to have been composed and interpolated during the Hellenistic period, which followed the conquest of Palestine by Alexander the Great in −332, and many sentences were supposed to have borne allusions to the events of the Maccabean war against the Seleucides, almost six hundred years after Isaiah. Now the same ideas and similar expressions were found on Ras Shamra tablets of a period six or seven hundred years before the time of the earlier prophets.

"With the present documents the history of the Hebrew language and of Syrian culture is pushed back toward the middle of the second pre-Christian millennium."[59] All proofs of late origin and all deductions based thereon become null and void before the evidence of the clay tablets.[60]

Bible criticism went to great pains to deny to Judaism of the pre-exilic period many of its achievements. By advancing the date of formulation of many social, moral, and religious imperatives of the Old Testament to the post-exilic period, Bible criticism had them originate in the Babylonian exile, referring some to the Seleucid period and to the influence of Greek thought.

The new view, predominant since the excavations at Ras Shamra, also regards many social, religious, and cultural elements of the Scriptures as copies, but of Canaanitic originals;[61] since they were already in existence some six hundred years before the time the Bible claims for them, they could not be of Jewish origin. The Canaanites paved the way to Jewish concepts in religion; their poetry had a high moral standard; their language, alphabet, style, and rhythm were inherited by the Jews; the ethos of social justice and the pathos of prophecy were Canaanitic hundreds of years before they became Israelite.[62] These and similar deductions were dictated by the age attributed to the tablets of Ras Shamra. In face of the striking parallels between the language, style, poetical forms, technical expressions, moral ideas, religious thought, temple ordinances, social institutions, treasury of legends and traditions, medical knowledge, apparel, and jewelry as reflected in the Ras Shamra tablets and in the pages of the Scriptures, the logical conclusion would have been that the tablets and the Books of the Scriptures containing these parallels are of the same age. But such a deduction was not thought of, owing to the obstacles of chronology already explained.

The revision of chronology requires the leveling of the time of Ras Shamra (Level I from the surface) to the time of the kings of Judea until Jehoshaphat. The presence of parallels in the life of Palestine and of a contemporary Syrian town, where the languages of neighboring peoples were learned, appears to be only natural.

If this reconstruction of world history by a correction of five to six hundred years puts a strain on the customary notions of history, how, then, can one's scientific conscience bridge a gap of double dimension and reconcile the results of the industrious efforts of Bible criticism with the archaeological finds of Ras Shamra? The span is twelve centuries.

Troglodytes or Carians?

The fourth language written in cuneiform on the tablets of the Ras Shamra library is called Khar. Words in Sumerian were accompanied by explanations in Khar. It appears to have been the local language, the language of the government and of a large part of the population. Despite the help of the bilingual syllabic dictionaries used by the scribes of Ras Shamra, the reading of Khar is not final. Had the words in Khar been explained in Sumerian, the task of the philologists would have been easier; but the translation and explanation of Sumerian words in Khar did not give all the necessary clues to the decipherers.

Before the excavations of Ras Shamra, frequent mention of "Khr" had already been encountered in various archaeological documents. Akkadian texts speak of "Khurri", and in Egyptian documents a part of Syria is often called "Kharu".

It had long been held that these Assyrian and Egyptian designations referred to the Horites or troglodytes of the early chapters of the Scriptures.[63]

With the discovery of the Tell el-Amarna archives in Egypt it was found that one of the letters of the archives was written, apart from the introduction, in an unknown tongue. This letter, written by Tushratta, king of Mitanni, dealt in its six hundred lines with some matters interpreted with the help of the other letters, and the language was deciphered. At first it was called Mitannian, but later changed to Subarean.

Then in the state archives of Boghazkeui in eastern Anatolia letters were found in a similar tongue, and its name was given as Khri. The people who spoke this language were called Khr. Scholars read the word differently—Khar and Khur—but finally they decided on Khur as the acceptable name, and accordingly the people are called Hurrians or Hurrites.

Despite the fact that the language of this people was found to have been put into writing, the identification of the Hurrians with the biblical Horites or troglodytes was maintained by a number of scholars.[64]

A definite vestige of the association of the Hurrians with Palestine has been discovered: on tablets from Tell Taannek, in the valley of Jezreel, Hurrian names were found.

With every new discovery it became increasingly obvious that the Hurrians exercised great influence on the civilization of the Near East. It was even stated that with the arrival of the Hurrians in this part of the world a new era in civilization had dawned.[65] In a sense they became the leading power, and "the story of their enormous expanse, from Armenia down to southern Palestine, and from the shores of the Mediterranean up to the borders of Persia, constitutes one of the most amazing chapters in the ancient history of the Near East".[66]

The language of this people has been studied by linguists in an endeavor to unriddle it,[67] but the historians know nothing of their history. "Hurrian" seemed therefore to be a tongue without a people. Those who spoke it were not Semitic, but neither were they Indo-Iranian.[68]

Then the writings in alphabetic Khar of Ras Shamra came to light. Translations from other languages into Khar proved that at least a part of the population used Khar as their daily speech. Who, then, were these Khar that impressed their name on Syria, their tongue on Asia Minor and on Mitanni, occupied a fortress in Palestine, were everywhere and nowhere in particular, were neither Semitic nor Indo-Iranian?

It became apparent not only that Khar was expressible in writing, but that the scribes who wrote in Khar were versed in a number of other languages as well, and wore themselves out in lexicographic study ("several rooms" in the library of Nikmed "contained only dictionaries and lexicons"[69]). Consequently the idea that the Khar were cave dwellers or troglodytes (the biblical Horites) appears wholly untenable.

Most probably the Hurrian people is but a creation of modern linguists. If we bring the scene five to six hundred years closer to our time we begin to wonder whether the Khar of the inscriptions are not the Carians often mentioned in classic literature. In Egyptian the Mediterranean Sea was called the Sea of Khar(u). Was it the sea of troglodytes or the sea of the Carians?

The Carians lived on the shore of the eastern Mediterranean and had settlements in many other parts of the world. They have left their traces in geographical names containing the syllable "car" or "cart" or "keret".[70] In very early times—that of the semilegendary King Minos of Crete—they manned his fleet. Herodotus says that at this time they were islanders and served as crews on the ships of Minos. "Minos had subdued much territory to himself and was victorious in war," and "this made the Carians too at that time to be very far the most regarded of all nations".[71]

"Then a long time afterwards, the Carians were driven from the islands by Dorians and Ionians and so came to the mainland."[72]

The mainland meant here is the southwestern corner of Asia Minor, where Halicarnassus, the native town of Herodotus, was situated.

Thucydides ascribed to King Minos the expulsion of the Carians: "Minos is the earliest of all those known to us by tradition who acquired a navy. . . . [He] became lord of the Cyclades islands and first colonizer of most of them, driving out the Carians and establishing his own sons in them as governors."[73]

The Carians are described by Thucydides as dispersed on many islands and engaging with the Phoenicians in piracy: "still more addicted to piracy were the islanders. These included Carians as well as Phoenicians, for Carians inhabited most of the islands." The close relations existing between the Carians and the Phoenicians are further attested to by the names of towns like Phoinix and Phoinikus in Caria.[74]

Knowledge of the whereabouts of Carian colonies outside southwest Asia Minor and the trails of Carian wandering was early lost to history. In the first century Strabo[75] wrote: "The emigrations of the Carians . . . are not likewise matters of off-hand knowledge to everybody."[76]

Cyprus was apparently included in "most of the islands", and the Carians survived there till a late date. Herodotus (V, 111) mentions a Carian shield-bearer on Cyprus in the early Persian days.

It does not require much ingenuity to come to the conclusion that the Khar of Ras Shamra were Carians.

The Carians settled not only in Cyprus but also on the shore of the mainland opposite, and similar graves in eastern Cyprus and in Ras Shamra are evidence of this. The peculiarity of the Carian graves was stressed by the early historian Thucydides, who wrote that Carians had inhabited most of the islands "as may be inferred from the fact that, when Delos was purified by the Athenians in this war [−426] and the graves of all who had ever died on the island were removed, over half were discovered to be Carians, being recognized by the fashion of the armour found buried with them, and by the mode of burial, which is that still in use among them".[77]

Modern archaeologists again point out the peculiar features of the graves of Ras Shamra and of eastern Cyprus.

Inasmuch as the Carians were inhabitants of northern Syria early in the first millennium before this era, it is only reasonable to look for a mention of them in the Scriptures. In the eighth century

Athaliah—daughter of Ahab and daughter-in-law of King Jehoshaphat of Jerusalem—the queen-mother who usurped the throne when her son Ahaziah was killed by Jehu on the road to Megiddo, had a bodyguard composed of "Cari". This bodyguard later participated in a revolt against Athaliah, when Jehoiada, the priest, made a covenant with "the rulers over hundreds, with the Cari, and the runners" (II Kings 11:4, 19)[78] and brought before them Jehoash, the boy who was secretly saved when Athaliah killed the royal family.

It is more than probable that the Kreti of the "Kreti and Pleti" (Cherethites and Pelethites) bodyguard of David (II Samuel 8:18), led by his marshal Benaiah, were the same Kari. In one place in the Scriptures (II Samuel 20:23) it is said that Benaiah was in command of Kari (or Kre) and Pleti. The Philistines, since days of old, have been considered the Kreti-Pleti. The word "Pleti" is generally regarded as a shortened form of "Philistines", and without sufficient ground they have been presumed to be the same people as the Kreti, and thus originated the theory that the Philistines came from Crete.[79] Pleti cannot be identical with Kreti or Kari, because whenever they are mentioned the two names are always connected by "and".[80]

The origin of the Kreti in Crete—heard even in the name—is also attested to by the Version of the Seventy, who translated "Kreti" by "Cretans". The Carians came from Crete. The Kreti also came from Crete and were identical with the Kari. It is obvious that Carians and Kari and Kreti were the same.

The Carians, royal bodyguard in the Jerusalem of Queen Athaliah, were employed in the same capacity in the Egypt of the seventh century, after they had arrived there together with the Ionians, driven by a gale.[81] The Carians occupied this position until Cambyses conquered Egypt.[82] They also formed the bodyguard of the kings of Lydia in the sixth century.

In this connection it is interesting to correlate Herodotus' statement (I, 171ff.) that the Carians invented and produced arms and were imitated by the Greeks with the fact that "in no other site in Syria or Palestine was found such a large quantity of arms as in Ras Shamra"[83] and with the Targum's translation of Kreti as bowmen.

At the end of the seventh century and the beginning of the sixth

Zephaniah (2:5) and Ezekiel (25:16) prophesied the end to come to the maritime shore of Keret.

When shortly thereafter Nebuchadnezzar subdued Tyre, the Phoenicians and the Carians fled to Carthage, and after that this colony grew to a metropolis.

The Carian Language

With the Ras Shamra tablets before the linguists, it would seem that the scholarly world is at last closer to the solution of the question, What was the language of the Carians? than was Strabo, who discussed it nineteen centuries ago.[84]

Homer, in his enumeration of the allies of Troy, included "the barbarously speaking Carians". Apollodorus understood these words as an allusion to the fact that the Carians spoke, not Greek, but some strange language.[85] Strabo concluded from Homer's reference that the Carians spoke Greek but pronounced it like barbarians. Strabo probably had in mind the Carians of Asia Minor, about whom Herodotus wrote that, with the settlement of the Carians there, the speech of the Caunians, the previous inhabitants, "has grown like to the Carians', or the Carians' to theirs".[86]

That the Carians used a speech not understandable to the Greeks is obvious from a story of Herodotus.[87] He narrates how a Carian came to a temple to hear the oracle; the Thebans stood amazed to hear the oracle use a language other than Greek. The stranger said that the words of the oracle were Carian, and he wrote them down.

The Carians used Greek too; Herodotus tells us that the Egyptians learned Greek from the Carians and Ionians who came to Egypt in the days of Psammetich in the seventh century.

A scattering of Carian words is found in the classic authors; also a few Carian personal names are preserved.[88] In Egypt, on various monuments, together with Greek autographs were found autographs not in Greek; they were written in Greek letters with the addition of a number of different characters so as to conform to the phonetics of the tongue. The conclusion was drawn that these were the names Carian mercenaries signed next to those of their Ionian comrades-in-arms.[89] The time was the seventh century. A number of epitaphs, some bilingual—in Carian and in Egyptian—were also discovered, but as the texts were apparently not exact translations from one language into the other, attempts to decipher the Carian have remained quite indecisive. The hypothesis that the language was Indo-German was disputed and rejected.[90] Neither is it Semitic.

The same diagnosis—that it is neither Indo-German nor Semitic—is made for the Hurrian (Khar) language of Mesopotamia. In both cases the characters, of the Khar and of the Carian inscriptions, are foreign to the language—in the one case they are borrowed from cuneiform, in the other from the Greek alphabet. It is of interest that a Mitanni element was discerned in Carian.[91] The Carian has not yet been read. In the summer of 1935 numerous Carian inscriptions were discovered near Mylasa in Caria (by Benveniste), but they have not yet been published. For reasons explained previously, it seems to me that it would be profitable to investigate the Khar of Ras Shamra, proceeding on the basis that it is the Carian language but in other characters, and to decipher Carian inscriptions with the help of the tablets of Ras Shamra.

The theory that the Ras Shamra tablets contain references to Ionians should not have been rejected; Ionians might have been mentioned in them, for it was not the fourteenth but the ninth century. It seems to me also permissible to conjecture that the word read as Ugarit[92] is the Carian-Ionian name of Euagoras. Kings of that name ruled on Cyprus, and the reign of one Euagoras in the fifth century before this era and of another early in the fourth century are known from Greek and Latin authors. The war of the second Euagoras against the Persians is mentioned in a later chapter of the present work. The influence of Cyprus on Ras Shamra was emphasized by the excavators; it is thought that in an early period this site on the Syrian shore might have been a colony of Cyprian rulers. This observation corroborates the surmise that an early king of Cyprus, a descendant of Carians who were driven to the east by the Ionians, built a city on the Syrian shore opposite the island of Cyprus on the ruins of a previous city and called it by his own name, Euagoras.

The name of the king Nikmedes (also written Nikmes, Nikmed) is the Ionian-Carian name Nikomedes. A similar name, Nikodamos, may have been a form more agreeable to a Semitic ear.[93]

The city of Didyma, whence came the Ionians to Ugarit, was in Ionia, but its name is Carian.[94] Tablets of the ninth century before the present era could contain this name, but not tablets of the fifteenth century.

It is curious, indeed, that the scholar who insisted that Ionians are referred to in the Ras Shamra texts nevertheless conservatively maintained that the Khar of Ras Shamra, mentioned together with the Ionians, were the Horites (troglodytes) of the Bible.[95] Carians and Ionians, not only in the Ras Shamra text, but in many texts of Greek authors, are mentioned together. Despite the fact that at an early date the Carians were driven from Crete and the Cyclades by the Greeks (Ionians), the two people intermingled, and the story of their first appearance together on the Egyptian shore shows that they became partners in their enterprises. In the classic literature the Carians usually appear either with the Ionians or with the Phoenicians.

If one day an Orphic hymn should be found in the dust of Ras Shamra it would be a lucky day for the excavators but no miracle.

According to Homer, the Carians participated in the defense of Troy. They might have had their own reminiscences and their own poems in which they sang the battles of Ilion. It is known that on Cyprus poems were composed dealing with the same subject as the *Iliad*,[96] but, except for twenty-five short fragments, nothing remains of them. Were these poems originally written in Carian?

The early relations between the Carians and Crete and between Crete and Cyprus should be remembered now that the Khar language of Ras Shamra is being read. Then a new attempt should be made to interpret the still unread characters of the Cyprian inscriptions, the linear script of Crete, and the pictorial script of that island, and thus to lift the veil that conceals the past of Crete and of the Minoan culture, the maritime adventures of the Carians in the second millennium, perhaps even the story of Atlantis.

Amenhotep II

Syria-Palestine of the period we are discussing was a region coveted by the pharaohs and striving for independence.

When the long and successful reign of Thutmose III came to its end, Amenhotep II (his royal name is usually read Okheperure) took the scepter. To the Asiatic provinces the death of Thutmose III was a signal for insurrection and the casting off of the Egyptian yoke. Amenhotep II marched at the head of a vast army of chariots, horsemen, and foot warriors to suppress the rebellion in Syria and Palestine. His Majesty "went against Retenu [Palestine] in his first victorious campaign, in order to extend his frontier. . . . His Majesty came to Shamash-Edom and devastated it. . . . His Majesty came to Ugarit and subdued all his adversaries. . . ."[97]

On the way to Syria Amenhotep II displayed his ability to use the bow in a demonstration before the local princes in order to impress and intimidate them.

He returned to Memphis with a few hundred nobles as war prisoners and a booty of some hundred horses and chariots or war carriages. On his return to Egypt he hanged some of the prisoners to the mast of his ship on the Nile with their heads down.

In his ninth year he repeated his expedition to Palestine, his goal being Aphek in lower Galilee. He plundered two villages "west of Socoh", and after pillaging other unimportant localities, he returned to Memphis with more prisoners. His harassing visits made him a common enemy of the kingdoms of Palestine and Syria. When he came again to Palestine, the main, and seemingly the only, battle was fought at a place called y-r'-s-t. Various solutions have been proposed for the identification of this locality.[98] However, it is an important fact that according to Amenhotep's annals he reached the place *one* day after his army left the Egyptian border.[99] Thus the place of the battle could have been only in southern Palestine.

Amenhotep called himself victorious, and it is accepted that this campaign was a victorious one. But was it really? What was the booty in the battle of y-r'-s-t?

> List of that which his majesty captured on this day: his horses 2, chariots 1, a coat of mail, 2 bows, a quiver full of arrows, a corselet and—[100]

some object the reading of which is no longer possible. But whatever may have been that last object, the complete spoils were pitiful indeed if all the king of Egypt could count after his victorious battle were one chariot, two horses, two bows, and one quiver "full of arrows". It was a defeat.[101]

After a victory an army usually marches deeper into the enemy's territory. But the lines directly following the enumeration of the spoils say that, "passing southward toward Egypt, his majesty proceeded by horse".[102] Immediately after the battle, the king turned toward Egypt.

When a king returns from a successful campaign of restoring order in the provinces, the cities located on his triumphal route home do not choose that moment for revolt. Vassal cities rebel on seeing their oppressor in flight, and this is just what happened, for the war annals relate that Asiatics of a city on the way to Egypt "plotted to make a plan for casting out the infantry of his majesty".[103]

During the remainder of his reign, for some decades, Amenhotep II did not return to Palestine, and there is no mention of any yearly tribute from there.[104]

To ascertain whether his expedition was a defeat, his subjective evaluation of the campaign must be compared with the scriptural record.

The son of Rehoboam, Abijah, king of Judah, succeeded in winning a decisive battle against Jeroboam, king of Israel (II Chronicles 13). This must mean that Egyptian domination was already declining.

After the short reign of Abijah, Asa, his son, followed him. "In his days the land was quiet ten years." He built fortified cities in Judah, constructed walls and towers, gates and bars. He said to Judah: "We have sought the Lord our God, and he hath given us rest on every side" (II Chronicles 14:7). So they built and prospered.

The destruction of the images of the pagan gods was in itself a rebellion (II Chronicles 14:5), for among them the first place surely belonged to the Egyptian gods, as the land since Shishak (Thutmose III) had been subject to the Egyptian crown. By fortifying the cities of Judah and recruiting his warriors, Asa clearly rejected Egyptian rule.

II CHRONICLES 14:8 And Asa had an army of men that bare targets and spears, out of Judah three hundred thousand; and out of Benjamin, that bare shields and drew bows, two hundred and fourscore thousand: all these were mighty men of valor.

The cities were fortified, the army stood ready.

II CHRONICLES 14:9–10 And there came out against them Zerah the Ethiopian with a host of a thousand thousand, and three hundred chariots; and came unto Mareshah.

Then Asa went out against him, and they set the battle in array in the valley of Zephathah at Mareshah.

Asa prayed to God for help.

II CHRONICLES 14:12–13 So the Lord smote the Ethiopians before Asa, and before Judah; and the Ethiopians fled.

And Asa and the people that were with him pursued them unto Gerar; and the Ethiopians were overthrown, that they could not recover themselves; for they were destroyed before the Lord, and before his host; and they carried away very much spoil.

Zerah the Ethiopian, who led an army of Ethiopians and Libyans (II Chronicles 16:8) from the southern and western borders of Egypt (like the army of the pharaoh Shishak), could be none other than a pharaoh. The way from Ethiopia to Palestine is along the valley of the Nile, and an Ethiopian army, in order to reach Palestine, would have had to conquer Egypt first. Moreover, the presence of Libyan soldiers in the army leaves little doubt that the king was the pharaoh of Egypt.

In the opinion of the exegetes (Graf, Erbt) the story of the Chronicles must have a historical basis in an Egyptian or an Arabian invasion.

The description of the battle of Mareshah or Moresheth[105] reveals why the pharaoh turned his back speedily on Palestine and his face toward Egypt, why from the field of this battle his army carried away "one bow and two horses", and why the population of the cities, presumably in Edomite southern Palestine, plotted against his garrisons.

It is a token of defeat when an Egyptian king recounts his own personal valor and fierceness on the battlefield, fighting himself against the soldiers of the enemy. It means that, when everyone had fled, His Majesty fought alone. In bombastic phrases, which do not refer to any special encounter, the inscription glorifies the ruler who battled alone: "Behold, he was like a fierce-eyed lion."

He was pursued only to Gerar. So he still had the satisfaction of taking with him on his return to Egypt a few chiefs of some villages, whom he burned alive in Egypt: his Memphis stele records this holocaust.

194

Amenhotep II was not a great man, but he was a large one. He was proud of his physical strength and boasted that no one could draw his bow. A large bow inscribed with his name was found a few decades ago in his sepulcher.

"There is not one who can draw his bow among his army, among the hill-country sheiks [or] among the princes of Retenu [Palestine] because his strength is so much greater than [that of] any king who has ever existed," says the Elephantine stele.[106]

"It is his story which furnished Herodotus with the legend that Cambyses was unable to draw the bow of the king of Ethiopia."[107] A modern scholar saw a common origin in this story, which survived in legendary form in Herodotus (Book III, 21ff.), and in the historical boast written on the stele of Elephantine by Amenhotep II, who lived many centuries earlier. The story of Herodotus has an Ethiopian king as the bragging bender of the bow of Amenhotep II. Was Amenhotep II an Ethiopian on the Egyptian throne?

In the veins of the Theban Dynasty there was Ethiopian blood.[108] Was the royal wife of Thutmose III a full-blooded Ethiopian and did she bear him a dark-skinned son? Or was Amenhotep II not the son of Thutmose III at all? He called himself son of Thutmose, but this claim need not have been literally true. He called his mother Hatshepsut.[109] Is it possible that before ascending the throne of Egypt he was a viceroy in Ethiopia?[110] Conventional chronology identifying Zerah with Osorkon of the Libyan Dynasty encounters difficulty in the biblical reference to Zerah as an Ethiopian.

It was a glorious accomplishment to carry away so decisive a victory from the battlefield, when the foe was not a petty Arabian prince—as some exegetes have thought[111]—or pharaoh of the ignominious Twenty-second Dynasty—as other exegetes have assumed—but Amenhotep II, the great pharaoh, the successor to Thutmose III, the greatest of all the pharaohs. It was a victory as sweeping as the defeat of the Hyksos-Amalekites by Saul, but, as we shall see, its effect on the subsequent period was not of equal importance. Politically, the victory was not sufficiently exploited, but this fact does not detract from its military value. Egypt, at the very zenith of its imperial might, was beaten by Asa, king of Judah, and this was not a victory over an Egyptian garrison or a detachment dispatched

195

to collect tribute, but over the multitude of the Egyptian-Ethiopian and Libyan hosts, at the head of whom stood the emperor-pharaoh himself.

With the rout of the Egyptian army in the south of Palestine, all of Syria-Palestine naturally was freed of the Egyptian yoke. The pharaoh had previously laid Ugarit waste and threatened all the kingdoms in this area; it is conceivable that the king of Judah had some help from the north, and the sympathy of the Syrian maritime peoples must certainly have been with Asa. The inscriptions of Amenhotep II reveal his ambition to dominate, in addition to the land of the Nile, the lands of the Jordan, Orontes, and Euphrates, which had rebelled after the death of Thutmose III. The great victory at Mareshah carried a message of freedom to all these peoples; the repercussions of the battle should have been heard in many countries and for many generations. But only once again does the Book of Chronicles pay tribute to this victory, and this in the words of the seer Hanani: "Were not the Ethiopians and the Lubim [Libyans] a huge host, with very many chariots and horsemen?" (II Chronicles 16:8.) It is also said that the population of the northern tribes went over to Judah because of the high esteem this country enjoyed after it had successfully repelled the pharaoh and his army (II Chronicles 15:9). Is no more material concerning the victory of Asa over Amenhotep II preserved? Such a great triumph should have had a greater echo.

It had this echo. A Phoenician poem sings of that victory over the host of Amenhotep II.

The Poem of Keret

Among the epics unearthed at Ras Shamra, one contains some historical material. The poem of Keret—the archaeologists call it by the name of its hero—has a historical setting. It was first translated and interpreted by Charles Virolleaud.[112] Later a very different meaning was given to the text. Virolleaud read in the text of the danger threatening the country of Keret, king of Sidon. The invasion of Negeb (south of Palestine) by the army of Terah aroused Keret's fears; he wept in the seclusion of his chamber. In his great distress he

196

was encouraged by a voice heard in a dream, and he went to meet the danger and joined the army of the defenders.

The names of Asher and Zebulun, two tribes, appear according to Virolleaud, in the poem. It is not clear from the poem whether the role of the tribe of Zebulun was that of an enemy or a friend.

Asher is mentioned repeatedly in a refrain, and the poem gives a vivid feeling of armed tribesmen hurrying to join the main army opposing Terah.

> Asher, two and two are gone, Asher, three and three are gone, shut the houses, marched together.

Volunteers joined the thousands of Hasis.

> Men of Hasis went by thousands, and by myriads, as a flood [yr].

They marched to meet the army of Terah. And Terah came into Negeb with a great force: "a great force of three hundred times ten thousand [rbt]", which would mean—if the translation is correct—three million men. Then the poem tells that the huge invading army, having been defeated, was in full retreat.

Who was Terah? asked Virolleaud. In Genesis the father of Abraham is called Terah. The theory was advanced[113] and found followers in France that patriarchal migrations and wars are described in the Phoenician poem of Keret. It was found to be a very illuminating addition to the legends about the sojourn and wandering of the patriarchs in Negeb—southern Palestine—as found in Genesis.

The patriarch Abraham came to Negeb; so did Terah of the poem. In the Scriptures it is said that Terah, Abraham's father, migrated from Ur of the Chaldees on the lower flow of the Euphrates to Harran in the northwest, and ended his days there (Genesis 11:32). A correction was introduced with the help of the poem, and it was agreed that Terah did not die in Harran but prepared the conquest of Canaan from the south and also accomplished it in part, and that Abraham capitulated when he met difficulties and left Canaan to seek refuge in Egypt.[114] Abraham and his two brothers, sons of the scriptural Terah, are not mentioned in the poem, and it was conjectured that this was because of the leading role played by Terah and the inconspicuous role of Abraham, the latter becoming an anonym in the multitude of the Terahites. And if the tale is very

197

different from the scriptural legend, still the combination—Negeb, the scene, and Terah, the invader—seemed to be a convincing parallel to Negeb, the scene, and Abraham, son of Terah, the invader. Consequently, the conclusion was drawn by Virolleaud that Terahites invaded the south of Canaan, meeting resistance on the part of the population, although in the Scriptures nothing is said about Abraham's war with the Canaanites, and in fact the peacefulness of his sojourn there is stressed.

An unexplained unconformity is the huge number of soldiers in the host of Terah: three hundred times ten thousand is very different from the number of persons in Abraham's household, servants included. The occurrence of the names Asher and Zebulun also presented a difficulty. Asher and Zebulun were sons of Israel of the Scriptures; these tribes were descendants of Abraham, son of Terah. How could Terah have battled with children of Asher and Zebulun, his descendants of many generations?

To meet this situation it was said that originally the names of Asher and Jebulun belonged to cantons inhabited by Canaanites. At a later date these places were conquered by the tribes of Israel, who did not give their names to, but received their names from, the cantons.[115]

Another translation and interpretation of the poem of Keret was presented. It rejects Terah, Asher, Zebulun, as proper names, finding for them the meanings: bridegroom (*terah*), after, behind (*atur*), sick man (*zebulun*).[116] It also denies the predominant martial theme of the poem, regarding it as a love romance. Thus are explained away the names of tribes not to be expected in Ugaritic times. Numerous other changes and corrections were offered.

It seems to us, however, that Virolleaud's translation was not far from the truth. Terah of the poem was, indeed, not the father of the patriarch, but the names of the tribes and the martial plot appear to be consistent with history.

It is, in fact, undisputed that Ugarit and the entire Phoenician coast were threatened by Amenhotep II in the period with which we are concerned.

Free from the limitations imposed by an incorrect estimate of the age of the Ras Shamra tablets, we pose this question: Is an unsuccessful invasion of southern Palestine by a large host known to us from the Scriptures?

Such an invasion occurred in the days of Asa, king of Judah, and it was headed by Zerah the Ethiopian. In the Second Book of Chronicles the invading army under Zerah is called a very large multitude, "thousand thousand", or a million.

> II CHRONICLES 14:9 And there came out against them Zerah the Ethiopian with a host of a thousand thousand. . . .

We have already recognized Zerah the Ethiopian as King Amenhotep II, and compared the scriptural and the Egyptian material.

We shall compare a few data from the Phoenician poem and from the inscriptions of Amenhotep II and from the Scriptures. If we again find parallels, obviously we shall strengthen the identification of Zerah as Amenhotep II.

In the Phoenician poem the army of Terah is described in these words:

> A great force of 300 times 10,000 [rbt] with harpes [hepeš] of copper, with daggers [šnn] of bronze.

It was noted that the poem used for copper scimitars the Egyptian word hepes,[117] and for the daggers of bronze, the Egyptian word snn.

In the tomb of Amenken,[118] King Amenhotep II is portrayed inspecting gifts to be distributed among his officers; these are swords described with the words "360 bronze hps [hepes]", and next to these swords are pictured one hundred and forty daggers. Inasmuch as the Egyptian name is given to the weapons in the Phoenician poem, their Egyptian origin would not be an arbitrary conclusion. The weapons are designated in the Phoenician poem by the word used on the Egyptian monument depicting Pharaoh Amenhotep II with weapons for his army. Accordingly, the forces of Terah were armed with weapons identical with those with which Amenhotep armed his soldiers. Examples of harpes (scimitars) have been found in Gezer in southern Palestine.[119]

The extant fragments of the poem describe the armies rushing to the battle. We see Keret marching across fields on which he finds arms thrown away by the fleeing soldiers of a beaten army; he comes to a spring tinged with blood.[120] He marches toward the cities of Edom with the intention of sharing in the loot.

According to the Scriptures, after the battle of Mareshah (Moresheth-gath) the Judean victors looted the cities of the south: ". . . the people . . . pursued them unto Gerar. . . . And they smote all the cities round about Gerar . . . and they spoiled all the cities; for there was exceedingly much spoil in them" (II Chronicles 14:13–14).

In the poem the march to the south to loot the cities after the victory is told in these words by the poet to Keret:

> Go a day and two days
> A third one and a fourth one,
> A fifth one and a sixth one,
> And on the seventh day you will meet Sapasites.
> And you shall come to Edom Rabbim
> And to Edom Serirot.

The poem is in the rhythm of the marching step of the hero and his men.

> Then he met the Sapasites
> And went to Edom Rabbot
> And to Edom Serirot.

The king of Edom begged Keret:

> Don't combat Edom Rabbot
> Nor Edom Serirot,
> Depart, king of Sidon,
> O, Keret, from my parvis![121]

And Keret asked from him his daughter instead of booty: she is "beautiful as Astarte".

We are intrigued to know whether Keret brought the daughter of the king of the Edomite city to Sidon as his consort, and a subsequently found fragment of the poem tells us that he did so. But we are more anxious to know who were the "Sapasites" before the walls of the cities of Edom, and what the name Serirot designates, and whether we are on the road we set ourselves to follow.

What does "people of Sapas" mean? A partial answer is at hand: "Sapas is the sun and the Sapasites got their name from it; Sapas of

200

Canaan was analogous to Shamash [sun] of the Assyrians and Babylonians."[122] The people who are repeatedly named as opposing Keret before the walls of the Edomite city were Sapas or Shamash men.

Cities in Edom were doomed to be looted, according to the poem and the Second Book of Chronicles, which also suggested that this fate was imposed on them because they apparently supported Zerah and his Ethiopian and Libyan host. We are therefore interested to know from what base Amenhotep II conducted his campaign against Palestine.

Amenhotep II's record of his campaign in Palestine-Syria begins verbatim:

> His majesty was in the city of Shamash Edom. His majesty furnished an example of bravery there.

The city of Shamash Edom is nowhere mentioned in extant Egyptian documents except in the registers of Thutmose III and in this monumental inscription of Amenhotep II. Is it a coincidence that, in the Phoenician poem about the invasion of Terah and his multitude, the name of the city is Edom and the people are called Sapas or Shamash? We have identified Terah of the poem as Zerah of the Scriptures. We have identified Zerah of the Scriptures as Amenhotep II. Now we find the same city with the same people as the military base of Terah of the poem and of Amenhotep II of the hieroglyphic annals.

Of the word "Sarira" or "Serirot" (plural) in the poem of Keret we have an explanation in the Septuagint, the Greek translation of the Scriptures. An addition to I Kings 12:24 gives some details concerning Jeroboam. His mother, called Zeruah in I Kings 11:26, is referred to there as Sarira. When he was appointed by Solomon to be over the northern part of the kingdom he built a town in Mount Ephraim and named it Sarira. When, after the death of Solomon, Jeroboam returned from his exile in Egypt he "came into the land of Sarira". The tribe of Ephraim assembled there and, according to the Septuagint, Jeroboam fortified Sarira.

Sarira was the name of a fortress built first about —920 and possibly of every fortress built following the plan of Jeroboam's fortress, which he called by the name of his mother. In the days of Keret, one generation later, the name Edom-Serirot (plural of Sarira)

201

is used. Half a millennium before Jeroboam, Serirot is an anachronism.

What was the role of Keret in the allied army? In II Samuel 8:18 it is written that Kreti and Pleti (Cherethites and Pelethites) were bodyguards of David. These were mercenaries in the service of the king of Jerusalem. Some sixty years later, in time of need, Asa called on the Sidonians for help; they themselves were alarmed by the invasion approaching from the south, as the first scene of the Keret poem vividly describes.

The defeat of the invading army under Terah, the weapons scattered over the fields, the wells stained red with blood, and the cities of Edom trembling before the victorious warriors are vivid pictures which fit the biblical story of the victory over Zerah and over the Edomite cities supporting him.

Analysis of the historical background of the Keret poem reveals that Virolleaud's translation was rejected without sufficient reason. It is a description of a war and a rout. The mention of the tribes of Asher and Zebulun in this setting is natural and does not call for a philological theory that gives other meanings to these names. These two tribes of the tribes of Israel were the closest neighbors of Sidon. Together with the Sidonians they left their homes and hastened to Mareshah to take part in the battle or to exploit Asa's victory. An allusion to this cooperation of the northern tribes is found in the Second Book of Chronicles (15:9) following the description of Asa's victory: "they fell to him [Asa] out of Israel in abundance, when they saw that the Lord his God was with him".

Comparing the record of the Second Book of Chronicles and the annals of Amenhotep II, who followed Thutmose III, we found that according to both sources a great army came from the borders of Egypt and invaded Palestine; that it reached the place Moresheth-Gath or Mu Areset, in Egyptian, one or two days' distance from the border (*nakhal mizraim*); and that it was turned back by defenders of the land. And that the domination of Egypt over Palestine ceased and was re-established only by the successor of Amenhotep II.

Comparing the annals of Amenhotep II with the poem of Keret, we find in both sources that the Phoenician coast was threatened (Ugarit is especially mentioned by Amenhotep II); that the invaders were armed with *hepes*, or copper scimitars; and that Shamash Edom played an important role in the campaign, being occupied first by Amenhotep II, then by Keret and his allies, who defeated the invader.

Comparing now Chronicles and the poem of Keret, we find that, according to the scriptural text, the army invading Negeb was composed of three hundred chariots and "a thousand thousand" of soldiers, and the defending army had three hundred thousand men of Judea; according to the poem, the invading army had three hundred times ten thousand men. At the head of the army was Zerah or (in the Virolleaud translation) Terah. The cities in Negeb were looted; Edom Serirot is mentioned, and has an explanation in the Translation of the Seventy. Northern tribes of Israel joined the army of the south in looting the cities of Negeb.

We would not attempt to derive an argument for the present reconstruction of history and to build our conclusions on Virolleaud's interpretation of the Keret poem, since this translation is a matter of controversy. On the other hand, we may offer a partial vindication of Virolleaud, and we find support for him in the historical setting as revealed by our reconstruction.

Some twenty years after the defeat at Mareshah, Thutmose IV, successor to Amenhotep II, re-established Egyptian hegemony over Syria and Palestine. There are no records of the campaign, but he bears the appellation "conqueror of Syria".[123] Thutmose IV met very little resistance, if any. The pressure of the Assyrians from the north made it desirable for the Syrians to submit to Egypt.[124]

The End of Ugarit

The excavator of Ras Shamra found that the city had been destroyed by violence and had not been rebuilt. Buildings were demolished; the library was burned and its walls fell on the tablets and crushed many of them. The last king whose name is mentioned on the documents that survived the fire is Nikmed. There was also found a proclamation which states that the city was captured and Nikmed and all the foreigners were expelled.

On the level of the structure destroyed by fire a seal impression of

Amenhotep III was found, as well as two letters of the type of the el-Amarna collection. Judged by these findings, Ugarit came to its end in the later days of Amenhotep III or in the early days of Akhnaton, a time known as the el-Amarna period.

In a letter written by Abimilki, king of Tyre, and found in the state archives of Akhet-Aton (el-Amarna), this vassal king informed the pharaoh of what had happened to Ugarit:

> And fire has consumed Ugarit, the city of the king; half of it is consumed, and its other half is not; and the people of the army of Hatti are not there.[125]

Half of Ugarit was burned, the other half was razed, and the invaders, soldiers of Hatti, retreated after they had wrought this destruction.

Mention of the destruction brought upon Ugarit and actual evidence of demolition by a violent hand convinced the investigators of Ras Shamra that the city ceased to exist in the very days to which the letter of Abimilki referred.

In the el-Amarna letters the city of Ugarit was referred to by name and her devastation was recorded, but the name of the king of Ugarit was not given. It could be established only inferentially. It is known that Nikmed lived during the el-Amarna period.[126] His name, on the latest tablets of the Ras Shamra library, and the two tablets of the el-Amarna type, bore evidence that the catastrophe of fire and destruction, as described by the king of Tyre and as found by the excavators, occurred in the days of Nikmed, in the period of el-Amarna.[127]

The proclamation found in Ras Shamra is directly related to the upheaval in the life of the city. Some invading king decreed that "the Jaman [Ionians], the people of Didyme, the Khar [Carians], the Cypriotes, all foreigners, together with the king Nikmed" were to be expelled from Ugarit, "all those who pillage you, all those who oppress you, all those who ruin you".[128] It appears that the proclamation was addressed to the Phoenician section of the city's population; it was written in ancient Hebrew with cuneiform characters.

The opening portion of the proclamation is missing; this is regrettable[129] because it might have revealed the name of the king who expelled Nikmed. Who was the king who conquered Ugarit,

204

burned it, expelled its population, and caused King Nikmed to flee? He will be identified in Chapter Eight.

Was Nikmed, his city burned, his fleet dispersed, a forerunner of the coming migration to Africa, where about the middle of the ninth century Phoenician refugees founded Carthage, or Keret the New,[130] in present-day Tunisia? Or did he direct his sails to Hellas, with which Ugarit traded in Mycenaean ware? Ugarit was also a market for the Ionians, who had their own colony there; they fled with Nikmed.

I believe Nikmed was not lost in the history of that century, so rich in events. I intend to trace his place of refuge later on.

Retarded Echoes

Three of the main conclusions of this chapter are: (1) The time-tables of Crete (Minoan ages) and of early Greece (Mycenaean ages) are displaced by the same stretch of time by which Egyptian dates are out of step with the revised chronology; (2) biblical criticism that ascribed the origin of many biblical texts to late centuries and to foreign influences is as erroneous on this issue as the reversal of it that assumes a borrowing of many biblical texts and institutions from the Canaanites of the fourteenth century; (3) the Hurrian language is but Carian; there existed no Hurrian nation.

We learned also that Amenhotep II is the scriptural Zerah, and that he lost the domination of Palestine and Syria in a battle at Mareshah; that the Keret poem actually refers to the harassing of Ugarit by Amenhotep II; and that Ugarit was laid waste in the ninth century.

The excavators and interpreters found themselves compelled to draw a number of inferences about the historical reappearance of one or another earlier cultural achievement. The sepulchral chambers of Ugarit influenced the architecture of sepulchral chambers on Cyprus—but not until more than half a millennium had elapsed. The naval catalogue of Ugarit reappeared in the epic creations of Homer after an interlude of several centuries. Jewels identical with those of Ugarit were worn by maidens of Jerusalem six or seven hundred years after the destruction of Ugarit. The poetic style and meter, the legal ordinances and sacerdotal practices, even the system of weights, re-emerged after an equally long period. Dividing strokes between written words were introduced into the script of Cyprus some seven hundred years after the script of Ras Shamra with the same characteristics had fallen into oblivion.

In western Asia Minor, where Homer lived, in the Jerusalem of the prophets, on Cyprus, all around Ugarit, the echo of its culture, its language, and its art returned only after a long period of latency.

Retarded echoes are never heard in Egypt; and how could it be otherwise? For the chronology of Ugarit-Ras Shamra is constructed to synchronize with the chronology of Egypt.

NOTES

1. Directed by Claude F. A. Schaeffer and reported in *Syria, Revue d'art oriental et d'archéologie*, 1929ff. Reprints of the first seven reports were published together under the title *Les Fouilles de Minet-el-Beida et de Ras Shamra*, 1929–36.

2. E. Forrer, *Syria, Revue d'art oriental et d'archéologie*, XIII (1932), 26.

3. Schaeffer, *Les Fouilles de Minet-el-Beida et de Ras Shamra, Campagne 1929* (Paris, 1929), p. 296 (*extrait de Syria, Revue d'art oriental et d'archéologie*); *La Deuxième Campagne de fouilles à Ras Shamra, 1930* (Paris, 1931), p. 4; *La Troisième Campagne de fouilles à Ras Shamra, 1931* (Paris, 1933), pp. 11, 24.

4. Schaeffer, *La Deuxième Campagne*, pp. 10–11.

5. The early Ras Shamra bibliography is given in Schaeffer's *Ugaritica I* (Paris, 1939). In the ten years following 1929, the number of publications exceeded five hundred.

6. Sir Arthur J. Evans, *The Palace of Minos* (London, 1921–35), II, 43, 101, 214, 286–89, 347; III, 12, 14, 348, 401–3.

7. "The chronology of prehistoric Greece is naturally far from certain although through connections with Egypt certain general dates can be given." A. J. B. Wace, "Prehistoric Greece" in *Cambridge Ancient History*, I (Cambridge, 1923), 173–80. "The difficulty comes when we attempt to fit these archaeological dates into any scheme of world chronology. . . . The one neighboring land where there is a fairly stable chronological system based on written documents and inscriptions is Egypt." Ibid., p. 174.

8. Ibid., p. 175.

9. Ibid., p. 177.

10. E. Gjerstad and others, *The Swedish Cyprus Expedition, 1927–1931* (Stockholm, 1934–37), I, 405.

11. Schaeffer, *The Cuneiform Texts of Ras Shamra-Ugarit* (London, 1939), p. 29.

12. Charles Virolleaud, "Les Inscriptions cunéiformes de Ras Shamra", *Syria, Revue d'art oriental et d'archéologie*, X (1929), 308.

13. In the ruins of the library of Ras-Shamra-Ugarit.

14. The semilegendary Aristomenes, who led the people of Messene in their battles against the Spartans in the years −684 and −683, was a son of Nikomedes—according to other sources, of a Pyrrhos (F. Hrozný, "Les Ioniens à Ras-Shamra", *Archiv Orientální*, IV [1932], 177). Aristotle mentions an Athenian archon of that name who flourished in −483. The name is also found later among the Spartans. In the third century Nikomedes I, king of Bíthynia on the eastern shore of the Bosphorus, built a new capital for himself, Nikomedeia.

15. Hrozný and E. Dhorme. See Hrozný, "Une Inscription de Ras-Shamra en langue Churrite", *Archiv Orientální*, IV (1932), 129, 176.

16. Schaeffer, *Cuneiform Texts*, p. 33.

17. E. Dhorme, "Première traduction des textes phéniciens de Ras Shamra", *Revue biblique*, XL (1931), 38. Also Hrozný, "Les Ioniens à Ras-Shamra", *Archiv Orientální*, IV (1932), 176.

18. Dhorme, *Revue biblique*, XL (1931); Hrozný, "Les Ioniens à Ras-Shamra", *Archiv Orientální*, IV (1932).

19. *Le ddmy est le gentilice d'un nom qui, sous la forme ddm, représente une divinité dans (text) 17, 6. Nous y verrions volontiers le Didyméen. La ville serait celle de Didyma et le dieu delui de Didyme, Apollon.*" Dhorme, *Revue biblique*, XL (1931); see also Hrozný, "Les Ioniens à Ras-Shamra", *Archiv Orientální*, IV (1932), 176.

20. In the British Museum, brought from Didyma (Didymaion) by C. T. Newton.

21. *La colonie égéenne d'Ugarit semble donc avoir été composée spécialement par les Ioniens originaires de Didyme près de Milet. . . . Nkmd . . . pourrait être considéré comme le roi des Ioniens qui s'emparèrent d'Ugarit au 13-ème siècle.*" Hrozný, "Les Ioniens à Ras Shamra", *Archiv Orientální*, IV (1932).

22. T. H. Gaster, "A Phoenician naval gazette; new light on Homer's Catalogue of ships", *Quarterly Statement of the Palestine Exploration Fund*, April 1938.

23. Schaeffer, *Cuneiform Texts*, p. 40.

24. Virolleaud, "Les Inscriptions cunéiformes", *Syria*, X (1929), 305.

25. H. Bauer and E. Dhorme, independently, in 1930.

26. H. L. Ginsberg, *Kitvei Ugarit*, Jerusalem, 1936.

27. *C'est un fait bien connu que les Chypriotes ont, à partir d'une épogue assez basse il est vrai, le VIe siècle, écrit leur langue au moyen d'une sorte de syllabaire comprenant soixante signes, dans lequels les mots sont séparés, comme à Ras Shamra, par un trait vertical, et dont on a précisément cherché jadis l'origine dans l'écriture accadienne. L'alphabet de Ras Shamra doit-il donc être considéré comme le prototype du syllabaire chypriote? Il peut sans doute*

paraître étrange qu'une écriture très simplifiée ait pu, à la longue, se compliquer à nouveau. . . ." Virolleaud, "Les Inscriptions cunéiformes", *Syria*, X (1929), 309.

28. This was already inferred from Semitic words met in the el-Amarna letters.
29. Some of the cuneiform texts in old Hebrew, found in Ras Shamra, bear reference to the south of Palestine-Canaan (Negeb), and for this reason Proto-Phoenician and Canaanite are applied *ad libitum* to the tongue.
30. Schaeffer, *Cuneiform Texts*, p. 35.
31. Ibid., p. 36.
32. Ibid., p. 59.
33. Attempts were made to find parallelism between the gods of the Ras Shamra texts and temples and the gods of the theological work of Sanchoniaton, an early Phoenician writer, quoted by Eusebius.
34. R. Dussaud, *Les Découvertes de Ras Shamra (Ugarit) et l'Ancien Testament* (Paris, 1937), p. 59.
35. Schaeffer, *Cuneiform Texts*, p. 60.
36. For instance in the name yw-il.
37. J. W. Jack, *The Ras Shamra Tablets* (Edinburgh, 1935): "A word of uncertain meaning, *mphrt* (community or family), which is found on two of the Ras Shamra tablets, occurs on the stele of Yehawmilk, king of Byblos (c.—650). Strange to say, the name Yehawmilk also appears on one of the Ras Shamra tablets." Cf. M. Dunand, "Nouvelle inscription phénicienne archaïque," *Revue biblique*, XXXIX (1930), 321ff. The same stele contains the phrase: "Baal Shamim and Baal Geval" (Byblos); the words "Baal Shamim" are also used in the treaty between Esarhaddon and the king of Tyre (seventh century). Ibid., p. 331.
38. Jack, *The Ras Shamra Tablets*, p. 9.
39. Dussaud, *Les Découvertes*, p. 61: "*Bien avant le récit du passage de la Mer Rouge par les Israélites, le folklore ou les mythes du sud de la Palestine connaissaient une légende ou le dieu El était représenté comme ayant fait surgir, d'entre les flots, le grand isthme désertique, que sépare la Mer Rouge de la Mediterranée. Il paraît, des lors, vraisemblable que cette légende est le prototype du récit concernant le passage de la Mer Rouge par les Israélites. . . .*
40. Jack, *The Ras Shamra Tablets*, p. 10.
41. Ibid.
42. Schaeffer, *Cuneiform Texts*, p. 58, quoting Dussaud, *Syria, revue d'art oriental et d'archéologie*, XVI (1935), 198.
43. Dussaud, *Les Découvertes*, p. 50; J. A. Montgomery and Z. S. Harris, *The Ras Shamra Mythological Texts* (Philadelphia, 1935), p. 16.
44. Jack, *The Ras Shamra Tablets*, p. 10.
45. Ibid., p. 7.
46. Dussaud, *Les Découvertes*, pp. 105–106.
47. Albright, *Archaeology and the Religion of Israel*, p. 38.
48. Schaeffer, *Cuneiform Texts*, p. 77.

49. Ibid., p. 47.

50. Ibid., p. 59.

51. Ibid., p. 41. M. B. Gordon (*Annals of Medical History*, IV [1942], 406–8) makes a point of the fact that *debelah* in Ugarit was used internally, not externally.

52. Schaeffer, *Cuneiform Tests*, p. 41.

53. Ibid.

54. "It is at Ras Shamra that one first meets with the system of weights later used by the Israelites and described in a certain passage of Exodus." Ibid., p. 27.

55. Ibid., Plate XXXII, Fig. 1.

56. Charles Virolleaud, "Un poème phénicien de Ras-Shamra", *La Deuxième Campagne de fouilles à Ras-Shamra*, pp. 209–210.

57. Schaeffer, *Cuneiform Texts*, p. 62.

58. "*C'est une révolution complète de l'exégèse des temps prémosaïques.*" Dussaud.

59. Montgomery and Harris, *Mythological Texts*, p. 1.

60. "*Reuss, Graf et Wellhausen ... on ne peut manquer de reviser leurs conclusions, en ce qui touche la basse époque et le peu de valeur des anciennes traditions israélites.*" Dussaud, *Les Découvertes*, p. 115.

61. Schaeffer, *Cuneiform Texts*, p. 59.

62. "*On reconnaîtra que si les Prophètes ont magnifiquement développé cette tendance pieuse, ils ne l'ont pas créé.*" Dussaud, *Les Découvertes*, p. 118.

63. Olmstead, *History of Palestine and Syria*, p. 140: "Kharu is doubtless to be connected in name with the Horites, who in Biblical times were remembered only as having been exterminated by the Edomites; it is also possible that there is some connection with the Hurrians."

64. See E. A. Speiser, *Mesopotamian Origins* (Philadelphia, 1930), p. 133; also his *Introduction to Hurrian* (New Haven, 1941), p. 3.

65. Speiser, *Mesopotamian Origins*, p. 152.

66. Ibid., p. 120.

67. See Speiser, *Introduction to Hurrian*.

68. Speiser, *Mesopotamian Origins*, p. 131.

69. Schaeffer, *Cuneiform Texts*, p. 37.

70. The south of Canaan, called in the Book of Joshua *Negeb-Kereti*, was, according to the opinion of various scholars, at an early age occupied by immigrants from Crete.

71. Herodotus (trans. A. D. Godley, 1921–24), I, 171.

72. Ibid.

73. Thucydides (trans. C. Foster Smith; London and New York, 1919), I, iv.

74. Georg Meyer, *Die Karier* (Göttingen, 1885), p. 3.

75. Strabo, *The Geography*, I, 3, 21.

76. Extensive studies were made, in which the name Car was tracked down all over the world in order to find traces of Carian and Phoenician navigation.

See Baron d'Eckstein, *Revue archéologique*, XIV (1857); XV (1858), and Brasseur de Bourbourg, *S'il existe des sources de l'histoire primitive du Mexique dans les monuments égyptiens* (Paris, 1864). Names such as Karkar or Carchemish (written also Gargemish) and the word "Kar" in names of cities, like Car Shalmaneser, may be mentioned in this connection.

77. Thucydides, I, viii.

78. The King James version has "the Captains and the Guard".

79. The Philistines came from the island of Caphtor (Deuteronomy 2:23; Amos 9:7; Jeremiah 47:4). Jeremiah speaks of the "Philistines, the remnants of the country of Caphtor". By identifying the Philistines with Kreti and Pleti, Caphtor was identified as Crete. It will be more in accord with historical evidence if we understand Caphtor to be Cyprus. If Caphtor was not Cyprus, then no name for Cyprus and no mention of the island would be found in the Scriptures, and that would be unlikely because Cyprus is very close to Syria. The islands of Khitiim (Jeremiah 2:10; Ezekiel 27:6), usually identified as Cyprus, signified all the islands and coastlands of the west, Macedonia, and even Italy. Cf. article "Cyprus" in the Jewish Encyclopedia.

80. The word "Pleti" was given still another explanation. The Targum translated "Kreti" as "bowmen", and "Pleti" as "slingers" from the word *palet*, "to cast" or "cast out". The same verb could be regarded as meaning "those who were cast" out by the sea, or "remnants of people escaped from some place on the sea": *iam polat* is "the sea threw out".

81. Herodotus, II, 152.

82. Ibid., II, 154.

83. Dussaud, *Les Découvertes*, p. 20.

84. Strabo, *The Geography*, XIV, ii, 27ff.

85. Ibid., with reference to Apollodorus, Athenian grammarian.

86. Herodotus, I, 172.

87. Ibid., VIII, 135.

88. See A. H. Sayce. "The Karian Language and Inscriptions", *Transactions of the Society of Biblical Archaeology*, IX (1886), 123–54. W. Brandenstein, "Karische Sprache", Pauly-Wissowa, *Real-Encyclopädie der classischen Altertumswissenschaft*, Supplement VI (1935), 140–46; F. Bork, "Die Sprache der Karer", *Archiv für Orientforschung*, VII (1931–32).

89. Lepsius noticed these signatures and drew this conclusion.

90. The hypothesis of the Iranian origin of Carian was put forth by P. de Lagarde. See P. Kretschmer, *Einleitung in die Geschichte der Griechischen Sprache* (Göttingen, 1896), pp. 376ff.

91. "*Neben dem elamoiden Kerne ist im Karischen ein starker Einschlag aus dem Mitanni deutlich zu erkennen.*" F. Bork, "Die Sprache der Karer", *Archiv für Orientforschung*, VII (1931), 23.

92. "*Ergt.*" See Hrozný, "Les Ioniens à Ras-Shamra", *Archiv Orientální*, IV (1932), 175. Compare Virolleaud, *Syria, Revue d'art oriental et d'archéologie*, XII, 351 and 557.

93. Nikodamos of Salamis on Cyprus minted coins ca —460 to —450 (Sir George F. Hill, *Catalogue of the Greek Coins of Cyprus* [London, 1904,] p. 52).

94. See the article "Didyma" by Bürchner in Pauly-Wissowa, *Real-Encyclopädie*.

95. "*Les inscriptions churrites de Ras Shamra démontrent une fois de plus la grande influence du peuple churrite ou horite sur la Syrie et la Palestine . Ce fait ne saurait nous surprendre, dès que nous connaissons le rôle joué par les Churrites en Syrie et en Palestine à la fin du troisième et dans la première moitié du second millénaire.*" Hrozný, *Archiv Orientální*, IV (1932), 127. See also Speiser, *Mesopotamian Origins*, p. 133.

96. Herodotus, II, 117.

97. A. M. Badawi, "Die neue historische Stele Amenophis' II", *Annales du service des antiquités de l'Egypte*, XLII (Cairo, 1943), 1–23.

98. Brugsch read the name "Arinath" and identified it with Orontes. Breasted and others accepted this view; F. W. von Bissing objected (*Die Statistische Tafel von Karnak* [Leipzig, 1897], p. 34); Petrie read "Arseth" and surmised it to be Haroshet on the Kishon (*History of Egypt*, II, 155). J. A. Wilson, however, verified the reading y-r-s-t on the Karnak fragment.

99. Amenhotep II started his campaign, according to the Memphis Stele, on the "First month of the third season, day 25". He reached y-r-s-t on the "first month of the third season, day 26" (Karnak variant), or only one day later. See "Egyptian Historical Texts" by J. A. Wilson in *Ancient Near Eastern Texts*, p. 245 and note 8 on the same page.

100. Breasted, *Records*, Vol. II, Sec. 785.

101. In a recent publication Sidney Smith arrived independently at the same conclusion that the expedition of Amenhotep II was a disastrous defeat. *Occasional Publications of the British School of Archaeology in Ankara* (London, 1949), Vol. I.

102. Breasted, *Records*, Vol. II, Sec. 786.

103. Ibid., Sec. 787.

104. That he lost Syria-Palestine may also be deduced from the fact that his successor, Thutmose IV, called himself "conqueror of Syria".

105. In Micah 1:14 the place is called Moresheth-Gath. The first syllable of Meresha or Me-reshet may possibly mean "the water of", as in Me-riba or Me-rom. The Egyptians, transcribing Moreshet, could write *mu-areset*, or the water of *areset*. Amenhotep II crossed "the arm of water [ford] of *arseth*" (Petrie, *History of Egypt*, II, 15).

106. Breasted, *Records*, Vol. II, Sec. 792.

107. Breasted, *A History of Egypt* (New York, 1912), p. 326.

108. Cf. Foucart in the *Bulletin de l'Institut Egyptien*, 5 série, II (1917), 268–269. Amenhotep I, an earlier king of the Eighteenth Dynasty, is pictured with a black face. I. Rosellini, *I Monumenti dell' Egitto e della Nubia* (Pisa, 1832–44).

109. Breasted, *Records*, Vol. II, Sec. 803. See Gauthier, *Le Livre des rois d'Egypte*, II, 287.

110. It is a curious circumstance that in the Abyssinian sacred tradition (*Kebra Nagast*) Menelik, king of Ethiopia and son of the Queen of the South, guest of Solomon, returned to Palestine to rob its Temple; after succeeding in stealing the holy Ark by a ruse, he fled to Ethiopia, pursued by Solomon, his father, to the borders of Egypt. Two historical elements are mingled in this legend. The sack of the Temple was the work of Shishak-Thutmose III, the successor of Queen Sheba-Hatshepsut. The one who fled from Palestine, pursued by the king of Jerusalem, was Zerah-Amenhotep II, the successor of Shishak-Thutmose III.

111. Who assumed a Kush in Arabia besides the Kush (Ethiopia) in Africa.

112. Charles Virolleaud, *La Légende de Keret, roi des Sidoniens* (Paris, 1936).

113. "*Comme nous l'avons indiqué déjà* (*Syria, Revue d'art oriental et d'archéologie*, XIV, 149), *ce nom de Trh est évidemment la même que celui du père d'Abraham.*" Ibid., p. 25.

114. "*Pour les Hébreux, Terah n'avait fait que préparer la conquête ou l'occupation de Canaan, et son fils Abram n'avait fait lui-même qu'ébaucher l'entreprise, puisque, devant les difficultés, il avait du abandonner la partie et s'était refugié en Egypte.*" Ibid., p. 32.

115. "*Il ne s'agit pas ici de deux des douze tribes* (*Asher et Zebulun*), *mais de deux de ces cantons canaanéens, dont les douze tribes prendront, un jour, le nom.*" Ibid., p. 18.

116. H. L. Ginsberg, in *Ancient Near Eastern Texts.* ed. Pritchard, pp. 142ff. Instead of "Asher, two and two are gone", he translated: "After two, two march."

117. Jack, *The Ras Shamra Tablets*, p. 38: "Three millions with copper scimitars or harpes (for which the Egyptian word *hepes* is used) and bronze daggers." Jack's is also the translation of these two lines from the poem.

118. Breasted, *Records*, Vol. II, Sec. 802.

119. See A. Lods, *Israel* (London, 1932), p. 64.

120. "*Mqr mmlat dm*". Ginsberg refers "*dm*" to the following sentence and gives it the meaning "Lo!"

121. Translated from Virolleaud, *La Légende de Keret*.

122. Ibid., p. 14.

123. Thutmose IV is twice called "conqueror of Syria" on the Stele Louvre C. 202. P. Pierret, *Recueil d'inscriptions inédites du Musée Egyptien du Louvre*, II partie (Paris, 1878), p. 35. Cf. *Journal of Egyptian Archaeology*, XXVII (1941), 18.

124. As was demonstrated in a previous chapter, Shishak is the scriptural name of Thutmose. Since the tablets of Ras Shamra belong to the period of the Amenhoteps and Thutmoses, we should expect to find in them, besides the biblical name of Zerah, that of Shishak. It was, in fact, among the first of the deciphered words and it caused considerable surprise. "*Le mot Swsk semble, un nom propre, à rapprocher peut-être de l'égyptien Sosenq, hebreu Sosaq, et Sisaq.*" Dhorme, *Revue biblique*, XL (1931), 55. The translator did

not dare to draw the correct conclusion, for what was this pharaoh of the ninth or tenth century doing in the middle of the second millennium?

125. El-Amarna Tablets, Letter 151.

126. Virolleaud, "Suppiluliuma et Niqmad d'Ugarit", *Revue hittite et asainique*, V (1940), 173f.; see Albright, *Archaeology and the Religion of Israel*, p. 38.

127. See Schaeffer, *Les Fouilles de Minet-el-Beida et de Ras Shamra, La Neuvième Campagne* (Syria, XIX [1938], 196), concerning the time when Ugarit was destroyed.

128. Hrozný, "Les Ioniens à Ras-Shamra", *Archiv Orientální*, IV (1932), 171; Dhorme, *Revue biblique*, XL (1931), 37–39.

129. "*Il est regrettable que le commencement de l'inscription n'ait pas été conservé; on peut supposer que les premières lignes de l'inscription contenaient des détails sur l'acteur principal des événements dépeints.*" Hrozný, "Les Ioniens à Ras-Shamra", *Archiv Orientální*, IV (1932), 176.

130. Keret, the city, and Keret, the personal name, have different spellings.

213

CHAPTER SIX

The el-Amarna Letters

The el-Amarna Letters and When They Were Written

A few small villages are scattered in the valley by the bank of the Nile where once stood Akhet-Aton, "the place where Aton rises". The site bears a name artificially composed by modern archaeologists, Tell el-Amarna. Ruins of temples, palaces, tombs, private dwellings, and workshops of craftsmen have been cleared of the desert sand that buried them for thousands of years.

In 1887 state archives were unearthed at Tell el-Amarna. A fellah woman digging in her yard turned up some clay tablets with cuneiform signs; the story runs that she sold her find for the equivalent of two shillings. Samples sent to the Louvre Museum were pronounced forgeries, but soon the scientific world recognized their genuineness.

During the ensuing years a hunt for the clay slabs was undertaken by many archaeologists on behalf of public and private collections. Meanwhile, the tablets had been scattered by unlicensed diggers and dealers in antiquities. Many were damaged, some broken into pieces by unskilled excavation, in the course of transportation, and, it is said, by partition among the clandestine excavators.

Up to the present over three hundred and sixty tablets have been recovered. With the exception of a few single tablets found in Palestine and in Syria, which obviously belong to the same collection, the entire lot is thought to have been found in or near the place recognized as the state archives of Akhet-Aton. Only a few tablets contain fragments of epic poems; all the others are letters exchanged between two successive kings of Egypt and their correspondents, the free kings of territories in the Middle East and Cyprus and various vassal kings and princes or officers in Syria and Palestine.

The kings in the north were not subject to the kings of Egypt, and they wrote "to my brother" and signed themselves "thy brother". The kings in Canaan and Syria, however, were under the scepter of the Nile dynasty and wrote "to my king, my lord" and signed themselves "thy servant". There are also letters addressed to certain dignitaries of the Egyptian court. Letters written by the pharaohs or in their names were obviously copies stored in the archives in order to preserve a record. The language of the tablets, with few exceptions, is Assyro-Babylonian (Akkadian), with many words in a Syrian dialect similar to Hebrew.[1]

The city of Akhet-Aton was built by the schismatic king Amenhotep IV, who abolished the cult of Amon of Thebes and introduced the cult of Aton, interpreted as the solar disk, and who changed his name to Akhnaton. But shortly after his reign the capital city of Akhet-Aton was abandoned, his religion was stamped as heresy, and his images were mutilated. His son-in-law, the young pharaoh Tutankhamen, reigned briefly in the old capital, Thebes. Then the dynasty became extinct. Akhet-Aton had a short history of only about twenty-five years before it was deserted by its inhabitants.

The time of the letters can be established with some precision. They were addressed to Nimmuria (Ni-ib-mu'-wa-ri-ia, Mi-im-mu-ri-ia, Im-mu-ri-ia), who was Amenhotep III, and to Naphuria (Na-ap-hu-ru-r-ia, Nam-hur-ia), who was Amenhotep IV (Akhnaton). The letters sent to Amenhotep III were presumably brought to Akhet-Aton from the archives at Thebes.

Long rows of shelves are filled with books and treatises dealing with these letters, which until recently were the most ancient exchange of state letters preserved. The science of history, it is thought, is in possession of historical testimony on a period previous to the entrance of the Israelites into Canaan. One of the main objects of investigation was to identify places and peoples named in the letters.

In the tablets written by the vassal king of Jerusalem (Urusalim) to the pharaoh, repeated mention is made of the "Habiru", who threatened the land from east of the Jordan. In letters written from other places, there is no reference to Habiru, but an invasion of sa-gaz-mesh (sa-gaz is also read ideographically habatu and translated "cutthroats", "pillagers") is mentioned over and over again. With the help of various letters it has been established that Habiru and *sagaz* (*habatu*) were identical. In letters from Syria the approach of the king of Hatti to the slopes of the mountains of Lebanon was reported with terror.

The impression received is that these invasions—of Habiru from the east and of the king of Hatti from the north—menaced Egyptian domination of Syria, a domination which, it was learned, actually ceased shortly after the reign of Akhnaton. In their letters the vassals incessantly ask for help against the invaders and often also against one another. King Akhnaton, "the first monotheist in world history",[2] did not care for his empire; he was immersed in his dream of "a religion of love". Little or no help was sent; the mastery of the pharaohs over Syria and Canaan was broken, and the control of Egypt over her Asiatic tributary provinces was swept away.

The name "king of Hatti" is generally understood as "king of the Hittites". In a later period—that of Seti I and Ramses II of the Nineteenth Dynasty—there were great wars between the kings of Hatti and the pharaohs. In a chapter dealing with that period the history of the "forgotten empire of the Hittites" will be analyzed. "The king of Hatti" of the el-Amarna period is supposed to have been one of the kings of that "forgotten empire".

The name Habiru, mentioned in the letters of the king of Jerusalem, became an important issue and gave rise to the following conjecture: these invaders could have been the Hebrews under Joshua drawing near to the borders of Canaan. The Habiru, too, emerged from the desert and approached the land from the other side of the Jordan. Arriving at the Promised Land during the time of Amenhotep III and Akhnaton, they were supposed to have left Egypt sometime in the days of Thutmose III or Amenhotep II.

This does not seem to be a well-founded construction, because these monarchs were conquerors and despots, too strong to allow the Israelites to put off the yoke of bondage. Other scholars have refused to identify the Israelites with the Habiru because the Hebrews were thought still to have been in Egypt at the time of Akhnaton, and the sole opportunity they would have had to leave would have been during a period of anarchy, when the once powerful dynasty was dying down, or during the interregnum following the extinction of that dynasty: these would have been the only times suitable for the rebellion of the slaves and their departure. In that view the Habiru came in one of the waves of nomads eager to settle in the land of

Canaan; other waves must have followed, one of them being the Israelites under Joshua ben Nun.

In the introductory chapter of this book I dwelt on the various theories relative to the time of the Exodus. There it was explained that a large group of scholars cannot compromise even on an Exodus at the end of the Eighteenth Dynasty, but have chosen Ramses II of the Nineteenth Dynasty to be the Pharaoh of the Exodus, holding the Exodus to be an insignificant event in the history of Egypt; or they select Ramses to be the Pharaoh of Oppression and his son Merneptah the Pharaoh of the Exodus.

As the el-Amarna tablets, in the opinion of these scholars, preceded the Exodus, the Habiru could not possibly have been the Israelites. Another link was sought to connect the narrative of the Scriptures with the details unfolded in the letters. A parallel to Joseph was found in the Egyptian courtier of Semitic name and obviously Semitic origin, Dudu,[3] whose memory is preserved not only as the addressee of a few el-Amarna letters but also as the owner of a rich sepulcher in Akhet-Aton. King Akhnaton, desirous of making a special gift to his favorites, presented each with a tomb built during the lifetime of the recipient and engraved with scenes from the life of the king and his family, the owner of the tomb—but usually not his family—being portrayed as a small figure receiving honors from the king. In the tomb of Dudu there are also figures of Semites rejoicing at Dudu's rewards. A letter written to this Dudu will be quoted later on.

Still another resemblance to Joseph was found in the person of Ianhama,[4] who according to the references in the el-Amarna letters, was an Egyptian chief in charge of the granaries of the state, where grain was bought by the Canaanite princes. There was a famine in the land, and it is an unceasing cry for grain that we hear from the letters of the king of Gubla and Sumur (Sumura).

One further conjecture may be recorded. The visit of Isaac and Rebecca or of Abraham and Sarah to Egypt[5] was linked to some references to a handmaid of the goddess of the city of Gubla and her husband, who were in Egypt. The king of Gubla and Sumur supported the couple in their desire to return to Canaan, if he did not actually ask for their extradition.

But the theory identifying the Habiru and the Hebrews was not abandoned; it seemed that otherwise the histories of these two

217

peoples of antiquity, the Egyptians and the Hebrews, who inhabited neighboring countries, would show no connecting link in the course of many hundreds of years of their early history. The other possible link—the Merneptah stele, with which I shall deal later—was also interpreted both in support of and against the Habiru-Hebrew theory.

The equation Habiru-Hebrews is still accepted by a large number of scholars: at the time when the el-Amarna letters were written the Israelite nomads of the desert were knocking at the gates of the land which they had come to conquer. Did this opinion contradict the scheme according to which the Israelites were still in bondage under Ramses II? If so, then the conception had to be formulated anew, and the migration of the Hebrews was then supposed to have proceeded in consecutive stages. Reconciling theories were put forth dividing the Exodus into several successive departures, the Rachel clan and the Leah clan leaving at different times; when the former reached Canaan the latter was still being oppressed in Egypt and followed later. Another theory had the Josephites coming from Egypt and the Jacobites directly from Mesopotamia; still another variation had the Jacobites coming from Egypt and the Abrahamites from the north.

A further difficulty presented by the equation Habiru-Hebrews arose from the fact that no person in the Book of Joshua could be identified in the el-Amarna letters. When the Israelites entered Canaan, Adonizedek was king of Jerusalem, Hoham king of Hebron, Piram king of Jarmuth, Japhia king of Lachish, and Debir king of Eglon (Joshua 10:3). Among the letters there are a number written by kings of some of these places but not by these kings. Much more important is the fact that there is little similarity in the events described in both sources. The episode of the siege of Jericho, the most remarkable occurrence in the first period of the conquest, is missing in the letters, and Jericho is not mentioned at all. This silence is strange, if the Habiru were the Hebrews under Joshua. No contemporaneous event can be traced in the letters.

The pharaohs of the Nineteenth Dynasty, Seti and Ramses II, left memorial monuments in Egypt and in Palestine regarding their passage through Palestine as conquerors of the land lost by the pharaohs of the el-Amarna period or by their successors. In the Books of Joshua and Judges, covering over four hundred years, nothing sug-

218

gests the hegemony of Egypt or her interference in the affairs of Canaan.

For all these reasons it seemed too bold to place the story of the conquest by the Hebrews so far back in time. The discussion is still in progress. Some of those who could not accept the theory that the Hebrews had already entered Canaan at the time of Amenhotep III and Akhnaton identified the Habiru as the Apiru, the workers in the Egyptian quarries on the Sinai Peninsula, on their seasonal journey from there to their homes in Lebanon; others identified them as migrants from the Babylonian district of Afiru.

How could the Hebrews have reached Canaan before they left Egypt? Or how could they have left their bondage in Egypt before it was weakened?

According to my chronological scheme, the letters of el-Amarna, sent and received by Amenhotep III and Akhnaton, were written, not in −1410 to −1370 as is generally accepted, but in −870 to −840, at the time of King Jehoshaphat in Jerusalem.[6] If this theory is correct, among the tablets of the el-Amarna collection we should expect to find letters written by the royal scribes, skilled in cuneiform, in the name of the Israelite kings of Jerusalem and of Samaria. The most prolific writer of letters among the princes and chiefs was the king of Sumur (Samaria). About sixty letters of his are preserved, fifty-four of them addressed to the king of Egypt. The pharaoh even wrote to him: "Thou writest to me more than all the regents."

The three hundred and sixty letters, linking the political past of the great and small nations of the Near East at an important period of remote antiquity, were objects of prolonged study in the interest of Egyptian, Babylonian, Hittite, Syrian, and Canaanite histories. The statement made above as to the time of the letters should not be accepted merely because it fits into a scheme built on other evidences of preceding or following periods. It should be demonstrated with respect to the letters themselves. Besides the Scriptures and the el-Amarna tablets, two other sources relate to the time of King Jehoshaphat: the stele of King Mesha of Moab and the inscriptions of the Assyrian king, Shalmaneser III. These relics, too, and not the Bible alone, must correspond to the contents of the el-Amarna letters, if it is true that Egyptian history must be revised and moved forward more than half a thousand years.

The letters of el-Amarna provide us with the names of princes and governors in Syria and Palestine and with the names of cities and walled places. Up to now not one of the personal names has been identified, and only several of the geographical names have been traced. I shall identify some important geographical locations and also a series of personal names.

Urusalim of the el-Amarna letters could not be misunderstood; there was no difficulty in recognizing it as Jerusalem. The difficulty arose only with respect to the passages in the Scriptures[7] according to which the pre-Israelite city was called Salem or Jebus, and not Jerusalem. It was decided that the el-Amarna letters exposed the error of these statements. If, however, the el-Amarna letters were written in the Israelite period, the above conflict does not exist.

Sumur (also Sumura) and Gubla are the most frequently named cities in the el-Amarna letters, each being mentioned more than one hundred times. Other cities are not mentioned even ten or fifteen times. "No king's or prince's name is given for this city Sumur, next to Gubla the most frequently referred to; despite the distress that came upon her, no letter from there is to be found in the entire el-Amarna correspondence."[8]

It is obvious from the content of the letters that Sumur was the "most important place" in Syria-Palestine and apparently also the seat of the deputy administering the region. It was a fortress.[9] A king's palace was there, and the frequent mention of this palace in the letters to the pharaoh gives the impression that it was a well-known building.

Sumur or Sumura was Samaria (Šemer, Šomron, in Hebrew). It could not be presumed that Sumur was Samaria because it was in the reign of Omri, father of Ahab, that this city was built, and in the period preceding the conquest of Joshua, of course, it had not existed.

I KINGS 16:24 And he [Omri] bought the hill Samaria of Shemer for two talents of silver, and built on the hill, and called the name of the city which he built, after the name of Shemer, owner of the hill, Samaria.

Since vowels were a late interpolation in the Hebrew Bible, inserted by the Masoretes ("carriers of the tradition") over a thousand years after the Old Testament had been completed, the name Semer can also be read Sumur.

Samaria was surrounded by a strong wall, and remnants of it have been unearthed. The city had a magnificent royal palace, and the ruins of it are seen today.

The identity of Sumur (Sumura) and Samaria will be shown in detail in the following pages, which describe the history of the period.

Together with Sumur, Gubla is named repeatedly. Evidence has been found in the letters that Gubla was considered heir to Sumur when this city was temporarily occupied by the Syrians. The king of Gubla wrote to the pharaoh in Egypt:

> LETTER 85: What was formerly given in [to] Sumura, should now be given in [to] Gubla.

There is scarcely a letter of the King of Gubla in which his uneasiness about Sumur does not find expression. He mentioned the name of Sumur or Sumura about eighty-five times in his letters, besides the numerous times that the city is referred to as "king's city" or "my city".

It was assumed that Gubla is Byblos, *Kpny* in Egyptian, but Gwal in Phoenician and Hebrew,[10] the Phoenician city north of Beirut. However, there must have been more than one Gwal ("border") in the region of Syria-Palestine; for instance, there is a reference in the Scriptures to Gwal in the south of Palestine.[11] It was also asked why the city of Gwal (Byblos) was changed to Gubla in the el-Amarna letters.

A few times the king of Gubla mentioned in his letters the city of Batruna, and it is identified as the ancient Botrys.[12] However, Menander, a Greek author, quoted by Josephus,[13] says of Ithobalos (Ethbaal), the king of Tyre in the ninth century, that "it was he who founded the city Botrys in Phoenicia". Having been built by the father-in-law of King Ahab, the city Botrys could be mentioned in the el-Amarna tablets only if the founding of the city preceded the el-Amarna age—which complies with the present reconstruction.

If Sumur is Samaria, then it is apparent that Gubla is the original name of Jezreel, the other capital in Israel. Omri built Samaria and its palace. Ahab, his son, built his palace in Jezreel, adjacent to the vineyard of Naboth. There Queen Jezebel, the wife of Ahab, later met a violent death. Jezebel (Izebel), in the memory of the people the most hated person of the period of the Kings, was a daughter of the Sidonian king Ethbaal (I Kings 16:31). She brought the evil of the heathen into the land, killed the prophets of Yahwe, and persecuted Elijah the Tishbite. Hundreds of prophets of Baal "eat at Jezebel's table" (I Kings 18:19). The king under her influence "went and served Baal, and worshipped him" (I Kings 16:31). Josephus Flavius says that Jezebel built a temple to the deity "whom they call Belias".[14]

The king of Gubla wrote in almost all his letters: "May Belit [Baalis] of Gubla give power. . . ." Belit of these letters seems to have been the deity Baaltis, or Belias, whose cult was brought from Phoenicia.[15]

Gubla thus appears to have been the original name of the royal residence city we know from the Scriptures as Jezreel. King Ahab must have had many wives, for he left seventy sons in Samaria (II Kings 10:1). But the daughter of the Sidonian king was Ahab's chief wife. She exercised charm and influence over him, and he built for her the new residence with her active participation, as the story of Naboth relates. It was not to be expected that Ahab, the apostate, would name the new residence in honor of the persecuted deity, the God of Israel ("the Lord will sow"). It is possible that the name of the new residence was derived from the name of the Phoenician city, dear to the Phoenician princess, now queen of Israel. It may be surmised, too, that the residence was named in honor of this wife, Jebel, or Gubla in cuneiform transcription, Izebel (Jezebel) in the biblical transcription.[16]

In the Scriptures there is a direct indication that Jezreel was previously called by the name of Queen Jezebel. When her life ended ignominiously, dogs tore her flesh "and the carcass of Jezebel [was] as dung upon the face of the field in the portion of Jezreel, about which they [should] not say, This is Jezebel" (II Kings 9:37)[17] The meaning of the sentence is that the name of the place should be blotted out with the death of the queen, by whose name it had been called. After that the city was given the name of the valley—Jezreel.[18]

We shall proceed on the assumption that Sumur and Gubla were

Samaria and Jezreel in Israel: the two cities were the two capitals of one state. The King of Gubla was worried about his other capital, alternately lost and recovered in incessant wars with the Syrians. When Sumur fell Gubla became its heir. We thus have the answer to the question of why no king of Sumur is mentioned in the el-Amarna letters, although the city is named so often; its king had his residence in Gubla. Possibly some of the letters which bear the usual blessing, "May Belit of Gubla give power", were written from Sumur (Samaria).

Having raised the curtain on the scene of the main action of the el-Amarna period, we must identify the persons on the stage.

The Five Kings

The kings of the ancient Orient usually bore many names. The letters of el-Amarna were addressed to the pharaoh Nimmuria and the pharaoh Naphuria. Nimmuria was Amenhotep III, and Naphuria, Amenhotep IV (Akhnaton). Egyptian kings regularly had as many as five names.[19] Not from the el-Amarna letters but from other Egyptian documents, it is known that Nimmuria was the throne name of Amenhotep III and Naphuria the throne name of Akhnaton. In the letters there is no mention of the names Amenhotep or Akhnaton.

The kings of Jerusalem, as well as the kings of Samaria and Damascus, also had more than one name. Five different names for Solomon are preserved.[20] King Hezekiah of Jerusalem had nine names.[21] In view of this practice, there is only a limited chance of finding in the el-Amarna letters the names of the kings of Palestine as we know them from the Scriptures.

But if the kings had many names, this does not mean that we are free in our choice of substitutes. No doors to indiscriminate identification are opened. The life and wars of the Syrian and Palestinian kings of this period are described with many details in the Scriptures and in the letters; all these details will be placed source against source.

At this point in the discussion, should my identification of Abdi-Hiba with Jehoshaphat, Rib-Addi with Ahab, and Ben-Hadad with Abdi-Ashirta seem arbitrary, I shall be pleased: in the hall of history, crowded with throngs of men from many centuries, I point straightway to certain figures bearing names entirely different from those of the persons we are looking for; they are even said to belong to an age separated by six centuries from the time of the persons we are seeking. Even before I have investigated the persons thus without apparent justification singled out, I shall insist on the identification.

The searching rod in my hand is the rod of time measurement: I reduce by six centuries the age of Thebes and el-Amarna, and I find King Jehoshaphat in Jerusalem, Ahab in Samaria, Ben-Hadad in Damascus. If my rod of time measurement does not mislead me, they are the kings who reigned in Jerusalem, Samaria, and Damascus in the el-Amarna period.

The el-Amarna letters were written in the days of Amenhotep III and his son Akhnaton some seventy-odd years after Thutmose III had conquered Palestine and sacked the Temple of Kadesh. Having established the contemporaneity of Solomon and Hatshepsut (Queen of Sheba), of Rehoboam and Thutmose III (Shishak), and of Asa and Amenhotep II (Zerah), we are compelled to conclude that the Jerusalem correspondent of Amenhotep III and Akhnaton was King Jehoshaphat. We are no longer free: either we have been wrong up to this point, or the contents of the el-Amarna letters will correspond to the scriptural information about the time of Jehoshaphat. We must be sure of this even before we open for the first time *The el-Amarna Tablets.*

Five kings—two successive kings of Damascus, one of Israel, one of Judah, and one of Moab—were the main characters on the stage of the political life of the Egyptian provinces of Syria and Palestine at the time under study. For two of them the Scriptures retained similar names: Hazael, the king of Damascus, is called Aziru, Azira, or Azaru in the el-Amarna letters. The king of Moab, Mesha, is called, as we shall see, Mesh in the letters.

The name of the king of Jerusalem of the el-Amarna letters is read Abdi-Hiba. However, the same characters, if regarded as ideographic, permit another reading; at first it was proposed to read it Ebed-Tov ("The Good Servant" in Hebrew)[22] and then Puti-Hiba;[23] others read it Aradhepa or Arthahepa.[24] From this fact we see that names written in cuneiform may be read in many ways, and the reading Abdi-Hiba is only one surmise among a number of others.[25] It would appear that the original reading Ebed-Tov is preferable.

The name of the king of Jerusalem, Jehoshaphat, may not even have been one of his several original names; it could have been a name given him by his people to commemorate his deeds. It means "Yahwe [Jahve] is the judge" or "one who judges in the name of Yahwe". This king sent Levites throughout the cities of Judah with a "book of the law of Yahwe" to teach the people (II Chronicles 17:9), and a high court for the "judgment of Yahwe" was established in Jerusalem (II Chronicles 19:8–10).

II CHRONICLES 19:5–6 And he set judges in the land throughout all the fenced cities of Judah, city by city.

And said to the judges, Take heed what ye do; for ye judge not for man, but for Yahwe, who is with you in the judgment.

He also built a new palace of justice in Jerusalem (II Chronicles 20:5). A king whose endeavor was dedicated to this work might receive in the memory of a nation the agnomen Jehoshaphat. The talmudic tradition asserts, for instance, that Solomon ("Peace") was a post-mortem name of the king, son of David.

The name Rib-Addi, written in ideograms, means "the elder [brother among the sons] of the father", the first part of the name signifying "the elder" brother or "the elder" son, and the second part "father". It is construed like the Hebrew name Ahab, the first part of which means "brother" (ah), the second part "father" (ab).

In the correspondence with Egypt the king of Amuru land was called Abdi-Ashirta (spelled also Abdu-Astarti, Adra-Astarti).

It may be gathered from the letters of the king of Sumur[26] that the seat of the kings of Amuru land, Abdi-Ashirta and Azaru, was in Dumaska (Damascus). Thus Amuru land of the letters is Aram (Syria) of the scriptural text. That Amuru land means Syria we learn also from an inscription of Shalmaneser III.[27]

According to Nicholas of Damascus, a historian of the first century before the present era, Ben-Hadad was a generic name for the kings of Damascus.[28] Many biblical scholars are also of the opinion that the name Ben-Hadad was a general designation for the kings of Damascus,[29] and historians assume that the king of Damascus who was Ahab's opponent was actually called Biridri; he was the leader of a coalition against Shalmaneser III, as we know from an inscription of that Assyrian king. "How the name came to be translated Ben-Hadad in the Scriptures is uncertain."[30]

We shall discover the real Biridri in the el-Amarna letters in the person of the commandant of Megiddo.

It was the custom to name kings in conformity with the religious worship practiced in their domains. The worship of Astarte and Asheroth in the time of Ben-Hadad is mentioned in the Scriptures.[31]

The name Abd' Astart was in use in the ninth century: Josephus Flavius, quoting a no longer extant work of Menander of Ephesus, gives a royal list of the Phoenician kings. Hiram, the contemporary of Solomon, had a grandson by the name of Abd' Astart, who at the age of thirty-nine was killed by four sons of his nurse.[32] This story may be an echo of the assassination of Ben-Hadad (II Kings 8).

The scriptural version of the name Hazael, *l* and *r* being interchangeable characters, differs from the name in the tablets (Aziru or Azaru) in the aspirate sound $\underline{h}$.[33] Scholars who derived Ivri (Hebrew) from Habiru should find no difficulty in deriving Hazael from Azaru. Josephus Flavius called Hazael by the name Azaelos.

The kings of Jerusalem and of Samaria were hereditary princes. In the el-Amarna letters they call themselves *rabiti sari*, princes or regents; from the context of the letters it is clear that each of them sat on the throne of his father. The regent king of Samaria recalled to the pharaoh the time when his father was helped by the pharaoh's father. And the king of Jerusalem wrote:

> LETTER 286: Behold, neither my father nor my mother has put me in this place. The mighty hand of the king had led me into the house of my father.

This means that the pharaoh of Egypt was wont to choose from among the royal princes one to succeed his father, the vassal king.

Governors were attached to "the regents", the vassal kings, as representatives of the Egyptian crown. One was for northern Syria, with his seat probably at Damascus. Another governor had his seat in Samaria (Sumur). We shall meet both of them not only in the letters but also in the Scriptures, and we shall recognize them there.

In Jerusalem there was no permanent deputy. In one of his letters the king of Jerusalem asked that the "deputy of the king", whose seat was in Gaza, be sent to visit Jerusalem.[34] Occasionally a deputy officer did visit Jerusalem.[35] These consular visits by invitation and the absence of a permanent representative of the Egyptian crown suggest that the king of Jerusalem was a vassal with greater independence than the other kings and chiefs.

Although the kings of Jerusalem and Samaria were themselves vassals, they had their own tributaries. The king of Jerusalem received homage and tribute in silver and cattle from the kings of Arabia and from the Philistines (II Chronicles 17:11). The king of Moab, Mesha, was tributary to Samaria (II Kings 3:4). In the el-Amarna letters we shall often meet "the rebel Mesh"; his name is mentioned so frequently that it was thought to be a grammatical form indicating a group. We shall readily see that *amel-gaz*-Mesh, in the *singular*, was "the rebel Mesh", the rebellious king of Moab, and the *amelut-gaz*-Mesh were "the people of the rebel Mesh" or the Moabites.

Unlike the kings, other persons on the historical scene generally had but one name. The names of the Egyptian military governors and those of the army chiefs in Judah, as well as several other names, are similar in the el-Amarna letters and in the Bible. The similarity or identity of the names is corroborated by the identity of the functions of their bearers, as presented in the Scriptures and in the letters.

Not only can the personages in Judah, Israel, Moab, and Syria be traced in the Books of Kings and Chronicles and in the letters of el-Amarna; the names of the rulers of the tiny kingdoms in Syria also may be compared in the el-Amarna letters and in the inscriptions of Shalmaneser, king of Assyria, who lived in the time of Jehoshaphat and Ahab. Both sources—the el-Amarna letters and the inscriptions of the Assyrian king—are written in the same characters, cuneiform. Later in this discourse on the el-Amarna letters the participants in the war of resistance against the invader from the north will be identified.

King Jehoshaphat had five captains at the head of his army.

> II CHRONICLES 17:14–19 And these are the numbers of them according [in obeisance] to the house of their fathers: Of Judah, the captains of thousands; Adnah the chief, and with him mighty men of valour three hundred thousand.
>
> And next to him was Jehohanan the captain and with him two hundred and fourscore thousand.
>
> And next to him was Amasiah the son of Zichri, who willingly offered himself unto the Lord; and with him two hundred thousand mighty men of valour.
>
> And of Benjamin; Eliada a mighty man of valour, and with him armed men with bow and shield two hundred thousand.
>
> And next to him was Jehozabad, and with him a hundred and fourscore thousand ready prepared for the war.
>
> These waited on the king, beside those whom the king put in the fenced cities throughout all Judah.

The phrases, "in obeisance to the house of their fathers", is of essential value. It points to the existence of a feudal system at that time, in which rank in the army was hereditary, passing from father to son. We shall find in the next generation the names of the captains Ishmael, the son of Jehohanan, and Elishaphat, the son of Zichri.[36] The number of men under each of the five chiefs might mean that in their feudal districts there were one, two, or three hundred thousand men fit for conscription; however, another explanation is possible, and it is presented in one of the following sections.

The el-Amarna letters offer rich material on the feudal system in Judah at that time.

We identify letters written by three of the five military chiefs of Jehoshaphat. Their position as military chiefs is the same in the letters as in the Book of Chronicles; their names are easily recognizable. A slight variation in one of the names has a connotation that will lead us to reflection on the development of religion in Judah and on the reform that took place shortly after the death of Jehoshaphat.

Addudani (also spelled Addadani) of the el-Amarna letters is called in the Scriptures Adna.[37] But the inscription of Shamshi-Ramman, who became the Assyrian king after Shalmaneser in −825, contains a reference to a gift he received from Ada-danu, prince of Gaza (Azati).[38] The "son of Zuchru" of the el-Amarna letters is called the "son of Zichri" in the Bible.[39] Iahzibada of the el-Amarna letters is Iehozabad (Jehozabad) in the Scriptures.

These officers were really important chiefs of the army, as the pharaoh corresponded directly with them; yet in their letters the expressions of obeisance disclose their more subordinate role and differ from those in the letters of the king of Jerusalem.

In keeping with his position as chief among the captains, Addudani carried on an extensive correspondence with the pharaoh, and four of his letters, written at length, are preserved. From them we learn the complicated system in which a chief was bound directly to the pharaoh, to the local deputy of the pharaoh, and to the king (regent) in Jerusalem. The pharaoh wrote to Addudani:

> LETTER 294: Hearken to thy deputy and protect the cities of the king, thy lord, which are in thy care.

Addudani replied with assurances of his loyalty:

> LETTER 292: Thus saith Addudani, thy servant . . . I have heard the words, which the king, my lord, has written to his servant: "Protect thy deputy and protect the cities of the king, thy lord." Behold, I protect, and, behold, I hearken day and night to the words of the king, my lord. And let the king, my lord, pay attention to his servant.

After this introduction he reported on local affairs, on preparations to meet the archers of the king, on caravans, on conflicts between him and a deputy, on a garrison he had placed in Jaffa, and so on.

The same words were written to the son "of Zuchru": "Protect the cities of the king which are in thy care." This order is repeated in the reply of the son of Zuchru. The custom of repeating in epistles whole sentences from the letter being answered resulted in the preservation of valuable items out of lost tablets of the el-Amarna period.

The first name of the writer is not preserved, only his second name, "of Zuchru".[40] In the scriptural list of the five chiefs of Jehoshaphat, only one is called by the name of his father: Amaziah, son of Zichri. It is interesting to note that in the el-Amarna letters also, only in the case of the son of Zuchru is the name of his father attached. The scriptural text explains the distinction: Zichri sacrificed himself willingly to God; his descendants were honored by the name "sons of Zichri" (II Chronicles 17:16; 23:1).

Zichri of the (Massorete) Bible is Zuchru of the el-Amarna letters; Amaziah, the son of Zichri, is the "son of Zuchru", the captain who wrote to the pharaoh on matters relating to the security of his district.[41]

Iehozabad (Jehozabad) of the Second Book of Chronicles is called Iahzibada in the letters he wrote to the pharaoh. These few short letters are acknowledgments of Pharaoh's orders. The place from which they were written is not indicated, but they were written from southern Palestine; in the Second Book of Chronicles (17:17–18) it is said that he was a chief in the land of Benjamin.

Iehozabad is mentioned as the last among the five military chiefs of Jehoshaphat's kingdom. He does not enter into discussions with the pharaoh, nor does he give advice like the first in rank among the captains, but as a good soldier he accepts and acknowledges the orders of the pharaoh. Here is one of his stereotyped letters:

> LETTER 275: To the king, my lord, my gods, my sun, say: Thus saith Iahzibada, thy servant, the dust of thy feet:
> At the feet of the king, my lord, my gods, my sun, seven times and seven times I fall down. The word which the king, my lord, my gods, my sun, has spoken to me, verily, I will execute it for the king, my lord.

The areas of administration of Judah and Benjamin, as they were divided among the chiefs and the *sari* of the cities, may be determined approximately by the combined information of the el-Amarna letters and the Scriptures. In Knudtzon's edition (and also in Mercer's) the el-Amarna letters are ordered according to the geographical location of their writers, the correspondents from the north preceding those from the south. It is to the credit of the industrious work of Knudtzon that the letters of Addadani, son of Zuchru, and Iahzibada are placed next to the letters of the king of Jerusalem; thus he correctly located these correspondents in the southern part of Palestine.

230

Adaia, the Deputy

We have in Chapter 23 of Second Chronicles a list of the chiefs in the sixteenth year after Jehoshaphat. Instead of Amaziah, son of Zichri, Elishaphat, son of Zichri, served; Ishmael, son of Jehohanan, replaced his father as one of the chiefs of the army. As previously stated, the post of chief passed from father to son and from one brother to another, clearly marking it as a feudal institution. In that sixteenth year after Jehoshaphat there was a chief named Maaseiah, son of Adaia (Adaja). Adaia must have lived in the days of Jehoshaphat and must have been in the service of the king. The el-Amarna letters give us a clue to his role. He was apparently the king's deputy in Edom, adjacent to Benjamin, and for some time also had charge over the gateway of Gaza through which traffic with Egypt was maintained.

In the Scriptures it is said that in the time of Jehoshaphat "there was then no king in Edom: a deputy was king" (I Kings 22:47). This land was under the control of the king of Jerusalem and was a dependency of Judah (II Chronicles 21:8).

Four times Adaia's name is mentioned in three passages in the letters of the king of Jerusalem:

LETTER 285: Addaia, the deputy of the king [pharaoh] . . .

LETTER 287: . . . Addaia has departed together with the garrison of the officers which the king had given. Let the king know that Addaia has said to me: "Verily, let me depart."[42]

LETTER 289: The garrison which thou hast sent. . . . Addaia has taken and placed in his house in Hazati.

These letters inform us that Adaia was a deputy of the pharaoh, and that he was subordinate to the king-regent in Jerusalem.[43] The deputy in Edom, a dependent land of Judah, was actually subordinate to the king of Jerusalem.

231

The list of the five chiefs of Jehoshaphat, as given in Second Chronicles, ended with the passage: "These waited on the king, beside those whom the king put in the fenced cities throughout all Judah." Jehoshaphat sent Levites to these city-princes to guide them spiritually:

> II CHRONICLES 17:7 Also in the third year of his reign he sent to his princes, even to Ben-Hail, and to Obadiah, and to Zechariah . . . to teach in the cities of Judah.

A city-prince whose name was Vidia wrote to the pharaoh from Askelon in southern Palestine. Seven tablets in the el-Amarna collection are signed with his name. His domain was confined to a single city. Accordingly, Vidia wrote:

> LETTER 320: I protect the place of the king, which is in my care.

> LETTER 326: I protect the city of the king, my lord.

Vidia received a deputy from the pharaoh and prepared a tribute for him.

In Egypt the monarch was deified as the incarnation of the god, the son of the sun, and the sun itself. In accordance with Egyptian religious beliefs the subordinate chiefs in the tributary lands addressed the pharaoh thus: "To the king, my lord, my god, my sun, the sun in heaven." So wrote also Vidia from Askelon.

The idolatry and pagan influence made the cities around Jerusalem "halt between two opinions", to use Elijah's expression. It was therefore necessary to undertake the task of enlightening the people: ". . . and he [Jehoshaphat] went out again through the people from Beer-sheba to mount Ephraim, and brought them back unto the Lord God of their fathers" (II Chronicles 19:4). These efforts met with only partial success, as the Scriptures acknowledge: ". . . yet the people had not prepared their hearts unto the God of their fathers" (II Chronicles 20:33).

In the Syrian and Palestinian realms the sovereign of Egypt kept his deputies at the side of the regent-kings. The deputy in Sumur during the first part of Rib-Addi's reign was, according to the el-Amarna letters, Aman-appa. In a letter from the king of Sumur to Aman-appa it is said:

> LETTER 73: Thou knowest my attitude: Whilst thou wast in Sumura, that I was thy faithful servant.

We meet this governor in the Second Book of Chronicles.

> II CHRONICLES 18:25 Then the king of Israel said, Take thou Micaiah [the prophet], and carry him back to Amon the governor of the city, and to Joash the king's son.

Because of his position the name of Amon, the governor, was placed before that of the prince of royal blood. That he was an Egyptian is implied by his name, which is the holy name of an Egyptian deity.[44]

After his return to Egypt Amon was regarded by the king of Sumur (Samaria) as a friend and as an advocate of his cause before the pharaoh, and was recommended as an expert in the military and political matters of Sumur. The king of Sumur wrote to the pharaoh:

> LETTER 74: Verily, Aman-appa is with thee. Ask him. He knows that and has seen the distress which oppresses me.

He asked Aman-appa to come again to Sumur and to bring archers with him:

> LETTER 77: Hast thou not said to thy lord that he sends thee at the head of the archers?

The king of Sumur was very intimate with the deputy of the pharaoh. In another letter he wrote:

> LETTER 93: "I come to thee," thou didst write to me. Hear me. Say to the king to give to thee three hundred people.

In all these matters the vassal king and the governor exchanged advice and expressed their common concern for Sumur (Samaria).

233

Aman-appa, as the letters show, was opposed to the policy of supporting the king of Damascus. In the Scriptures, too, Amon, the governor of Samaria, by imprisoning the prophet who warned against making war on the king of Damascus,[45] showed his endorsement of the policy of regaining the lost cities by arms.

In the time of the el-Amarna correspondence Aman-appa was an aged man; he did not live until the end of that period.[46] The king of Sumur (Samaria), who for some time had heard nothing from Aman-appa in Egypt, wrote to him affectionately: "If thou art dead, I shall die, too."[47]

"To Aman-appa, my father, thus said Rib-Addi, thy son," wrote the king of Sumur (Samaria). This expression of respect is preserved in dialogues of the same period, as recorded in the Scriptures. The king of Samaria spoke similarly to Elisha: "My father" (II Kings 6:21).

The title by which the governor, Amon, is known in the Scriptures, *sar*, is often applied to dignitaries in the el-Amarna letters.

In the Scriptures the name of the governor of Samaria is given as Amon. In the el-Amarna letters the governor of Sumur was Aman-appa. They were the same person.[48]

The First Siege of Samaria by the King of Damascus

The kings of Judah and Israel were loyal to the Egyptian crown; but the king of the Syrian kingdom used the balance of strength in the north and the south to increase his domain.

The letters of Abdi-Ashirta (Ben-Hadad) of Damascus were humble despite or because of his treacherous intentions. The usual form of respectful address toward a potentate was: "I fall down seven and seven times to the feet of my lord"; to this the king of Damascus usually added when writing to the pharaoh: "Thy servant and the mud of thy feet, thy dog."

According to the Scriptures, Ben-Hadad, the king of Damascus, was a descendant of Rezon, who "fled from his lord" and "gathered men unto him, and became captain over a band; . . . and they went to Damascus . . . and reigned in Damascus".[49]

The king of Sumur (Samaria) and other vassal kings called the king of Damascus "the slave".

LETTER 71: What is Abdi-Ashirta, the servant, the dog, that he should take the land of the king to himself? What is his family?

From the days of Rezon on, in consequence of Damascus' politics, which encouraged a spirit of rivalry between Israel and Judah, these two countries continued to be hostile to each other. Baasha, king of Israel, built Ramah against Judah and threatened her. Asa, king of Judah, sent presents to Ben-Hadad, and Ben-Hadad turned against Baasha and "smote Ijon, and Dan, and Abel-beth-maachah, and all Cinneroth, with all the land of Naphtali" (I Kings 15:20). Judah took Mount Ephraim (II Chronicles 17:2). This happened two generations after Solomon and a few decades before the el-Amarna period.

The new dynasty of Omri at its beginning strengthened the kingdom of Israel. It was the time when the Egyptian hegemony over Palestine had been re-established by Thutmose IV, the father of Amenhotep III.

In the time of Ahab, the son of Omri, Ben-Hadad[50] renewed hostilities and arranged a coalition of chieftains depending on him:

> I KINGS 20:1 And Ben-Hadad the king of Syria gathered all his host together: and there were thirty and two kings with him, and horses and chariots.

In the letter of the king of Sumur (Samaria) we find a complaint:

> LETTER 90: All the majors [chieftains] are one with Abdi-Ashirta.

The kings of Damascus laid siege to Samaria.

> I KINGS 20:1 ... and he [Ben-Hadad] went up and besieged Samaria, and warred against it.

This campaign opened a long series of sieges, battles, short truces, and renewed oppressions that occupied the period described in the last six chapters of the First Book of Kings and the first nine chapters of the Second Book.

> Hostility against Sumur has become very great

235

is repeated in several letters from the king of Sumur (Samaria).

The time was opportune. The land of Israel was visited by drought; national feeling in the Ten Tribes was dwindling. The worship of pagan images laid the northern kingdom open to the spiritual influences of the surrounding nations. Religious ties between Samaria, Sidon, and Damascus melted away the frontiers. Even the prophets—Elijah from Gilead and Elisha, his successor—intervened in the state affairs of Damascus, visited there and were visited by people from there.

In this state of spiritual and material decline the land of Samaria was a prey to the shrewd warrior and politician of Damascus. All the lands as far north as the Orontes flows were "the lands of the king, the lord", the pharaoh of Egypt; but this circumstance did not prevent the Damascene from seeking to extend his domain. He was aware that the approach of the king of Assyria along the valley of the Orontes provided him with a chance to play a double game. He knew that the Egyptian king would not like to see him desert and openly go over to the war-minded Assyrians, the conquerors who pounded at the strongholds of northern Syria without declaring war on Egypt. The hazardous policy of the king of Damascus looked for a victim; Samaria was marked to be the first, and it was put under siege.

King Ahab asked a prophet whence help would come. "The young men of the governors of the provinces" would put the Syrians to flight, was the answer (I Kings 20:14). What did this mean? Why should the Syrian host be afraid of the governors' guard if they were not afraid of the king's army? In the battle at the wall of Samaria "two hundred and thirty-two young men of the governors of the provinces", leading Samaria's small garrison, put the Syrians to flight. The answer may be found in the letters.

The king of Sumur and Gubla (Jezreel), in his letters to the pharaoh and also to the governors, repeatedly asked that small detachments of archers be sent him. One such letter was quoted in the previous section: the governor was asked to send three hundred people to relieve the city.

The bearers of the emblem of the Egyptian state (the young men of the governors of the provinces) were a kind of gendarmerie attached to the governors of the pharaoh. These small detachments numbered tens, seldom hundreds, of men. In executing their duty, they were backed by the regular troops of Egypt, and their appearance at the place of dispute between the vassals of the Egyptian crown heralded a definite decision on the part of the pharaoh to support one of the rivals with arms. The impatience with which such a detachment for the relief of Samaria was awaited is reflected in the following passage from a letter of the king of Sumur to Haia (?), a dignitary in Egypt:

LETTER 71: Why hast thou held back and not said to the king that he should send archers that they may take Sumura? What is Abdi-Ashirta, the servant, the dog, that he should take the land of the king to himself? . . . send me fifty pairs of horses and two hundred infantry . . . till the archers go forth. . . .

Ben-Hadad boasted of his troops that the dust of Samaria would not suffice for handfuls for all the people that followed him (I Kings 20:10). The victory over this army was accomplished when the representatives of the suzerainty appeared at the head of the defenders of Samaria.

I KINGS 20:19 So these young men of the governors of the provinces came out of the city, and the army which followed them. 21 And the king of Israel went out, and smote the horses and chariots, and slew the Syrians with a great slaughter.

The young men of the governors were the soldiers of the pharaoh.

LETTER 129: Who can stand against the soldiers of the king [pharaoh]?

Later on, when again pressed by the army of Damascus, the king of Sumur (Samaria) recalled this or a similar event:

LETTER 121: I have written to the palace: "Send archers." Have they not formerly regained the lands for the king?

and similarly he wrote in another letter of that later period:

LETTER 132: Formerly Abdi-Ashirta opposed me, and I wrote to thy father: "Send royal archers. And the whole land will be taken in (a few) days."

237

Once more he recalled this memorable event, and his words accord with the story in the Scriptures:

LETTER 138: When Abdi-Ashrati conquered Sumurri, I protected the city by my own hand. I had no garrison. But I wrote to the king, my lord, and soldiers came and they took Sumuri.

The Capture and Release of the King of Damascus by the King of Samaria

One year after "the young men of the governors of the provinces" freed Samaria from the siege, Ben-Hadad came once more against the king of Israel and was met by the defenders on the plain of Aphek. "The children of Israel pitched before them like two little flocks of kids; but the Syrians filled the country" (I Kings 20:27).

The dread this army inspired arose from its multitude; we read in the el-Amarna letters that it was composed of wild and irregular troops. In the battle the Syrians were once more defeated, and Ben-Hadad fled to Aphek and hid. His servants said to him: "The kings of the house of Israel are merciful kings; ... go out to the king of Israel." Then they "came to the king of Israel, and said, Thy servant Ben-Hadad saith, I pray thee, let me live. And he said, Is he yet alive? he is my brother. Now the men did diligently observe ... and did hastily catch it: and they said, Thy brother Ben-Hadad. Then he said, Go ye, bring him. Then Ben-Hadad came forth to him; and he caused him to come up into the chariot. And Ben-Hadad said unto him, The cities, which my father took from thy father, I will restore; and thou shalt make streets for thee in Damascus, as my father made in Samaria. Then said Ahab, I will send thee away with this covenant. So he made a covenant with him, and sent him away" (I Kings 20:31–34).

From this story we learn that Ben-Hadad was defeated and captured but released; a covenant was concluded with him; the Syrians had built a part of Samaria; there was a dispute between the Syrians and the Israelites about a number of cities.

It appears that in the letters of the king of Sumur we have references to all these events, some of which happened only a few years before the date of the correspondence.

In the times of distress and oppression that came later, the king of Sumur, remembering these better days, wrote to the pharaoh:

LETTER 127: When, formerly, Abdi-Ashratu marched up against me, I was mighty; but, behold, now my people are crushed . . .

Looking for help from the pharaoh against Azaru (Hazael), the son of Abdi-Ashirta (Ben-Hadad[51]), he recalled the capture of the king of Damascus:

LETTER 117: If my words were regarded, then Azaru would truly be captured even as his father.

Ben-Hadad was captured and released:

LETTER 117: Abdi-Ashirta, with all that belongs to him, was not (then) taken, as I have said.

The treaty ("the covenant") with Ben-Hadad was made at a time when the Syrian was defeated. A prophet, distressed by the credulity of the king of Samaria, disguised himself as a wounded warrior, stopped the king on the road, and said to him: ". . . thy life shall go for his [Ben-Hadad's] life, and thy people for his people" (I Kings 20:42).

This scriptural prophecy was fulfilled. We read the complaints of the king of Sumur that he is oppressed by the king of Damascus, whom he once released. He even contemplates a new covenant with the king of Damascus, this time himself the humble partner:

LETTER 83: Why hast thou held back, so that thy land is taken? . . . I have written for a garrison and for horses and they were not given. Send an answer to me. Otherwise I will make a treaty with Abdi-Ashirta. . . . Then should I be rescued.

He was in the position of the once wretched king of Damascus, who, when vanquished, asked for a treaty of peace.

The covenant, concluded after the battle of Aphek, which was made in favor of Israel, endured for three years: "And they continued three years without war between Syria and Israel" (I Kings 22:1); then hostilities between Samaria and Damascus were renewed.

The war, which started with the siege of Samaria, was continued at Aphek, and after a truce of three years, broke out again at Ramoth-Gilead. Accordingly, the king of Sumur (Samaria) wrote: "Three times, these years, has he [Abdi-Ashirta] opposed me."[52]

Ships, Chieftains, or Legions?

In the el-Amarna letters a word recurs frequently which in some places fits in the text and in others does not: it is the word *elippe*, translated "a ship".

This translation is appropriate in the letter of the king of Tyre, who wrote that the king of Beirut left in a ship, and in a letter of the king of Sumur (Samaria), who asked that supplies be sent to him in ships.

In the ninth century Palestine tried to maintain the maritime traditions of the preceding century. Jehoshaphat ventured to repeat Solomon's undertaking and built ships in Ezion-Geber on the Aqaba Gulf of the Red Sea in order to send them "to Tarshish" (II Chronicles 20:35f.), an enterprise that ended disastrously before it had scarcely begun when a storm, so sudden in this gulf, smashed the fleet. The Phoenician cities of Tyre, Sidon, and Beirut maintained their maritime traffic down to a much later age. To translate *elippe* as "ship" in the above examples is undoubtedly correct. The reading "ships" might even be correct in connection with an inland region of Palestine if there was at that time navigation on the Jordan (as Strabo [XVI, 2] described it later), or on the Sea of Galilee. But in some cases one finds ships traveling on dry land, and performing other acts for which a vehicle of transportation on water is not suited. For example, the *elippe* are said to have penetrated into Amuru land and conspired with the killers of Abdi-Ashirta.

In the Hebrew language there is a word *ilpha* (*aleph, lamed, phe*) derived from the Syrian and meaning a "ship".[53] An old Hebrew word *aluph* (also *aleph, lamed, phe*) means "a prince, lord of a clan, head of a family".[54] It seems to me that *elippe* in the el-Amarna letters must sometimes signify chieftain or the head of a small tribe. One or another city mentioned in the el-Amarna letters and presumed to be a harbor need not necessarily be a maritime city or a lake city just because *elippe* arrived there and did something. In

240

these instances the word is misunderstood by modern scholars, but they can be comforted by the fact that a similar mistake was committed by the penman, who, in writing the annals of this period, used old chronicles.

When Ben-Hadad with his captains fled from the battlefield of Aphek into that city, a part of the city wall collapsed and "fell upon twenty and seven thousand" men (I Kings 20:30). This sounds like sheer exaggeration. Aphek was not such a large city, and tens of thousands of persons would hardly be killed by a falling wall. Now *eleph* (again *aleph, lamed, phe*) is "thousand",[55] and *aluph* is, as I have mentioned, a "chieftain".

It will be remembered that in the description of the feudal domains of the five chiefs of Jehoshaphat it was said that one had under him three hundred thousand men of valor, a second, two hundred and fourscore thousand, and so on. Although at the time of David men available for conscription in Judah numbered five hundred thousand, and in Israel eight hundred thousand,[56] and one century later Asa commanded over three hundred thousand from Judah and two hundred and fourscore thousand from Benjamin, nevertheless, it seems that the passage about the chiefs of Jehoshaphat would better reflect the military might of the Palestine princes of that age if it were read as follows: "Adnah the chief, and with him mighty men of valour three hundred chieftains. And next to him was Jehohanan the captain, and with him two hundred and fourscore chieftains," and so on.

There is contemporaneous evidence that feudal domains were reckoned by the number of chieftains or heads of communities. Mesha wrote on his stele: "And the chiefs of Daibon were fifty, for all Daibon was obedient (to me). And I reigned over a hundred (chiefs) in the cities which I added to the land."

"Thousands" in the biblical story of the disaster at Aphek and probably also in the passage about the chiefs of Jehoshaphat, and "ships" in a number of the el-Amarna letters, must be revised, and "chieftains" or "heads of communities" substituted, thus making the passages in the letters and the Scriptures reasonable.

The King of Samaria Seeks an Ally
against the King of Damascus

Because of one of the cities in dispute, the truce was broken.

> I KINGS 22:3 And the king of Israel said unto his servants, Know ye that Ramoth in Gilead is ours, and we be still, and take it not out of the hand of the king of Syria?

The dispute over the cities of Israel, captured by the king of Damascus, is recorded scores of times in the el-Amarna letters.

According to the Scriptures, in the beginning these cities were "Ijon, and Dan, and Abel-beth-maachah, and all Cinneroth, with all the land of Naphtali" (I Kings 15:20), "cities of Israel", which the king of Damascus smote in days gone by. Later, other territory was added to that gained by the king of Damascus.

"Abdi-Ashirta, the dog, he seeks to take all cities," wrote the king of Sumur (Samaria).

> LETTER 81: Let the king, my lord, know that powerful is the hostility of Abdi-Ashirta, and that he has taken all my cities to himself.

The king of Sumur (Samaria) looked for an ally to recover these lost cities. He thought that if one of the king-regents would side with him he would be able to return blow for blow and expel the bands of Syrians.

> LETTER 85: If one regent would make common cause with me, then I would drive Abdi-Ashirta out of Amurri.

> I KINGS 22:4 And he said unto Jehoshaphat, Wilt thou go with me to battle to Ramoth-Gilead?

At that time Governor Aman-appa was in Samaria. A prophet admonished the kings of Jerusalem and Samaria not to go to war against the king of Damascus. The prophet was turned over to the governor (I Kings 22:36).

King Jehoshaphat, at the beginning of his reign, strengthened himself against Israel: "And he placed forces in all the fenced cities of Judah ... and in the cities of Ephraim, which Asa his father had taken" (II Chronicles 17:2). Later he concluded peace with Israel and agreed to join Ahab in his campaign against the king of Damascus at Ramoth-Gilead. He felt that the Syrian was growing too strong and that the day might come when he would threaten Jerusalem too. Possibly, also, he wished to make amends for the sin of his father Asa, who had called the king of Damascus to his assistance when defending his land against Baasha of Israel.

The two kings joined forces and met their enemy at Ramoth-Gilead. In the course of the battle a chance arrow struck King Ahab.

Ahab or Jehoram: Two Versions of the Scriptures

With the story of this first battle of Ramoth-Gilead we reach the period during which the el-Amarna letters were written. The king who wrote more than sixty of the extant letters called himself—if the reading is correct—Rib-Addi. Was their author the scriptural Ahab or Jehoram, Ahab's son? They were written in the latter part of Jehoshaphat's reign in Jerusalem, and their contents accord with the events of that time.

According to the more expanded version of the story in the Scriptures, Ahab died from the wound he received in the battle at Ramoth-Gilead; Ahaziah, his son, followed him on the throne of Israel, reigning for two years; after the death of Ahaziah, Jehoram, his brother, reigned. The less amplified but probably older version in the Scriptures implies that Ahab was apparently only wounded at Ramoth-Gilead and reigned nine years longer.

The beginning of the reign of Jehoram in Israel is recorded in two contradictory statements:

> II KINGS 1:17 And Jehoram reigned in his [Ahaziah's] stead in the second year of Jehoram the son of Jehoshaphat king of Judah.

> II KINGS 3:1 Now Jehoram the son of Ahab began to reign over Israel in Samaria in the eighteenth year of Jehoshaphat of Judah.[57]

Jehoshaphat ruled for twenty-five years.[58] The difference in these two statements consists of nine years: the last seven years of Jehoshaphat and the first two years of his son's reign.[59] The inconsistent records of course presented difficulties to chronologists and exegetes. The question under discussion, which was only a chronological difficulty, becomes a problem of importance for the study of Palestinian history in the period of the el-Amarna letters, for it was during these nine years that the major part of the letters was written.

The problem, it should be made clear, does not result from a comparison of the Scriptures with the el-Amarna letters, but from the divergencies in the scriptural texts. The el-Amarna letters will have to help in clarifying the question at issue.

If, during the last seven years of the reign of King Jehoshaphat, Jehoram reigned in Israel, it would also be he who wrote the sixty-five letters preserved in the el-Amarna archives. But if the other version of the Second Book of Kings is correct, and during the last seven years of King Jehoshaphat's reign King Ahab still reigned in Israel, then he must have been the author of the letters and the events of these seven or nine years must have occurred in his time. Ahab, dying one or two years after Jehoshaphat, would have met with the experiences ascribed to his son Jehoram. He would not have died at the hand of one of Ben-Hadad's archers but would only have been wounded and would have survived Ben-Hadad. Similarly the rebellion of Mesha, king of Moab, would not have taken place after the death of Ahab but after his defeat at Ramoth-Gilead.

The description of the battle at Ramoth-Gilead gives one the feeling that the hand of a later scribe tried to mingle different sources. Having been wounded by an archer, Ahab "said unto the driver of his chariot: Turn thine hand, and carry me out of the host; for I am wounded". This indicates that the wounded Ahab abandoned the field of battle. But the next passage contradicts this: "And the battle increased that day: and the king was stayed up in his chariot against the Syrians, and died at even: and the blood ran out of the wound into the midst of the chariot" (I Kings 22).

This story closes a chapter of the drama known as the crime of Ahab, whose wife, Jezebel, gave him the vineyard of Naboth adjacent to his palace in Jezreel to build there a garden of herbs. When Ahab rose up to go down to the vineyard to take possession of it, Elijah the Tishbite came to meet him there. And Ahab said to

244

Elijah: "Hast thou found me, O mine enemy?" And the feared man answered: "I have found thee. . . . Hast thou killed and also taken possession? . . . In the place where dogs licked the blood of Naboth shall dogs lick thy blood, even thine. The dogs shall eat Jezebel by the wall of Jezreel" (I Kings 21).

In accordance with this curse, after the battle of Ramoth-Gilead "one washed the chariot in the pool of Samaria; and the dogs licked up his [Ahab's] blood; and they washed his armour; according unto the word of the Lord which he spake" (I Kings 22).

But the story of the meeting of the seer with the king had a sequel. When Ahab heard the words of the prophet he rent his clothes and put sackcloth on his body, and fasted and went softly. And Elijah the Tishbite heard the word of the Lord: "Seest thou how Ahab humbleth himself before me? because he humbleth himself before me, I will not bring the evil in his days: but in his son's days will I bring the evil upon his house."

And what about the pardon? Ahab humbled himself, and the misfortune fated for his lifetime was postponed to the next generation. But in spite of this did the evil overtake him?

The effort of the scribe to unify the diverse elements was unsuccessful, as there are inconsistencies in the text. This editor of the Books of Kings was helpless in the face of two different versions, and while giving preference to the tradition that Jehoram was king in Israel during the last seven years of Jehoshaphat, he did not suppress the other version, but in describing the history of the period, he evaded the issue by writing in a number of chapters the impersonal and indefinite "king of Israel".[60]

Already in the story of the battle at Ramoth-Gilead, every time the king of Jerusalem is referred to, he is called by name, Jehoshaphat, but the king of Samaria is referred to fifteen times as "king of Israel": "And the king of Israel and Jehoshaphat", "And the king of Israel said unto Jehoshaphat"; in the body of the story the name of Ahab is not mentioned.

Again, in the war against Moab, the king of Jerusalem is repeatedly called by name, but not the "king of Israel": "So the king of Israel and Jehoshaphat . . ." Only the introductory remark informs us that Jehoram is intended. In the story of Naaman's healing there is but the recurrent "king of Israel" and no name; neither does the story of attempts against the life of the "king of Israel" contain his name. In the long story of the second siege of Samaria, when the king met the mother who killed her child, and in the story of the rescue from this siege, not once does the name of the king accompany the phrase "king of Israel". This is singular in the Books of Kings and Chronicles.

A contemporaneous source, which may shed light on the matter, is the stele of King Mesha of Moab. On the stele it is engraved that Omri, king of Israel, oppressed Moab many days, and "his son succeeded him; and he also said, 'I will afflict Moab.' In my days said he thus." Then follows:

> Omri took possession of the land of Medeba and (Israel) dwelt therein, during his days and half his son's days, forty years; but Chemosh restored it in my days.[61]

Here is an event of importance ascribed to the reign of Ahab, son of Omri, according to one source (the stele of Mesha), and to the reign of Jehoram, son of Ahab, according to another source, one of the two scriptural versions.

The Second Book of Kings opens with these words: "Then Moab rebelled against Israel after the death of Ahab." This contradicts the stele of Mesha, which says that Mesha rebelled when Omri's son had reigned only half his years.

If Ahab was not killed, but only wounded, at Ramoth-Gilead, the defeat having been a signal for the rebellion of Moab, the expression, "and half the days of his [Omri's] son", would accord with the version that Ahab reigned during the time otherwise allotted to Jehoram.

Besides the fact that the rebellion of Mesha took place not after the death of Ahab but in the middle of his reign, the figure of forty years of the reign of Omri and half of the reign of his son conflicts with the scriptural account. Forty years of oppression of Moab may be taken as a rough figure, forty being the measure of a generation, or a period of time, similar to our counting by centuries; but the account must correspond approximately to the time passed; it would suggest a longer reign for Omri and Ahab or for one of them.[62]

It is recorded that Omri reigned twelve years over Israel, six of them in Tirzah (I Kings 16:23), and that Ahab reigned twenty-two years in Samaria (I Kings 16:29). Either the reign of Ahab started later than recorded or it lasted longer. Since it is said of his father,

Omri, that "six years reigned he in Tirzah" but for twelve years he reigned over Israel, it is clear that the last six years he reigned in his new capital Samaria (Shemer). Similarly the twenty-two years of Ahab's reign in Samaria may refer to his reign in that capital alone, the years of his reign in his new capital Jezreel not being mentioned.

In one of his last letters the king of Sumur (Samaria) wrote of himself to Pharaoh Akhnaton:

> LETTER 137: Behold, I cannot come to the lands of Egypt. I am old, and my body is afflicted with a severe disease.

Ahab could have said that he was old; not so a second son of his in the first part of his reign.

The inscription of King Shalmaneser III of Assyria-Babylonia, like the stele of Mesha, conflicts with the same version of the Scriptures. Shalmaneser wrote that in his sixth year he battled a coalition of Syrian and Palestinian princes at Karkar. Among the princes Ahab is mentioned; he delivered an army of ten thousand soldiers and two thousand chariots to the allied host.[63] In the eighteenth year of his reign Shalmaneser wrote that he received "a tribute of the men of Tyre, Sidon, and of Jehu, of the house of Omri".[64]

During the twelve years between the sixth and the eighteenth year of Shalmaneser's reign, the reign of Ahab had to come to an end, Ahaziah had to reign two years, Jehoram twelve years, and Jehu had to reign for some undefined period. But even if Ahab had died right after the battle of Karkar,[65] and Jehu's tribute mentioned in the inscription were paid immediately after he seized the throne, there would still not be twelve years left for Jehoram's reign.

The Mesha stele thus requires the extension of Ahab's reign, and the inscription of Shalmaneser requires shortening of Jehoram's reign so as to bring Ahab closer to Jehu.

Under the circumstances described, a hypothesis may be formulated, and the very existence of Jehoram, king of Israel, may be questioned.[66]

What could have misled the annalist into calling the son of Ahab by the name of Jehoram? Jehoram was a son of Jehoshaphat and the son-in-law of Ahab (II Chronicles 21:6). Obviously it was Jehoshaphat's policy to establish good relations with Israel through this alliance; he probably hoped to unite the kingdom under his son

Jehoram and Athaliah, the daughter of Ahab and Jezebel. Jehoshaphat visited Samaria and allied himself with Ahab in his military undertakings.

Ahab is said to have been killed at Ramoth-Gilead. Jehoram, his son, is said to have been wounded also at Ramoth-Gilead, under similar circumstances, with the king of Judah again an ally in the battle (II Kings 8:28). Many other details ascribed to the reigns of these two kings of Israel bear the mark of confusing similarity.

The annalist placed an Ahaziah and a Jehoram in Israel and another Jehoram and another Ahaziah in Judah. The state of things as he found them in older chronicles was probably confusing.

Possibly Jehoram, son of Jehoshaphat, acted as a regent, succeeding Ahab, his father-in-law, and Ahaziah, son of Ahab, who died after a short reign and a long illness. Of Jehoram it is said: "And he walked in the way of the kings of Israel, like as did the house of Ahab: for he had the daughter of Ahab to wife."[67] Thereafter the throne of Israel was seized by Jehu, and Jehoram was killed together with Ahaziah of Judah. Ahab's daughter Athaliah, wife of Jehoram, seized the throne of Jerusalem when her husband and son were killed.

Following the description of the assassination of Jehoram and Ahaziah by Jehu, the Second Book of Kings relates: "And Ahab had seventy sons in Samaria." Jehu sent letters to the elders of Samaria with the challenge: "Look even out the best and meetest of your master's sons, and set him on his father's throne." The king who is called here "your master" was Ahab, not Jehoram, son of Ahab, and one of his (Ahab's) sons should have succeeded him, had it not been for the massacre in which all his sons perished. This also would explain Ahab's reign extending almost until the rebellion of Jehu.

On subsequent pages it will be shown that rumors of Ahab's death were spread while he was still alive. A later annalist might well be misled if contemporaries were mistaken.

Regardless of whether this hypothesis concerning Jehoram is correct or incorrect, the documents of Mesha, of Shalmaneser, and of el-Amarna unanimously support the version of the Scriptures according to which Ahab was alive during the last seven years of Jehoshaphant's reign.

In the contest between the two versions of the Book of Kings, all three non-scriptural sources testify in favor of the minor version and

248

against the prevailing text; the rendering of II Kings 1:17 must be looked upon as the correct one. This would mean that Ahab died not before but after Jehoshaphat, which would signify that the king-author of more than sixty of the preserved clay tablets was Ahab and no one else.

NOTES

1. The translations into German are by Hugo Winckler and by J. A. Knudtzon. The work of the last-named Scandinavian scientist is of classical value for the study of the Tell el-Amarna tablets. The translation into English is by S. A. B. Mercer (1939). Twelve letters found since the publication by Knudtzon are included in Mercer's English edition. The letters are similarly numbered in Knudtzon's and Mercer's editions. In this chapter quotations from the letters are taken from the English version of Mercer (*The Tell el-Amarna Tablets* [Toronto, 1939]). However, the translations have been checked in Knudtzon's version.

2. Breasted, Weigall, Freud.

3. Mercer, *The Tell el-Amarna Tablets*, pp. 510ff.; Barton, *Archaeology and the Bible*, p. 368; H. Ranke, in *Zeitschrift für Aegyptische Sprache*, LVI (1920), 69–71. Albright, "Cuneiform Material for Egyptian Prosopography", *Journal of Near Eastern Studies*, V (1946), 22, n. 62.

4. Cf. Marquart, *Chronologische Untersuchungen*. pp. 35ff., and Jeremias, *Das Alte Testament im Lichte des Alten Orients* (2nd ed.; Leipzig, 1906), pp. 390ff.

5. O. Weber in J. A. Knudtzon, *Die El-Amarna-Tafeln* (Leipzig, 1915), p. 1172.

6. The readers of this chapter are advised to read beforehand I Kings 16-22; II Kings 1–10; and II Chronicles 16–22.

7. See Genesis 14:18; Joshua 15:8; 18:28; Judges 19:10–11; I Chronicles 11:4–5.

8. Weber in Knudtzon, *Die El-Amarna-Tafeln*, p. 1135.

9. Letter 81.

10. Joshua 13:5; Ezekiel 27:9.

11. Psalms 83:7. Cf. I Kings 5:18 (the Hebrew text). See R. Dussaud, *Syria, revue d'art oriental et d'archéologie*, IV (1923), 300f.

12. Dhorme, *Revue biblique* (1908), 509f.; Weber, in Knudtzon, *Die El-Amarna-Tafeln*, p. 1165.

13. *Against Apion*, I, 116; *Jewish Antiquities,* VIII, 1.

14. Josephus, *Jewish Antiquities*, VIII, xiii, 1.

15. Philo of Byblos, *Fragments*, 2, 25.

16. It is possible that the name Jezebel (Izebel) is a later form of Zebel; addition of the *I* lends to the name an ignominious character of denial or curse, as in the name I-chabod (I Samuel 3:21).

249

17. The King James translation is: "so that they shall not say, This is Jezebel".

18. The site of the residence city of Jezreel has not been established. Its traditional site in the east of the valley disclosed no antiquities. It is probable, rather, that Jezreel is to be looked for in the west of the valley. Ahab, taking a daughter of the Sidonian king to wife, might have been anxious also to have a share in the maritime trade of the Phoenicians. Elijah ran without stopping from Carmel to Jezreel (I Kings 18:46).

19. A throne name, a personal name, and epithets; some of them could have been changed during the lifetime of a monarch.

20. Sources are brought together by Ginzberg, *Legends*, VI, 277.

21. Tractate Sanhedrin 94,a; Jerome on Isaiah 20:1 and 36:1; Ginzberg, *Legends*, VI, 370.

22. By H. Winckler. See A. H. Sayce, *Records of the Past* (New Series, 6 Vols.; London, 1889–93).

23. By A. Gustavs, *Die Personennamen in den Tontafeln von Tell Taanek* (Leipzig, 1928), p. 10.

24. By Dhorme. Hiba is presumably the Hurrian form of the name of a Hittite deity, Hepa. Cf. B. Maisler, *Untersuchungen zur alten Geschichte und Ethnographie Syriens und Palästinas*, I (Giessen, 1930), 37.

25. A name containing the part *hiba* is known among the officers of King David: Eliahba (Elihiba) in II Samuel 23:32.

26. Letter 107.

27. D. D. Luckenbill, *Ancient Records of Assyria* (Chicago, 1926–27), I, Sec. 601. The Amorites were a tribe of Syria and Canaan.

28. Josephus, *Jewish Antiquities*, VI, 5.

29. See Jack, *Samaria in Ahab's Time*, p. 119, note 3. Compare Jeremiah 49:27 and Amos 1:4.

30. Jack, *Samaria in Ahab's Time*. See Meyer, *Geschichte des Altertums*, II, Pt. 2 (2nd ed.; 1931), p. 274, note 2; p. 332, note 1.

31. Asheroth is usually translated "groves", as in I Kings 18:19 ("the prophets of the groves"). On Ashera and Astarte, see M. Ohnefalsch-Richter, *Kypros: The Bible and Homer* (London, 1893), pp. 141ff.

32. *Against Apion*, I, 122.

33. We have other examples in the Scriptures as well as in the letters, where *h* or *kh* was often freely added or deleted. Hadoram of the Second Book of Chronicles (II Chronicles 10:18) is called Adoram in the First Book of Kings (12:18). Another example is Adad and Hadad, two transcriptions of the same name (I Kings 11:14ff.). Ammunira, king of Beirut, in some letters, is Hamuniri of other letters. "The sound h in the biblical name Hazael happens to occur in Akkadian as Haza-ilu but the spelling Aza-ilu, if it occurred, would be quite in accordance with the facts observable in other cases." (Professor I. J. Gelb, written communication of May 15, 1951.)

34. Letter 287.

35. Letter 289.

36. II Chronicles 23:1.

37. In the form "Adna" the divine name "Addu" (Addu of Dan) is mutilated; this mutilation was probably the work of the holy penman, who would not admit that a man close to the pious Jehoshaphat had borne the name of Addu-Dani.

38. Luckenbill, *Records of Assyria*, I, Sec. 722. Mercer (*Tell el-Amarna Tablets*, p. 375, note) relates Azzati to Gaza (Aza in Hebrew).

39. II Chronicles 17:16.

40. Letter 334.

41. Letter 335.

42. In his transliteration, Knudtzon gives these varying spellings of the name.

43. In Letter 254 it is said that Dumuia was entrusted to Adaia. Does this name mean an individual, or could it stand for Dumah in Seir, or Edom (Isaiah 21:11)?

44. Amon was also the name of the son of Manasseh, king of Jerusalem, in the seventh century (II Chronicles 33:20–25). Of Manasseh it is said that he "made Judah and the inhabitants of Jerusalem to err" (II Chronicles 33:9).

45. I Kings 22:26–27.

46. Letter 106: "There is hostility against Sumur. And verily, its deputy is now dead."

47. Letter 87.

48. Early in this century in Tell Taannek, the biblical Taanach, on the hills in the region of the Esdraelon Valley, a few tablets written in cuneiform were found; they are very similar to those of the el-Amarna collection. In one of them a governor by the name of Aman-hasir exacted tribute from the local mayor (E. Sellin, 'Tell Ta'annek", *Denkschriften der Akademie der Wissenschaften*, Philosophisch-Historische Klasse, Vol. 50 [Vienna, 1904]). The reading of Aman-hasir was revised by Albright to Aman-hatpe: "Aman-hatpe, Governor of Palestine", *Zeitschrift für ägyptische Sprache und Altertumskunde*, LXII (1927), 63f. His reading was accepted by A. Gustavs, *Die Personennamen in den Tontafeln von Tell Taanek*, p. 26. Albright made a surmise that this governor was the future pharaoh Amenhotep II.

49. I Kings 11:23–24.

50. The opponent of Ahab is generally regarded as a son of Ben-Hadad I, the adversary of Baasha, and therefore is named Ben-Hadad II.

51. That Hazael was a son of Ben-Hadad, see infra.

52. Letter 85.

53. See Levy, *Wörterbuch über die Talmudim und Midrashim*.

54. Ibid.

55. Sar-ha-eleph is a captain over a thousand. It might have been the origin of *aluph*, "chieftain".

56. II Samuel 24:9.

57. See also II Kings 8:16.

251

58. II Chronicles 20:31.

59. The same discrepancy of nine years exists in the records of the reigns of Baasha and Asa.

60. The annals (the Books of Kings and Chronicles) were composed during and after the Exile in Babylon, since the Exile is narrated in Chronicles and the return from the Exile in Kings. The editor of Kings indicated that his work was a compilation by referring to "the book of the chronicles of the kings of Israel", which seems to have been a larger work than the canonical Chronicles. Also, the books of the prophets Nathan, Iddo, and others (not extant at the time when the Scriptures were revised and canonized) are referred to in the annals.

61. Translation by S. R. Driver.

62. Compare I Kings 16:23 and 16:29.

63. Luckenbill, *Records of Assyria*, I, Sec. 610.

64. Cf. ibid., Sec. 672. Jehu was a son of Jehoshaphat, son of Nimshi. Was he a son of a daughter of Omri?

65. "Within these thirteen years, 854–842, must fall the death of Ahab, the reigns of Ahaziah and Jehoram, and the accession of Jehu. There appears to be no time left for Ahab after 854. The death of Ahab, however, cannot be assigned to so early a date as 854." K. Marti in Encyclopaedia Biblica, I (New York, 1899), "Chronology".

66. It seems problematic that Ahab, who persecuted the cult of Yahwe, would have called his son Jehoram (Jahwe is exalted). This would be a forceful argument but for the fact that the scriptural names of Ahab's other children—Ahaziah, Joash, and Athaliah—invite the same question.

67. II Chronicles 21:6. Cf. II Kings 8:16–18.

252

CHAPTER SEVEN

The el-Amarna Letters (Continued)

Famine

Tribes of the desert, driven by famine, came to Trans-Jordan and crossed the river, only to find the land of Israel in even greater distress and misery than that from which they had fled. The fields in Israel yielded nothing. The meadows were scorched by the sun; only thorns sprang from the barren land.

The first forecast of the drought came from Elijah the Tishbite, who said to the king:

I KINGS 17:1 . . . there shall not be dew nor rain these years.

The heavens breathed heat, the trees in the field withered. The brook Cherith, where the prophet looked for a little water, dried up like other brooks.

I KINGS 17:7 And it came to pass after a while, that the brook dried up, because there had been no rain in the land.

Eager eyes were turned skyward, looking to see whether a cloud would appear. The prophet said that his blessing on the poor widow should last "until the day that the Lord sendeth rain upon the earth" (I Kings 17:14).

Unlike Egypt, which is dependent on a river,[1] Palestine's harvest depends solely on rain; the drought brought the famine.

I KINGS 17:2. And there was a sore famine in Samaria.

It was a time of hunger and starvation unparalleled in the history of the period of the kings of Israel. The want lasted seven years.

II KINGS 8:1 . . . the Lord hath called for a famine; and it shall also come upon the land seven years.

This famine overshadowed all other events of that time; the mark

253

of hunger was impressed on the entire period. A succession of chapters in the Scriptures describes the famine. During those years the el-Amarna letters were written, and it would be inconceivable that they should not reflect these conditions. The fact is that the letters of el-Amarna written by the king of Sumur (Samaria) are as eloquent as the Book of Kings.

The famine was so great that children were sold for bread.

> LETTER 74: Our sons and our daughters have come to an end, together with ourselves, because they are given in Iarimuta for the saving of our lives. My field is a wife, who is without a husband, deficient in cultivation.

After the sons and the daughters had been sold into slavery to save them and the remnants of the population from starvation, the implements of the households also went in exchange for food.

> LETTER 75: The sons, the daughters, have come to an end, and the wooden implements of the houses, because they are given in Iarimuta for the saving of our lives.

The king repeated his comparison of a land without seed to a woman deserted by her husband; he was to repeat it in many letters. In more than thirty places in his letters the king of Sumur (Samaria) writes of the distress of the famine or pleads for provisions for the population or the army.

> LETTER 79: Give me something to feed them [the archers], I have nothing.

> LETTER 83: ... Give grain for my provision.

> LETTER 85: There is no grain for our support. What shall I say to my peasants? Their sons, their daughters have come to an end. ... Send grain in ships and preserve the life of his servant and his city. ... May it seem good to the king, my lord, that grain be given, produce of the land of Iarimuta.

> LETTER 86: There is nothing to give for deliverance. ... And from the land of Iarimuta should grain be given for our nourishment.

Grain received from Iarimuta in exchange for the freedom of a part of the people was rationed and divided in scanty portions among the peasants. (As to the whereabouts of Iarimuta, the reader will be instructed in a later section.)

We see the turn of the years:

LETTER 85: Two years I measure my grain.

LETTER 86: I measure our grain three years already.

LETTER 90: My field is deficient in cultivation . . . and I measure my grain.

LETTER 91: I measure my grain.

The indication in the Scriptures that the famine endured for seven years (II Kings 8:1) corresponds to the impression received from these letters; after the third year of the drought the famine still was constantly recalled, although on the basis of the letters alone we cannot calculate the time precisely. The clue of the seven years helps to establish a timetable for some of the letters of el-Amarna.

The springs and wells dried up; a prince from the north, an ally of the king of Egypt, intended to come to the help of Sumur (Samaria).

LETTER 85: . . . but there was no water for him to drink, and so he has returned to his land.

The lack of water is described in the story of the brook that dried up after rainless seasons (I Kings 17:7); the passage was quoted above. The king of Israel, who undertook an expedition against Mesha, was in a bad plight, because "there was no water for the host, and for the cattle that followed them" (II Kings 3:9). "And there was a dearth in the land" (II Kings 4:38). In the letter of the king of Babylon we read: "The road [to Egypt] is very long, the water supply cut off, and the weather hot."[2]

This drought and the anxiety of the king of Israel are equally reflected in that part of the story where he said to Obadiah, the governor of his house:

I KINGS 18: 5–6 . . . Go into the land, unto all fountains of water, and unto all brooks: peradventure we may find grass to save the horses and mules alive, that we lose not all the beasts.

So they divided the land between them to pass throughout it . . .

A quotation from a letter written to the pharaoh by a man of Gubla, but not by the king himself, may lead us to a conjecture about the author of the letter. After reporting on political affairs, the writer referred to his care of the asses:

LETTER 94: Formerly, in respect to the asses, the king has commanded that they be given to his faithful servant. But, verily, he has now nothing . . .

The name of the author at the beginning of the letter is destroyed. It might have been Obadiah, a man who would have the prerogative to write to the pharaoh, and who was charged with the care of the animals. The man who, together with Ahab, king of Samaria, was concerned that "we lose not all the beasts", reported to the Egyptian sovereign that no asses survived the famine and the drought.

The plague among the animals in Samaria is mentioned in a letter written from Egypt by a chief; he had heard that the people of Sumura were not allowed to enter Gubla, since in Sumura there was a plague.

LETTER 96: What sort of a plague is among the asses?

Outside of Samaria, as in the capital, the population was starving. The king brought the riches of the realm, together with the children sold into slavery "for the saving of the life", to a place called Iarimuta, in order to take grain from there; the list of the treasures exists in part. But the king did not get the grain, for the ruler of the region from which it was to be sent apparently allied himself with Damascus.

The dreadful seven years' famine left deep marks. A thousand years later the rabbinical Haggada related: "In the first year everything stored in the houses was eaten up. In the second, the people supported themselves with what they could scrape together in the fields. The flesh of the clean animals sufficed for the third year; in the

256

fourth the sufferers resorted to the unclean animals; in the fifth, to the reptiles and insects; and in the sixth the monstrous thing happened that women, crazed by hunger, consumed their own children as food. ... In the seventh year, men sought to gnaw the flesh from their own bones."[3]

The contemporaneous letters of the king of Sumur (Samaria) reflect the agony of those years. He wrote of his futile efforts to obtain grain for sustenance from Iarimuta "that I die not".[4] Again he wrote: "Everything is consumed",[5] and again: 'Everything is gone",[6] and the king had nothing to give to his people, no grain for bread and no seed to try for a crop the next year.

The king wrote that he had no provisions for his peasants,[7] that he was afraid of them,[8] that they would rebel,[9] that they intended to desert: "My peasants intend to desert",[10] and that his land was waste and the peasants were departing for places where grain was to be found:

> LETTER 125: And there is no grain for my provisions, and the peasants have departed for the cities where there is grain for their provisions.

This emigration is reflected in II Kings 8:1:

> Arise, and go thou and thine household, and sojourn wheresoever thou canst sojourn: for the Lord hath called for a famine.

It is significant that in the Scriptures and in the letters the years-long famine is localized in the land of Samaria (Sumur).

The shortest record—short because only a few words remain of it on a mutilated tablet—speaks most eloquently:

> dust ... yield ... of the lands.

Mesha's Rebellion

Mesha, king of Moab, paid tribute to the king of Israel: "And Mesha king of Moab was a sheepmaster, and rendered unto the king of Israel a hundred thousand lambs, and a hundred thousand rams, with the wool" (II Kings 3:4).

257

After the defeat at Ramoth-Gilead, Moab rebelled against the king of Israel (II Kings 1:1 and 3:5). The allied armies of Israel, Judah, and Edom suffered greatly from lack of water on their march around the Dead Sea to bring the king of Moab to obedience. They reached Moab from the south and wrought destruction upon the land. The king of Moab tried to break through the siege but could not. "Then he took his eldest son that should have reigned in his stead, and offered him for a burnt offering upon the wall. And there was great indignation against Israel: and they departed from him, and returned to their own land" (II Kings 3:27).

It is evident that Israel was defeated and that its allies could not help. We are not given any details of that "great indignation".

In the sixties of the last century Arabs of Dhiban in Trans-Jordan, the ancient Dibon, showed a traveler a black basalt stele on which were cut ancient Hebrew characters. After the stone had been sold to a museum the Arabs regretted the transaction. They began to think that a treasure was hidden inside the stone and that the strange writing told of it, and so they decided to open the stone. Furthermore, they reasoned that by breaking it up they would have more objects for sale. And if the stone possessed a charm they had to destroy it first. They heated it in fire and poured cold water over it, and it broke into many pieces.

A young scholar, however, had previously succeeded in secretly having a cast of the stone made by an Arab. Later the broken basalt in a very defective condition was purchased for the Louvre Museum. The cast supplies the missing parts, but all together it is only the upper part of the stele. The stone, which at the time of its discovery was estimated to be the oldest inscription in Hebrew characters in the hands of the archaeologists, established the fact that the Moabites used Hebrew.

The text gives an account of Mesha's victory over Israel. It opens with these words:

"I am Mesha, son of Chemosh, king of Moab, the Daibonite [Dibonite].[11] My father reigned over Moab for thirty years, and I reigned after my father. And I made this high place for Chemosh in KRKHH, a high place of salvation, because he had saved me from all the assailants, and because he had let me see my desire upon all them that hated me.

"Omri, king of Israel, afflicted Moab for many days, because Chemosh was angry with his land. And his son succeeded him; and he also said, I will afflict Moab. In my days said he thus; but I saw my desire upon him and upon his house, and Israel perished with an everlasting destruction."[12]

Mesha recorded that he restored Medeba; built Baal-Meon and Kiryathen; fought against Ataroth, the dwelling place of Gad from of old, fortified by the king of Israel, and slew all the people of that city. The oracle of Chemosh told him, "Go, take Nebo against Israel," and he took it and slew all that were in it, seven thousand men and boys and women and girls and maidservants. "And the king of Israel had built Yahas, and abode in it, while he fought against me. But Chemosh drave him out from before me; and I took . . . it to add it unto Daibon." Then Mesha built and repaired the walls and palace in Karkhah with the slave labor of the prisoners of Israel, and it is recounted in more detail; he also built other places. He continued with the war and went against the Israelite city of Horonen, and here the record is interrupted; the lines are defaced, but it may be presumed that other war exploits against Israel were recorded. "And I" can be read a few more times. The lower part of the stele is missing altogether. The cities mentioned on the stele are in Trans-Jordan, but Karkhah is not known.

Every word and every letter and every obliterated portion were objects of careful examination: the stele of Mesha is regarded as the greatest single discovery in biblical archaeology, especially as it provides a parallel record to a scriptural narrative.

The Book of Kings relates that "there was great indignation against Israel" at the attempt to subjugate Moab, which rebelled; on what form this indignation took, the record is silent. The Book of Chronicles discloses that the Moabites, together with the Ammonites, invaded Palestine with the help of the Syrians. From the Mesha stele we know that Mesha, king of Moab, took his revenge by an "everlasting destruction" of Israel, and that Mehedeba, Ataroth, Nebo, and Yahas were retaken. The "everlasting destruction" is not to be seen merely in the recapture of four or five not very significant places in Trans-Jordan.

The cuneiform letters of el-Amarna, found twenty years after the discovery of the Mesha stele, are regarded as historical documents preceding the stele of Mesha by some five hundred and fifty years, and relating to the Canaanite period. The present research shows

259

that the documents are contemporaneous, having been written in the middle of the ninth century.

In letters from Palestine, and especially in those written by the king of Samaria, we should expect to find direct reference to the rebellion of Mesha. We should also expect to find mention of events, the record of which was carved on the lower, missing part of the stele of Mesha. And, in fact, the letters of el-Amarna bear extensive witness to the history of that war, and also give us material with which to reconstruct roughly the missing part of the stele.

Already in the earliest preserved letter of the king of Sumur (Samaria) he had written to the pharaoh (Amenhotep III):

> LETTER 68: Let the king, my lord, know ... the hostility of the *sa.gaz.Mesh* troops is very great against me. So let not the king, my lord, hold back from Sumur that it be not quite annexed to the *sa.gaz.Mesh* troops.

In the translations of the el-Amarna letters we find *sa-gaz-Mesh* rendered as "*sa-gaz*-people". Putting these words into English, *sa-gaz*, which ideographically can also be read "*habatu*", is translated "plunderers" or "cutthroats" or "rebellious bandits". "Mesh" is understood as the suffix of the plural. It is met with in the letters scores of times, always in connection with the plundering rebels. Sometimes the intruding pillagers are called *amelut sa-gaz-Mesh*, *amelut* being the men or the people; sometimes the text speaks of *gaz-Mesh* as of a single person, and the translators again neglect "Mesh" and translate "robber". "He takes thy cities the *amel-gaz-Mesh*, the dog," and the words are rendered "the *gaz* man", *Mesh* being dropped. But the text speaks in these cases of a single person, and therefore Mesh cannot here be the suffix for the plural.

I shall not translate Mesh either, because it is the personal name of King Mesha, but I shall not omit it from the text of the translation. Thus we shall read: "The hostility of the rebels [pillagers] of Mesh is very great against me"; and in the other sentence: "He takes thy cities, the rebel Mesh, the dog."[13]

According to the inscription on the stele of Mesha, the revolt took place in the middle of Ahab's reign. As the earlier letters of the king of Sumur mention this rebellion, we are already in the second part of the reign of Ahab.

In another letter of the early years of the el-Amarna correspondence, the king of Sumur (Samaria) wrote again:

LETTER 69: Behold, now they rise up day and night in rebellions against me.

Ambi, a geographical name referred to in connection with the rebellion of Mesh,[14] seems to be the name of the land of Moab or its capital, or its people. Possibly Ammi, repeatedly mentioned in connection with Ambi, is Ammon;[15] in Genesis (19:38) the Ammonites are called Ammi. A city named Rubute also appears; this must be Rabbath-Ammon, the capital.[16]

The king of Sumur (Samaria) wrote to a dignitary in Egypt:

LETTER 73: When he [Abdi-Ashirta] wrote to the people of Ammia: "Kill your lord," and they joined the *amelut-gaz* [pillagers], then said the regents: "Thus will he do to us." And so all lands will join with the *amelut-gaz* [pillagers].

The Ammonites, their king having been killed, joined the herdsmen on the fields of Moab.

"There cometh a great multitude" of "the children of Moab, and the children of Ammon, and with them other[s] beside the Ammonites", was reported to King Jehoshaphat (II Chronicles 20: 1–2).

The king of Sumur wrote:

LETTER 79: Know that since the arrival of Aman-appa to me all *amelut-gaz-Mesh* [the people of the bandit Mesh] have directed their face against me, in accordance with the demand of Abdi-Ashirta. Let my lord listen to the words of his servant, and let him send me a garrison to defend the city of the king, until the march of the archers. And if there are no archers then all lands will unite with *amelut-gaz-Mesh* [the people of the bandit Mesh].

What is generally surmised from the records of the Books of Kings and Chronicles—the hidden hand of the king of Damascus in the rebellion of Moab[17] and in the turmoil among the tribes of the desert—grows to certainty in face of these letters. The king of Sumur (Samaria) asked that horses and infantry be sent him "so that he [Abdi-Ashirta] might not assemble all the bandits of Mesh".[18]

Letters are rare in which the king of Sumur (Samaria) does not mention the revolt of Mesha and the role which the ruler of Damascus played in it, whether openly or secretly. The role of Damascus in the war between Samaria and Moab may be read also in the Second Book of Chronicles (20:1–2).

The "Great Indignation": A Reconstruction of the Obscure and Missing Portions of the Stele of Mesha

"The city of the king, my lord," Sumur (Samaria), was threatened by the "bandits" or "cutthroats" of Mesh (Mesha). Already in the earliest extant letters of the king of Sumur the pharaoh was warned that the rebels were endangering the capital. Then he was warned that "the land of the king and Sumur, your garrison town", would unite with the people of the rebel Mesh (*amelu-gaz-Mesh*), and "thou held back".[19] The king of Sumur asked for archers, making it clear that he was unable to defend his land against the troops of the rebel Mesh.[20] The help did not come. So he wrote again:

> LETTER 83: Listen to me. Why hast thou held back, so that thy land is taken? Let it not be said: "In the days of the princes *amelut-gas-Mesh* [the people of the rebel Mesh] have taken all lands." Let not such things be said in future days. "And thou wast not able to rescue it" . . . if Sumura and Bit-Arkha also are now lost . . .

In the days when this letter was written Sumur (Samaria) was only threatened. Another stern warning came in a letter from the king of Sumur: If the pharaoh will not listen to the words he writes to him, then

> LETTER 88: . . . all the lands of the king, even as far as Egypt, will unite with *amelut-gaz-Mesh* [people of the rebel Mesh].

Finally it becomes clear that with the help of the king of Damascus the troops of Mesha entered Samaria (Sumur).

> LETTER 91: Why dost thou sit and hold back, so that he takes thy cities *amel-gaz-Mesh* [the rebel Mesh], the dog? When he had taken Sumura . . .

The king of Sumur, his capital lost, looked desperately for help from Egypt and asked the pharaoh in the same letter:

LETTER 91: So mayest thou give a thousand minas of silver and one hundred minas of gold. Then will he [the rebel Mesh] depart from me. And he has taken all my cities ...

We know from the Scriptures that the king of Samaria (Sumur) had negotiated to free his capital from a siege by the payment of silver and gold; it was during the first siege of Samaria described in I Kings 20. He was told then: "Thou shalt deliver me thy silver and thy gold. ..."

With all the information that the el-Amarna letters provide now in hand and properly read, we are faced with the surprising fact that the rebel Mesha of Moab succeeded in entering Samaria.

The first thought that presents itself is this: we have only the upper portion of the stele of Mesha; in the lower lost portion "the everlasting destruction" or the "great indignation" (II Kings 3:27) were probably described. It is also possible that the Mesha stele may have been but one part of a twin inscription on two monoliths, each of which carried half the story.

That we have only the top portion of the stele is evident; that there were two steles is not impossible. In any event we have to re-examine the stele of Mesha, and in doing this, it may be that we shall see things to which we paid insufficient attention before.

After the capture of the city of Yahas the king of Moab, at war with Israel, turned to building activities in an unidentified place:

"I built KRKHH [Karkhah], the wall of Yearim [or of the Forests], and the wall of Ophel. And I built its gates, and I built its towers. And I built the king's palace, and I made the two reservoirs for water in the midst of the city. And there was no cistern in the midst of the city, in Karkhah. And I said to all the people: 'Make you every man a cistern in his house." And I cut out the cuttings for Karkhah with (the help of) prisoners of Israel."

There is no city known by the name of Karkhah. Some scholars have thought that it was a portion of the city of Dibon. Another conjecture was that it represents the city of Kir-ha-Kharoshet. For some reason mention of this building activity in Karkhah is the central theme of this stele, and thus, together with the references to Karkhah in the introductory passage, lends emphasis to the importance of the activities there during the war against the Israelites.

With the help of the el-Amarna letters it may be assumed that KRKHH was the capital of the entire Egyptian domain in Palestine, or Sumur (Samaria), which Mesha, the rebel, entered after his success in overcoming the resistance of the fortified cities to the east. Samaria—on a straight line—is about twenty miles from the Jordan, the border of Ammon.

Kerakh or Karkha is in Hebrew "a very large city, encircled by a wall, to which foreigners come yearly for trade purposes",[21] that is, a metropolis.[22]

According to the inscription of Mesha, in Karkhah there was "Ophel", the wall of which he built or repaired. In the Scriptures a part of the city called "Ophel" is mentioned only with respect to Jerusalem and Samaria. Gehazi, the servant of Elisha, "the prophet that is in Samaria", returning to the house of his master with the presents from Naaman, parted from the servants of the *sar* when they reached Ophel in the city of Samaria (II Kings 5:24). Reference to Ophel in Karkhah on the stele of Mesha is of unquestionable value.

The king's palace (*beth-melech*), so often mentioned in the el-Amarna letters as being in Sumur, and unearthed in the Samaria of Omri and Ahab (Sebastieh of today), is also spoken of as being in Karkhah or the metropolis of the Mesha stele.

The attitude of Mesha in conquering the capital, "the city of the king [pharaoh]", was characteristic indeed. He repaired the city, its palace, and its walls. In his estimation it did not belong to the king of Israel. It belonged to the pharaoh. Whoever built in it showed his peaceful attitude toward Egypt, and also acquired the right to be regarded as the senior among the vassal princes of Syria–Palestine.

In other letters, written from northern Syria, it is said that the following places, possibly outside the domain of Israel, were despoiled by the rebels of Mesh (Mesha): Mahzibti, Giluni, Magdali, and Uste, also Tahsi and Ubi.[23]

Like a turbulent wind, the Arabs attacked the land from all directions. The peasants on the plains of the coast, hungry and thirsty, joined the troops of rebellious tribes driving across the land.

A ransom was probably paid to Mesh (Mesha) for releasing Sumur (Samaria), and if so, it was apparently paid out of the Egyptian treasury, as the Egyptian king saw in Samaria a palace-city of his own.

The stele was prepared to be erected in Karkhah-Samaria, but was left in Dibon. The vassal triumphed over his lord and the lord of his lord and probably received a ransom. This is what was meant by the "great indignation" of the biblical text and the "everlasting destruction" of the Mesha stele.

Or did the help dispatched by governor Aman-appa arrive in time and the oppressors of Samaria depart without ransom, as in the story of I Kings 20?

Arza, the Courtier

The king of Damascus followed the troops of the rebel Mesh and marched once again toward the city of Samaria (Sumur), left in ruins by wars and sieges and deserted by most of her inhabitants because of the famine, and entered it. He wrote to the pharaoh:

> LETTER 60: Behold, I am a servant of the king and dog of his house, and the whole land of Amurri I guard for the king, my lord. . . . Behold, all kings of the king seek selected troops to snatch the lands from my hand . . . If my plenipotentiary brings life from the king, the sun, then I shall reap the grain of Sumur, and I shall guard all lands for the king, my sun, my lord.

He had an excuse for entering Sumur, the excuse of one who insists on protecting a city that does not want his protection: he "delivered it from the hands of the troops of Sehlal". "There were no people in Sumur to protect it." If it had not been for him, "the troops of Sehlal would now have burned with fire Sumur and her palace". That these troops were incited by him he did not say.

He wrote to a chief in Egypt:

> LETTER 62: But when I hurried up here . . . and came to Sumur, the people were not there, who had dwelt in her palace; then, behold, the people, who dwelt in her palace, were Sabi-ilu, Bi-sitanu, Maia, and Arzaia. Behold, only four people there were, who dwelt in her palace, and they said to me: "Deliver us out of the hand of the troops of Sehlal," and I delivered them out of

the hand of the troops of Sehlal. . . . What lies did the regents tell thee? And thou believest them.

One of the four persons who remained in the palace in Samaria, when it was entered by the king of Damascus, was called Arzaia. Probably it was the same old courtier, the dweller in palaces, who is familiar to us from the First Book of Kings. There he is called Arza.

It was a few decades earlier that Elah, the son of Baasha, after a reign of two years, was killed by Zimri, a captain, when the young king was "drinking himself drunk in the house of Arza steward of his house" (I Kings 16:9). Since then twelve years of Omri's reign and a number of years of Ahab's reign had passed. Arza, as always, enjoyed the air of palace life.

From the Scriptures we know only of two sieges of Samaria and nothing of her fall. But we read that when the king of Samaria had allowed the vanquished Ben-Hadad to go in peace with a covenant of brotherhood, a prophet told the king: "Because thou hast let go out of thy hand a man appointed to utter destruction, therefore thy life shall go for his life, and thy people for his people" (I Kings 20:42). In these words is disguised the story we read in the letters of the king of Sumur.

> LETTER 74: Behold, now, the king [the pharaoh] has let his faithful city go out of his hand . . . Mighty is the hostility of the *gaz*-people [bandits] against me.

In passing, it may be noted that the turn of speech, "let go out of the hand", used in the scriptural text, is repeated in this letter.

Jerusalem in Peril

From the hills surrounding Jerusalem the mountains of Moab are distinctly seen in the clear air over the Jordan and the Dead Sea. But no movement of troops can be discerned from that distance, unless by their multitude they color the slopes and the ravines.

> II CHRONICLES 20:1–3 It came to pass . . . that the children of Moab, and the children of Ammon, and with them other beside the Ammonites, came against Jehoshaphat to battle.

Then there came some that told Jehoshaphat, saying There cometh a great multitude against thee from beyond the sea, from Syria . . .[24]

And Jehoshaphat feared.

His prayer before the congregation of Judah and Jerusalem is preserved in the Book of Chronicles. It begins with these words:

II CHRONICLES 20:6 . . . O Lord God of our fathers, are not thou God in heaven? and rulest not thou over all the kingdoms of the heathen? and in thine hand is there not power and might, so that none is able to withstand thee?

Then he reminded the Lord that the land had been given to the people of Israel forever:

II CHRONICLES 20:7 Art not thou our God who . . . gavest it to the seed of Abraham thy friend for ever?

He expressed his belief that the Lord would not abandon the place where a sanctuary was built to His name, and would come with help.

II CHRONICLES 20:8–9 And they [thy people Israel] dwelt therein, and have built thee a sanctuary therein for thy name, saying,

If, when evil cometh upon us . . . and [we] cry unto thee in our affliction, then thou wilt hear and help.

Then Jehoshaphat explained the distress of his people.

II CHRONICLES 20:10–11 And now, behold, the children of Ammon and Moab and mount Seir . . .

. . . come to cast us out of thy possession, which thou hast given us to inherit.

He concluded with an invocation declaring his and his people's helplessness because of the great number of the invading hordes.

II CHRONICLES 20:12 . . . We have no might against this great company that cometh against us; neither know we what to do; but our eyes are upon thee.

The feelings of the king of Jerusalem at the sight of the immense hordes converging on his kingdom are expressed both in his prayer

267

and in his letters. On earth his house was in vassalage to Egypt, and the pharaoh was obliged to protect him.

The king of Jerusalem wrote to the pharaoh:

> LETTER 288: Let the king care for his land. The land of the king will be lost. All of it will be taken from me; there is hostility to me as far as the lands of Seeri and even to Gintikirmil.[25] There is peace to all the regents, but to me there is hostility.

The same region—Mount Seir (Seeri)—is distinctly named in both sources as the far land from which a part of the invaders came.

The king expressed the belief that his lord would not abandon the place on which he had set his name forever:

> LETTER 287: Verily, the king has set his name upon the land of Urusalim for ever. Therefore he cannot abandon the lands of Urusalim [Jerusalem].

At the same time he voiced his fear that if no help arrived the invaders would cast them out of his lord's possession.

> LETTER 287: If there are no archers [this year], then there will also remain to the king no lands and no regents.

There exists a similarity between his appeal to his Lord in heaven and to his lord on earth, but observing that help was not arriving from the latter, he wrote impatient words he would not have addressed to the first:

> LETTER 288: Although a man sees the facts, yet the two eyes of the king, my lord, do not see. .... The Habiru are taking the cities of the king.

The king of Jerusalem, unlike other vassal kings, omits expressions of respect for the gods of Egypt; he does not call the pharaoh "my sun, my god", as all other vassal correspondents did; in distinction to other writers of the letters, he does not mention his God; he may be recognized as a servant of a Lord whose name he would not profane in his letters to his pagan protector.

That the same type of appeal should issue from the mouth and from the hand of the same human being is natural. In this case the similarity casts light on the genuineness of this prayer in Chronicles; it implies also the religious purity of the king-monotheist.

The letters contain details about this invasion and the threat to Jerusalem. The shepherds of Moab and Seir took Rabbath of Ammon in Trans-Jordan with the help of those chieftains who went over to the invaders, and the population of Ammon joined the nomads.

LETTER 289: After they have taken Rubuda, they seek now to take Urusalim.

Rubuda, written also Rubute,[26] we recognized as "Rabbath of the children of Ammon", a city still existing today. Separate bands crossed the borders of the kingdom of Judah simultaneously at a number of places. This is what is meant by the designation, "from beyond the sea [the Dead Sea], from Syria".

That Syria offered a through way for them we learn from the letters of the king of Sumur (Samaria). A letter from Palestine reports that the bands appeared even in the valley of Ajalon.[27] This explains the retreat of the population to the stronghold of Jerusalem.

II CHRONICLES 20:4. And Judah gathered themselves together, to ask help of the Lord: even out of all the cities of Judah they came to seek the Lord.

And the king of Jerusalem wrote to the pharaoh:

LETTER 289: The whole land of the king has deserted.

An unexpected turn of events spared Jerusalem from humiliation. The approach of the "multitude" of "the children of Ammon, Moab, and mount Seir" toward Jerusalem was interrupted by disagreement that flared up among the allies.

II CHRONICLES 20:23 For the children of Ammon and Moab stood up against the inhabitants of mount Seir, utterly to slay and destroy them: and when they had made an end of the inhabitants of Seir, every one helped to destroy another.

Something of these happenings is reflected in a letter of the king of Sumur:

LETTER 76: Behold, he [Abdi-Ashirta] has now mustered all *amelut gaz* [bandits] against Sigata and Ambi.

These cities were in the area of Ammon and Moab respectively. Sigata appears to be Succoth on the river Jordan.

The king of Jerusalem pointed to the roving tribes penetrating from the wastes of Trans-Jordan, and called them Habiru. Habiru is derived from the Hebrew root *haber*, a member of a band, and *habiru* means "bandits", and is used for "companions of thieves" in Isaiah 1:23, "troops of robbers" in Hosea 6:9, and "companion of a destroyer" in Proverbs 28:24.

This meaning of the word "Habiru" should have been suggested by the fact that *sa-gaz*, which is translated "bandits", "cutthroats", is interchanged with the term "Habiru".[28] The various theories about Habiru (Khabiru) of the el-Amarna letters—that it signifies "Ivri" (Hebrew), or "apiru" (miners), or "Afiru" (from the Babylonian region of Afiru—are thus found to be without foundation.

The Revolt of the Sodomites

During the uneasy period through which Jerusalem passed, the king of Judah apparently reached the conviction that a small and ancient colony on the road from the Jordan to Jerusalem was inclined to take the side of the enemies. Suwardata, the prince of Kelti, changed his policy according to the way the wind blew. At one time he wrote that he followed the king of Jerusalem against the bandits (*sa-gaz*).[29] At another time he opposed the king of Jerusalem and complained that the king had taken Kelti. In his turn he was accused by the king of Jerusalem of disloyalty to Egyptian interests.

Kelti[30] may be identified as Wadi Kelt on the way from Jerusalem to Jericho.

A campaign of Jehoshaphat is briefly recorded in the First Book of Kings:

> I KINGS 22:46 And the remnants of the Sodomites, which remained in the days of his father Asa, he took out of the land.

Wadi Kelt, where now a copious stream of water comes out of the ground and flows toward the Dead Sea and a few anchorites hide themselves among the rocks, was obviously the last abode of the remnants of the valley which became a dead lake. They proved to be disloyal at the time of the invasion of the desert tribes. The king of

Jerusalem wrote to them then, "Follow me", but they did not. More details may be found in the letters of the king of Jerusalem and Suwardata.

The name of this prince of the Sodomites may cause the philologists to reflect on their racial origin.[31]

The Second Siege of Samaria

The king of Damascus again laid siege to Samaria.

> II KINGS 6:24 ... King of Syria gathered all his host, and went up, and besieged Samaria.

The king of Sumur (Samaria) wrote:

> LETTER 92: He [Abdi-Ashirta] has now come forth against me.

"Hostility against Sumur has become very great" is repeated in many letters of the king of that city. In a number of letters he complains of the distress of that city plagued by siege and famine. This famine in the besieged city is described in the Book of Kings.

> II KINGS 6:25 And there was a great famine in Samaria: and, behold, they besieged it, until an ass's head was sold for fourscore pieces of silver, and the fourth part of a cab of dove's dung for five pieces of silver.

The population of the city and its garrison fainted from hunger, but the king of Samaria continued to defend the city and supervise the bastions.

> II KINGS 6:26 ... the king of Israel was passing by upon the wall ...

This was a wall of a fortress. In one of the el-Amarna letters the king wrote of the city:

> LETTER 81: Formerly Sumura and its people were a fortress and garrison for us.

When, during one of these inspections, the king learned that a case of cannibalism had occurred in the city and that hunger had over-

271

come the mother instinct, he rent his clothes and the people saw him in sackcloth (II Kings 6:29).

The king of Sumur (Samaria) dispatched the following message to the pharaoh:

> LETTER 74: May the king hear the words of his servant, and give life [provisions] to his servant, that his servant may live. Then will I defend his faithful city. ... Art thou kindly disposed toward me? What shall I do in my solitude? Behold, thus I ask day and night.

So wrote the king who, in hours of despair, put sackcloth on himself.

His only hope lay in the quick arrival of help from without, but time was passing and it seemed that no help would come to him.

> LETTER 74: If there is not a man to deliver me out of the hand of the enemy [Abdi-Ashirta], and we—the regents—are put out of the lands, then all the lands will unite with the *amelut-gaz* [pillagers] ... And if the king should (then) march forth all lands would be hostile to him, and what could he do for us then? Thus have they formed a conspiracy with one another, and thus have I great fear that there is no man to rescue me out of their hand.

Anxious to have help in time, he appealed to Aman-appa.[32] Aman-appa was the dignitary to whom the king of Samaria (Sumur) wrote: "Thou knowest my attitude, whilst thou wast in Sumura, that I was thy faithful servant." To this plenipotentiary and ex-governor, Amon of Samaria (I Kings 22:26), the king put this question:

> LETTER 73: Why hast thou held back and not spoken to the king thy lord, in order that thou mayest march forth with the archers and that thou mayest fall upon the land of Amurri? If they perceive that the archers have gone forth, they will leave even their cities and depart.

This quotation is of peculiar interest for comparison with the story of the end of the second siege of Samaria, according to the Book of Kings.

> II KINGS 7:6–7 For the Lord had made the host of the Syrians to hear a noise of chariots, and a noise of horses, even the noise o

a great host: and they said one to another, Lo, the king of Israel hath hired against us the kings of the Hittites, and the kings of the Egyptians, to come upon us.

Wherefore they arose and fled in the twilight, and left their tents, and their horses, and their asses, even the camp as it was, and fled for their life.

We know now that the fear of the Syrians at the walls of Samaria was not groundless. The king of Israel really negotiated for the departure of archers from Egypt. He knew in advance that merely on hearing that "the archers have gone forth" the Syrians would leave their tents and be put to flight.

The letters of el-Amarna and the Book of Kings present two records which supplement each other.

The city was oppressed but not conquered. How this happened is told in the Second Book of Kings, in Chapters 6–7. In accordance with that record, we find this statement by the king of Sumur (Samaria):

> LETTER 106: They have been able to oppress it [Sumur] but they have not been able to conquer it.

When gathering his host for the campaign that ended in flight, the king of Damascus wrote to his warriors: "Assemble yourselves in the house of Ninib."[33] This might mean: "Make an alliance with the king of Assyria."[34] Behind the rebellion of the Moabite Mesha was the intrigue of the king of Damascus, and he in his turn was incited by the king of Assyria.

But when the king of Damascus fled from the walls of Samaria and returned home from his unsuccessful campaign, he wrote to the pharaoh:

> LETTER 64: To the king, my lord, say. Thus saith Abdi-Ashtarti, the servant of the king: At the feet of the king, my lord, I have fallen seven times ... and seven times in addition, upon breast as well as back. May the king, my lord, learn that enmity is mighty against me. So may it seem good to the king, my lord, to send a powerful man to protect me.

The commander of the army of the king of Damascus bore the name of Naaman. "Now Naaman, captain [*sar*] of the host of the king of Syria [Aram], was a great man with his master, and honourable, because by him the Lord had given deliverance unto Syria: he was also a mighty man in valour" (II Kings 5:1).

The fifth chapter of the Second Book of Kings narrates the story of the healing of this captain by Elisha the prophet. It informs us in passing that the Syrians were prowling about in bands (as often mentioned in the el-Amarna letters) and carrying away captives out of the land of Israel, and that among these captives was an Israelite girl who became a handmaid in the household of the captain. The members of the household advised him to try the treatment of the prophet in Samaria. And the king of Syria said to Naaman: "Go to, go, and I will send a letter unto the king of Israel."

When the king of Israel received the letter, he rent his clothes. "Am I God," he asked, "that this man doth send unto me to recover a man of his leprosy? wherefore consider, I pray you, and see how he seeketh a quarrel against me."

Elisha the seer intervened when he heard that the king had rent his clothes. The story of how Elisha healed Naaman the captain is well known.[35]

In the quoted portion of the story, two facts are somewhat strange. First, inasmuch as Ben-Hadad himself was at the head of the thirty-two captains of his army,[36] why, in this story of the wondrous healing, is the deliverance of Syria credited to a captain Naaman? Second, the king of Israel was a lifelong rival of the king of Damascus. Why, then, did this request to cure a sick captain inspire in the king of Israel such dread that he rent his clothes?

For an explanation of the real role of this captain Naaman we shall look to the contemporaneous letters. A man by whom Syria received deliverance must be identifiable in the letters. We recognize him in the person of Ianhama, called also Iaanhamu.

Ianhama, the pharaoh's deputy in Syria, was sent to the king of Damascus with prerogatives similar to those which Aman-appa had when he was with the king of Samaria. Naaman's title in the Scriptures—sar—is also used in the letters. He was a plenipotentiary of the king of Egypt, in charge of the army and walled cities of Amuru land (Syria), later also the overseer of stores of grain. He had great influence in all matters of Syrian administration. Judged by his name, he was of Syrian origin, as were some other dignitaries at the court of Thebes.[37] Ianhama is a Semitic name. "Ianhamu was a powerful Egyptian agent in Syria, where he was respected as a good and wise man, and where he proved himself to be the most faithful of the pharaoh's servants."[38]

The servant of Elisha said: "Behold, my master hath spared Naaman this Syrian [Arami], in not receiving at his hand that which he brought." When healed, Naaman asked Elisha to give him two mules' burden of earth, "for thy servant will henceforth offer neither burnt offering nor sacrifice unto other gods, but unto the Lord. In this thing the Lord pardon thy servant, that when my master goes into the house of Rimmon to worship there, and he leaneth on my hand, and I bow myself in the house of Rimmon: ... the Lord pardon thy servant in this thing." Here the god Rimmon was probably the god Ra-Amon, the chief god of the Egyptians. Ianhama's master was Amenhotep III, and later Akhnaton. Ianhama, as may be seen from the context of the letters, protected the king of Damascus and helped him in his rise to power. At various times he is reported to have been in Syria and in Egypt, favoring a policy of balanced relations in Syria in view of the necessity to secure the position of Damascus in the Syrian bloc against the "king of Hatti".

Accordingly it is said in the Second Book of Kings that through Naaman, "a great man with his master, and honourable ... the Lord had given deliverance unto Syria".

In the early letters of the king of Sumur (Samaria) his fear of the mighty deputy of the pharaoh is plainly expressed. In one letter he wrote to the pharaoh: "Thou must rescue me out of the hand of Iaanhamu."[39] He asked the pharaoh to inform his deputy that he, Ianhama, would be responsible if anything should happen to the person of the king of Samaria. "Say to Ianhamu: 'Rib-Addi is even in thy hands, and all that will be done to him rests upon thee'."[40]

Later on, when Aman-appa left Samaria and died in Egypt,[41] the king of Samaria wrote to the pharaoh asking him to appoint Ianhama governor in Samaria. He asked the pharaoh: "May it seem right to my lord to send Ianhama as his deputy. I hear from the mouth of the people that he is a wise man and all people love him."[42] We recall the scriptural words about Naaman, that he was an "honourable" man.

What happened that the king of Samaria, who once feared Ian-hama and asked to be rescued from his hand, should now himself recommend that very man for the governorship of Samaria? In another letter he again asks the pharaoh to send Ianhama and in the next one he praises him in these words: "There is no servant like Ianhama, a faithful servant to the king."[43]

Ianhama wrote from Egypt to the king of Samaria: "Go and occupy Sumur until I come . . . enter, fear not,"[44] and the king of Sumur (Samaria) wrote him, in the hope that the pharaoh would appoint Ianhama: "Hasten very quickly thy arrival," and explained that because of the hostility of the people of Ambi he was unable to enter Samaria.

The letters do not show why the fear of the king of Samaria changed into confidence with respect to the Syrian deputy. The Scriptures provide the explanation in the story of the healing of Naaman by the prophet of Samaria. Naaman was very grateful to the prophet, and also to the king of Samaria, to whom he came with a letter from the king of Damascus. "Behold, now I know that there is no God in all the earth, but in Israel" (II Kings 5:15). Elisha even declared that he would heal Naaman in order to help the king of Israel politically.

So he became a friend. When the king of Damascus was killed, as will be narrated on a subsequent page, Ianhama (Naaman) was apparently in Egypt. He did not support the next Syrian king; he corresponded with the king of Samaria and favored him.

Certain other features of the role and character of Ianhama, reflected in the letters, are shown also in the Scriptures. He was a generous man. This appears in the story of the healing: he gave to the servant of the prophet two talents of silver and two changes of garments, more than the servant had asked for, when the prophet refused to take ten talents of silver, six thousand pieces of gold, and ten changes of raiment.

It is of interest to find that, according to the letters, Ianhama was in charge of the pharaoh's treasury in Syria, being over "money and clothing".[45]

Paying with a combination of money and clothing is a true custom of that time. In one letter the king of Sumur wrote that he paid "thirteen silvers and one pair of garments" for someone to go on an errand.[46]

According to the Scriptures, the Syrian governor had in his house an Israelite girl, a captive carried off by a Syrian band, who waited on his wife.[47] In an early letter of the king of Sumur (Samaria) to the pharaoh there is a complaint that two people from his domain are detained in the private home of Ianhama.[48]

The el-Amarna letters also speak of him as the generous patron of a Palestinian youth, who was educated in Egypt at his expense.[49]

The man "by whom the Lord had given deliverance unto Syria" and who was feared by the king of Samaria was Ianhama.[50] How this captain changed his attitude and became a supporter of the king of Samaria is recorded in the letters and is explained by the Scriptures.

The Letters of the "Great Woman of Shunem"

Informative letters were sent to the pharaoh by a woman correspondent called Baalat Neše. Two of them are preserved. Who might this female person be who was at that time in Palestine and of high enough rank to write to the sovereign in Egypt? She is the only lady correspondent from Palestine in the el-Amarna collection.

In her first letter[51] she informed the pharaoh that the pillagers (sa-gaz people) had sent bands to Ajaluna (Ajalon). She wrote about "two sons of Milkili" in connection with a raid that apparently threatened her native town. The menace was not averted because she wrote again:

> LETTER 274: Thus said Baalat-Nese, thy handmaid, the dust of thy feet. . . . Let the king, my lord, deliver his land out of the hands of the *amelut sa-gaz-Mesh* [people of the bandit Mesh], that it may not be destroyed. . . .

And she advised the pharaoh that the invaders were coming nearer and that another town had fallen: "Verily, this is for the information of the king, my lord."

In Letter 250 from the hand of another correspondent we read that Milkili "took stand against Shunama and Burkuna". If we put this statement together with the fact that the lady correspondent complained to the pharaoh of the bands of Milkili that threatened her domicile, it becomes quite apparent that her home was either at Shunama or Burkuna.

If this very simple deduction is not fallacious, then not only would the lady of Shunama or Burkuna be one of the el-Amarna correspondents, but she would also have a page in the Old Testament. And in fact she has. It begins with the verse:

> II KINGS 4:8 And it fell on a day, that Elisha passed to Shunem, where was a great woman.

Shunem is doubtless Shunama. To suppose that there was another "great woman" at the same time in the same town or in the neighboring town of Burkuna would be a forced conjecture.

The name of the "great woman", Baalat-Neše, might be understood as "a woman to whom occurred a wonder" (Baalat-Nes).[52] The story of Elisha breathing life into the son of the lady of Shunem would have been a matter of talk in the palaces of that time; it was the subject of the royal audience granted by the king of Israel to the servant of Elisha (II Kings 8:4). The woman could also have been known in the palace of Egypt as "a woman to whom was a wonder". Her high position was enhanced by the wide circulation of the story about the revival of her child.

The very existence of a "great lady" by the name of Baalat-Neše in the city or district of Shunem (Shunama) at the time of a famine and of the revolt of Mesh (Mesha) seems to throw a sidelight on the coming and going of a healer and holy man venerated in folk traditions, Elisha.

The King of Damascus Conspires against the Life of the King of Samaria

Famine, lack of a consistent policy in dealing with the Syrian vassals, and frequent and prolonged absences of the Egyptian governors contributed to the reduction of the area to a state of anarchy. The king of Sumur (Samaria) warned the pharaoh:

> LETTER 75: Aduna king of Irqata, mercenaries have killed [him] and there is no one who has said anything to Abdi-Ashirta, al-

though thou didst know it. . . . The people of Ammi have killed its lord. So I am afraid.

He was afraid and not without reason, for he himself was chosen to be a victim and to be killed from ambush.

> II KINGS 6:8 Then the king of Syria warred against Israel, and took counsel with his servants, saying, In such and such a place shall be my camp.
> 10 And the king of Israel sent to the place which the man of God told him and warned him of, and saved himself there, not once nor twice.

The corresponding passage in the el-Amarna tablets is found in the letters of the king of Sumur to the pharaoh and also to his former governor, Aman-appa:

> LETTER 81: A stranger stood with drawn dagger against me; but I killed him. . . . Abdi-Ashirta, at his command was committed this deed against me. Behold, thus I remain shut up in the midst of my city. I cannot go out of the gate. . . . I have been wounded nine times, and so I have feared for my life.

The letters describe the terror which the king of Damascus inspired in the king of Samaria (Sumur), then at his residence in Jezreel (Gubla), by setting an ambush at the places where the hunted rival was supposed to pass. The king of Sumur wrote:

> LETTER 88: He [Abdi-Ashirta, king of Damascus] slinks around about every gate of Gubla. . . . We cannot go out of the gates.

Josephus Flavius, who related the story following the Scriptures, wrote of "secret attempts on the life of the Israelite king"[53]: "There were some Syrians lying in wait to kill him"[54] and the king did not dare to go out of the city; but "Adados [Ben-Hadad] was unsuccessful in his plot" and "decided to fight openly".

The grove of Jezebel (Gubla) provided the bands of the king of Damascus with hiding places. The grove is mentioned in the Scriptures: "And Ahab made a grove" (I Kings 16:33). The garden of herbs in the vineyard of Naboth was part of the grove. Probably because of the scarcity of water during a number of years, the grove had withered. The king decided to cut it down for reasons of safety, being afraid for his life. He wrote to the pharaoh:

> LETTER 91: He [Abdi-Ashirta] sought to take Gubla, and I myself have felled my groves.

It may be that we have here an explanation of why the vineyard of Naboth, planted by Ahab and Jezebel to serve as a garden, is called only "the field" in the fatal epilogue to the drama of Naboth (II Kings 9:37).

The King of Damascus Is Killed While Lying Ill

After years of sieges and battles the day came when the adversary of the king of Samaria was afflicted with a grave disease.

> II KINGS 8:7-9 And Elisha came to Damascus; and Ben-Hadad the king of Syria was sick; and it was told him, saying, The man of God is come hither.
> And the king said unto Hazel . . . go . . . enquire of the Lord by him, saying, Shall I recover of this disease?
> So Hazael went to meet him . . . and stood before him, and said, Thy son Ben-Hadad king of Syria hath sent me to thee, saying, Shall I recover of this disease?

The question of whether the king of Damascus would recover from his sickness or die is repeated in the letter of the king of Sumur (Samaria) to "a chief" in Egypt:

> LETTER 95: Abdi-Ashirta is very sick, who knows but that he will die?

He died on his sickbed, but not from his disease; he was killed.[55]

A man in Gubla reported to the pharaoh:

> LETTER 101: They have indeed killed Abdi-Ashirta, whom the king [pharaoh] had placed over them.

A more detailed record is preserved in the Hebrew sources, where Hazael is named as the assassin.

II KINGS 8:15 And it came to pass on the morrow, that he took a thick cloth, and dipped it in water and spread it on his face, so that he died: and Hazael reigned in his stead.

Both records agree that Ben-Hadad (Abdi-Ashirta) was gravely ill, but he did not die from his disease; he met a violent death.

The el-Amarna letters provide us with additional information. Hazael (Aziru, Azaru) was a son of Ben-Hadad: "Aziru, a son of Abdi-Ashirta, is with his brothers in Dumasqa."[56] This information is not entirely new. Nicholas of Damascus, an author of the first century before the present era, wrote: "After the death of Adad (Hadad), his descendants reigned for ten generations, each of whom inherited from his father the name and the crown."[57] It was suggested that Nicholas of Damascus had been mistaken.[58] Now we have authentic documents proving that he was correct in declaring Hazael to be a son of Ben-Hadad. It is told in the Scriptures that Hazael "returned to his lord" in Damascus after greeting Elisha. Calling his father, the king, "lord" is entirely in keeping with the customs and the language.[59]

Hazael was not the lawful heir to the throne. Like Ahab, who had one queen but many wives and seventy sons by them in Samaria, Ben-Hadad probably had many sons. From an inscription of Shalmaneser it may be inferred that Hazael was born of a concubine: "Hazael, son of nobody, seized the throne."[60] In the el-Amarna letters he is persistently called slave.

But if Menander's story about Abd' Astartus, grandson of Hiram, is the story of Abdi-Ashirta, also called Abdu-Astarti,[61] the scriptural Ben-Hadad, then the king of Damascus was killed by the sons of his nurse. Sons of the royal nurse were held in honor, and occasionally could even lay claim to the throne.[62]

Aziru, also called Azaru[63] (Hazael or Azaelos of Josephus), was anxious to secure the consent of the pharaoh to his seizure of the throne of the murdered Abdi-Ashirta (Ben-Hadad). The assassination had not been accomplished in public, so Hazael could try to exculpate himself, pleading that he was wrongly accused of the crime. He wrote: "I have not sinned. Not the least have I done against the king, my lord. The king, my lord, knows the people who have committed crime."[64]

These words imply that the accusation had reached the Egyptian capital.

Hazael, "the Dog", Burns the Strongholds of Israel

The reign of Hazael (Azaru, Aziru) proved to be even more disastrous for the realm of Israel than the reign of the slain king of Damascus. The famine still plagued the people of Samaria in the days of King Hazael as in the days of his father. Samaria was in a state of almost constant seige and changed hands over a period of more than five years.[65] The king of Sumur (Samaria) wrote:

> LETTER 125: Aziru has again oppressed me. . . . My cities belong to Aziru, and he seeks after me. . . . What are the dogs, the sons of Abdi-Ashirta, that they act according to their heart's wish, and cause the cities of the king to go up in smoke?

This is almost exactly what Elisha said when, before the gates of Damascus, he announced to Hazael that he would be king over Syria:

> II KINGS 8:11–13 . . . and the man of God wept.
> And Hazael said, Why weepeth my lord? and he answered, Because I know the evil that thou wilt do unto the children of Israel: their strongholds wilt thou set on fire . . .
> And Hazael said, But what, is thy servant a dog, that he should do this great thing?

His expression, "is thy servant a dog . . .?" which accidentally escaped oblivion, was a typical figure of speech at the time of the el-Amarna letters. Many chieftains and governors concluded their letters with this sentence: "Is thy servant a dog that he shall not hear the words of the king, the lord?"

In referring to Aziru (Hazael), in many a letter the king of Sumur (Samaria) and other dispatchers of messages also use the word "dog". Sometimes the name of Aziru is not mentioned at all, only the word "dog". For example, in Letter 108 it is written, "because of the dog", and the pharaoh knew who was meant.

Another figure of speech out of the mouth of Hazael in the Scriptures is repeated in his letters. When Hazael went to meet the prophet Elisha he said:

> II KINGS 8:9 Thy son Ben-Hadad king of Syria hath sent me to thee.

In his letters to Dudu in Egypt Aziru wrote:

> LETTER 158: To Dudu, my lord, my father. Thus saith Aziri, thy son, thy servant.

He employed the expressions, "thy son", "thy servant", in letters and in conversation to show respect. In the only dialogue preserved in the Scriptures in which Hazael participates, there are three turns of speech that also appear in his letters. The context of the dialogue—the question of whether the king of Damascus would survive, and the statement that he, Hazael, the new king, would cause the cities (of Israel) to go up in smoke—is also preserved in the el-Amarna letters. It is therefore a precious example of the authenticity of the scriptural orations and dialogues.

At this point it is appropriate to quote one of the letters of Aziru (Hazael) to this mighty man in Egypt who is called Dudu. It will give us an idea of who was the secret force in the colonial office of Egypt which supported Abdi-Ashirta and his son in their adventurous campaigns in Syria and Palestine.

> LETTER 158: To Dudu, my lord, my father, Thus saith Aziri, thy son, thy servant: At the feet of my father I will fall down. May my father be well. . . . Behold, thou are there, my father, and whatever is the wish of Dudu, my father, write, and I will indeed perform it. Behold, thou art my father and my lord, and I am thy son. The lands of Amurri are thy lands, and my house is thy house, and all thy wish write. And, behold, I will perform thy wish. And, behold, thou sittest before the king, my lord, —— enemies have spoken words of slander to my father before the king, my lord. But do thou not admit them. And behold, mayest thou sit before the king, my lord, when I arise and words of slander against me not admit. . . . But if the king, my lord, does not love me but hates me, what shall I then say?

Who knows whether all the letters of the king of Sumur (Samaria) reached the eyes of Akhnaton? The king of Samaria complained that many of his messages went unanswered and did not receive due attention.

This Dudu, whose splendid sepulcher is preserved among the tombs of the courtiers of Akhnaton in Akhet-Aton (Tell el-Amarna), is thought to have been of Semitic origin. His reward at the hand of the Pharaoh was celebrated by some Asiatics, as the murals in the tomb witness.[66] These Asiatics were Syrians. The name Dudu is a Semitic name of the Kings period in Palestine. It appears in the Scriptures as Dodo (II Samuel 23:9, 24).

It is conceivable that this Dudu was a descendant of Hadad the Edomite, who was an adversary of Solomon and who married into the royal house of Egypt (I Kings 11:19).

Dudu, the chamberlain and the "chief mouthpiece of all the foreign lands", as he described himself in an inscription in his sepulchral hall, made matters easier for Aziru (Hazael), and some of the tablets of the king of Sumur (Samaria) waited twenty-seven centuries before receiving attention. Their author, desperate at the thought that his messages might not have been reported to the pharaoh, and fearing that he was carrying on a soliloquy, closed a letter with a bitter plea:

> LETTER 122: I in my solitude protect my right. . . . What should I do? Hear! I beg: refuse not. There are people in the presence of the king, or there are not? Hear me! Behold, so have I written to the palace; but thou hast not hearkened.

The words of Elisha announcing to Hazael his accession to the throne and his wars against Israel were fulfilled. Hazael (Aziru) oppressed Israel even more than did Ben-Hadad (Abdi-Ashirta). A tablet was sent to Pharaoh from Gubla (Jezreel):

> LETTER 127: When, formerly, Abdi-Ashratu [Abdi-Ashirta] marched up against me, I was mighty; but, behold, now my people are crushed.

In a series of raids Hazael depopulated the land.

> LETTER 109: But, verily, the sons of Abdi-Ashirta, the slave, the dog, have taken the cities of the king and the cities of his regents, according to their pleasure. . . . Under such circum-

stances my heart burned. ... And they strive for crime, outrage ...

The outrages of Hazael are described in the Scriptures: not only did he kill the men, but he crushed their children and ripped open their women with child (II Kings 8:12).

Another tablet was sent to the pharaoh from Gubla:

> LETTER 124: Aziru has taken all my cities, Gubla in her solitude is left to me. ... Behold, soldiers have gone up against Gubla ... And if he takes it, where shall I stand?

This oppression by Aziru (Hazael) lasted many years; the last letters of el-Amarna still tell of it; it endured as long as he lived and reigned.

A parallel account is in the following passage of the Scriptures:

> II KINGS 10:32 In those days the Lord began to cut Israel short: and Hazael smote them in all the coasts of Israel.

"All my cities" in the language of the king of Sumur (Samaria) of the el-Amarna letters are 'all the coasts of Israel" in the language of the Second Book of Kings.

The Last Letters of Ahab

"The people of Gubla and my house and my wife said to me: 'Attach yourself to the son of Abdi-Ashirta and let us make peace between us.' But I refused."[67]

So wrote the king "whom Jezebel his wife stirred up" (I Kings 21:25). The protests of the population he met with reprisals. "And they said: 'How long wilt thou kill us? Where wilt thou get people to live in the city?' " he himself reported. "When I was discouraged I wrenched from my heart a decision,"[68] and he went to Beirut to make friends with the king of that city and prepare a refuge in case Aziru (Hazael) drove him out of his city. The king of Samaria was a son-in-law of the king of Sidon (I Kings 16:31), and the king of Beirut was probably a relative of this family. When he departed his city closed its gates behind him. From his refuge he complained to the pharaoh. He reported that he had massacred his opponents, which was not the first time he had done so: one such instance is also

285

recorded in the Scriptures in the story of the condemning to death of all the members of the opposing party of Yahwists (I Kings 18:3ff.). He feared an insurrection of his people, and this he had already expressed repeatedly in his early and later letters: "I am afraid that the peasants will slay me", "I fear my peasants", "my peasants will rebel".[69]

He wrote as though he feared the curse of Elijah pronounced in the field of Naboth: "In the place where dogs licked the blood of Naboth shall dogs lick thy blood, even thine."

> LETTER 138: And behold, the people of Gubla have now written to me: ... Behold their hostile words: "Give it, his flowing blood."

This detail of his refuge in Beirut and later in Sidon,[70] apparently in the house of Jezebel's relatives for more than one year, is not recorded in the Scriptures, but it was this absence that gave rise to the rumors that he was dead. "They said: 'Our lord is indeed dead?' " "They had said: 'Rib-Addi is dead, and we are out of his power.' "[71] This episode and the belief that the king had died may have contributed to the confusion of a later chronicler, and may have been one reason for the many chronological and dynastic contradictions between the different versions of the Scriptures.

Here the letters of Rib-Addi, the scriptural Ahab, cease. He, "an old man", asked, with very little hope of gaining the attention of the Egyptian suzerain, whether the pharaoh would assign "Buruzilim" to him as a residence.[72] Did he mean Jerusalem, and did his scribe merely show his ignorance, as in many other letters of the el-Amarna collection where the names of cities and persons are spelled arbitrarily? Or is it that the Hebrew preposition "in" (b' in Hebrew) remained attached to the name of the city written in cuneiform?

Jehoshaphat had died only shortly before. Apparently the pharaoh gave the kingdom of Israel to Jehoram, son of Jehoshaphat and son-in-law of Ahab; it seems that Jehoram took it upon himself to restore order in the northern kingdom, leaving his young son Ahaziah in Jerusalem.[73]

The last letter of Rib-Addi ends with the admonition: "When I die, my sons, the servants of the king, will still live, and they will write to the king: 'O bring us back into our city.' "

The feared revolt was led by the long-suppressed Yahwist party.

The conspiracy in the army, which opposed Hazael, broke out, and the revolt struck Jezreel. Jehu drove furiously to Jezreel and killed Jehoram, previously wounded in a battle with the Syrians, and Ahaziah, who had come from Jerusalem to visit him. Jezebel was thrown out of the window, and Jehu's horses trod upon her, and her flesh was torn by dogs in the field of Naboth, leaving for burial only the skull and the feet and the palms of her hands.

The seventy sons of Ahab who dwelt in Samaria were also put to death, and their heads were carried in baskets to Jezreel (II Kings 10:7).

The prince of Beirut informed the pharaoh that the children of the king of Gubla and Sumur (Letter 142) were given by his brother into the hands of offenders. From the Scriptures we know that Jehu killed the kin of Ahab in Samaria and Jezreel. "So Jehu slew all the remainder of the house of Ahab in Jezreel" (II Kings 10:11).

The sixty-five letters of Rib-Addi, king of Sumur and Gubla (Samaria and Jezebel-Jezreel), written to the pharaoh Amenhotep III, the pharaoh Akhnaton, and the governor Aman-appa, disclose the human nature of their author; he was a man with a heavy heart, sad and worried. There is not one cheerful line in his letters. It is true that the events of his time in his land justified this state of mind. No one else among all the correspondents of the el-Amarna period wrote such distressed letters. "What shall I do in my solitude? Behold, this I ask day and night." He impressed his people, too, by his melancholy.

"And Ahab came into his house heavy and displeased. . . . And he laid him down upon his bed, and turned away his face, and would eat no bread"—so when Naboth refused him the vineyard (I Kings 21:4). "He rent his clothes, and put sackcloth upon his flesh, and fasted, and lay in sackcloth, and went softly"—so when Elijah cursed him and Jezebel for what had been done to Naboth, who was sentenced to death by stoning on the testimony of false witnesses (I Kings 21:27). "And the king of Israel went to his house heavy and displeased, and came to Samaria"—so when he heard that he had failed by letting Ben-Hadad go away with a covenant (I Kings 20:43).

The Scriptures condemned this idolator who was possessed by religious zeal: "And Ahab the son of Omri did evil in the sight of the Lord above all that were before him" (I Kings 16:30).

The rabbinical tradition condemned Ahab for his idolatry and the persecution of the prophets of Yahweh, but could not close its eyes to his patriotism and the deep emotions of his perturbed soul. "In the heavenly court of justice, at Ahab's trial, the accusing witnesses and his defenders exactly balanced each other in number and statements, until the spirit of Naboth appeared and turned the scale against Ahab."[74]

A minute study of this period might create the false impression that Ben-Hadad and Hazael were the principal adversaries in the long Jewish history, and not just two villains who might be found in any generation. However, they became the national heroes of the Syrian history. It is a fact that in Syria, in the time of Josephus Flavius, nine hundred years after the time I describe, these two kings were honored as great builders and great conquerors and great characters, and the people of Damascus cherished the memory of these national heroes and saints. Josephus, after describing the act of murder, wrote: "Adados and Azaelos who ruled after him are to this day honored as gods because of their benefactions and the building of temples with which they adorned the city of Damascus. And they [the Syrians and the people of Damascus] have processions every day in honour of these kings and glory in their antiquity. . . ."[75] The Arabs of Damascus feasted in memory of Ben-Hadad and his murderous son. Their rival, Ahab, went down in the memory of his people as a great sinner in Israel.

The attention given to the historical events and even to some minor happenings of the period with which this chapter deals is due mainly to the fact that the priceless el-Amarna letters were written at that time, and they provided rich material for parallels. From the standpoint of the biblical narrative, the importance lent to the period is enhanced, for the letters deal with the mist-shrouded time of Elijah and Elisha, the seers.

NOTES

1. The Blue Nile is fed by tropical rains that fall in Ethiopia, and by melted snow from its mountains.
2. Letter 7.
3. Ginzberg, *Legends*, IV, 190–191.
4. Letter 105.

5. Letter 112.

6. Letter 117.

7. Letter 118.

8. Letter 117.

9. Letter 130.

10. Letter 114.

11. On Dibon, cf. Numbers 21:30; Joshua 13:9.

12. Translation by S. R. Driver. A modern translation by W. F. Albright differs in a few details ("Palestine Inscriptions" in *Ancient Near Eastern Texts*, ed. Pritchard).

13. Letter 91.

14. Letters 72, 102.

15. Ambi and Ammia, not far from Sumur, were placed by historians close to the coast and identified with Enfe near Tripolis. Both names, Ambi and Ammia were assumed to be two names for one place. See Mercer, *Tell el-Amarna Tablets*, p. 269.

16. "Rabbath of the children of Ammon" (Deuteronomy 3:11).

17. Cf. E. Dussaud, *Les Monuments palestiniens et judaïques* (Musée du Louvre, Paris, 1912), p. 13: *Mésa ne nous dit pas, mais cela résulte nettement des renseignements bibliques, que le secret de sa fortune tint à l'habilité avec laquelle il sut profiter des revers qu'éprouva Israel après la mort d'Achab et dont l'agent le plus actif fut le roi de Damas. Il n'est pas douteux que la région, au nord de Dibon, fut occupée de nouveau par Mésa en accord avec Hazaël, roi de Damas, et peut-être sous la suzeraineté de ce dernier.*"

18. Letter 71.

19. Letter 76.

20. Letter 79.

21. Levy, *Wörterbuch über die Talmudim und Midrashim*.

22. This well-known word is written with two *khaf* letters; in the inscription of Mesha the letters are *khuf* and *heth*, the other characters for *k* and *kh*. But in the same inscription of Mesha the word "city" is *kar*, written also with *khuf*, and it is probable that the original writing of Kerakh and Karkhah with *khuf* is correct, being derived from *kar*, the city. *Kar* for city is of Carian origin. Similarly, today we write both "Carians" and "Karians".

23. Letters 185 and 189.

24. *Me'ever haiam me'aram* gives no support to the King James version, "from beyond the sea on this side Syria".

25. In the translation of Knudtzon, and likewise in that of Mercer, the period is placed differently: "There is hostility to me. As far as the lands of Seeri and even to Gintikirmil there is peace to all regents, and to me there is hostility." Comparing this text with the biblical text, we see that the new period precedes the words: "There is peace."

26. Letter 290.

27. Letter 273.

28. Habiru "is also written with an ideogram signifying 'cutthroats'," C. J. Gadd, *The Fall of Nineveh* (London, 1923).

29. This letter of Suwardata confirms what has been established on the basis of other considerations—that *sa-gaz* ("bandits", "pillagers") and Habiru were the same, or if a difference is to be drawn between these two denominations, the invaders themselves were not different.

30. The usual identification is with Kila, about eight miles northwest of Hebron "but in the letter 289 it seems more closely associated with Bethshan and Shechem" (Mercer, *Tell el-Amarna Tablets*, p. 694).

31. After the death of Jehoshaphat "Libnah revolted" (II Chronicles 21:10) from under the hand of his successor. It appears that the disturbing activities of a certain Labaia at the time this king of Jerusalem was still alive were a preliminary phase of that revolt. It is supposed that Labaia was not only the name of a chief but also of a group of the inhabitants (Weber in Knudtzon *Die El-Amarna-Tafeln*, p. 1558). Libnah, also called Labina (Josephus), was situated between Makkedah and Lachish (Joshua 10:28f.). Labia (Joseph), was situated between Makkedah and Lachish (Joshua 10:28f.). Labaia at one time approached Makkedah (Letter 244) and another time attacked Gezer north of Lachish (Letter 254). When Labaia sacked Gezer in the south he wrote to the pharaoh that this was "his only crime" and that it was not true that he disobeyed the deputy or refused tribute.

32. He wrote also to another dignitary in Egypt (Haia), explaining his plight and asking that horses and infantry be sent (Letter 71).

33. Letter 74.

34. *Beth-Ninib* was a town in Palestine (Letter 290); Letter 74 may refer to this town. However, see Weber, in Knudtzon, p. 1160.

35. He ordered seven baths in the Jordan. The Jordan is rich in sulfur, potassium, and magnesium, which enter the river from the springs at the Sea of Galilee and form the deposits of the Dead Sea, where the water evaporates and the salts remain.

36. Or "thirty and two kings that helped him" (I Kings 20:16).

37. As, for example, Dudu, referred to later in this chapter.

38. Mercer, *Tell el-Almarna Tablets*, p. 297. See also Weber, in Knudtzon, *Die El-Amarna-Tafeln*, p. 1068.

39. Letter 83.

40. Letter 83.

41. Letter 106.

42. Letter 106.

43. Letter 118.

44. Letter 102.

45. Letter 85.

46. Letter 112.

47. II Kings 5:2.

48. Letter 83.

49. Letter 296.

50. The biblical Naaman could be a cognomen; it means "truthful".

51. Letter 273.

52. This translation should be substituted for "Mistress of Lions" (see Mercer, *Tell el-Amarna Tablets*, note to Letter 273); the ideogram for "lion" is *neše*, but this ideogram could have been used to represent the phonetically similar word in Hebrew which means "sign" or "miracle".

53. *Jewish Antiquities*, IX, 60 (trans. R. Marcus).

54. Ibid., IX, 51.

55. "*Abda-Ashirta ist aber nicht eines natürlichen Todes infolge dieser Erkrankung gestorben, sondern ermordet worden, und zwar offenbar von Amurru-Leuten selbst. . . . Die wahren Umstände . . . sind leider infolge der Lückenhaftigkeit des Textes nicht deutlich zu erkennen.*" Weber, in Knudtzon, *Die El-Amarna-Tafeln*, p. 1132. "*Aus 105,25f. ist wohl zu entnehmen, dass die Söhne des Abdi-Ashirta bei seinem gewaltsamen Ende kaum ganz unbeteiligt sein können.*" Ibid., p. 1198.

56. Letter 107.

57. Josephus (*Jewish Antiquities*, VII, 102), who cited Nicholas of Damascus.

58. "*En tout cas, il paraît difficile d'admettre que la dynastie des Hadad ait duré dix générations, car, en 845, Hazaël assassina Ben-Hadad II et fonda une dynastie nouvelle.*" Th. Reinach, *Textes*, p. 80.

59. It was also an Egyptian usage. See Erman, A., and Blackman, A. M., *The Literature of the Ancient Egyptians* (London, 1927), p. 42.

60. Luckenbill, *Ancient Records of Assyria*, I, Sec. 681.

61. Letter 64. See section "The Five Kings", above.

62. Cf. H. W. Helck, "Der Einfluss der Militärführer in der 18. ägyptischen Dynastie", *Untersuchungen zur Geschichte und Altertumskunde Aegyptens*, 14 (1939), 66–70. In Egypt this role of the milk brothers is noticeable only during the Eighteenth Dynasty.

63. Letter 117.

64. Letter 157.

65. Letter 106.

66. N. de Garis Davies, *The Rock Tombs of El-Amarna*, Vol. VI, *The Tomb of Tutu*. J. D. S. Pendlebury, *Tell el-Amarna*, p. 51.

67. Letter 136.

68. Ibid.

69. Letters 77, 117, 130.

70. Letter 162, by the pharaoh. Jezebel was from Sidon.

71. Letter 138.

72. Letter 137.

73. See section "Ahab or Jehoram", above.

74. Ginzberg, *Legends*, IV, 187.

75. Josephus, *Jewish Antiquities*, IX, 92–94.

CHAPTER EIGHT

The el-Amarna Letters (Concluded)

Iarimuta

On reading the letters of the king of Sumur (Samaria), we are impressed by the frequent mention of a place called Iarimuta, also Rimuta,[1] and by the role it played in the aspirations of the king. This place is referred to only in his letters, eighteen times in thirteen letters. From there his people obtained grain in the years of the famine in return for heavy payments, trading of household implements, and even giving children into slavery. In previous times grain from Iarimuta had belonged by right to Sumur (Samaria). The king wrote to the pharaoh asking his command: "May it seem good to the king, my lord, that grain be given, produce of the land Iarimuta."[2] In like manner he wrote to the governor Aman-appa: "Say to thy lord that there should be given to his servant the produce of the land of Iarimuta, as was formerly given to Sumura."[3]

The king of Sumur (Samaria) had a claim on this place, and he submitted the claim to three Egyptian deputies, two of them being Aman-appa and Ianhama, "and they have recognized my right". The dispute was with the king of Damascus over Iarimuta.

> LETTER 105: Because of that which belongs to me ... he has become hostile to me. ... He oppressed —— and he has taken —— he has become hostile to me on account of —— to take grain for nourishment from Iarimuta that I die not.

For a time Iarimuta was held by the king of Sumur and Gubla, and his army was stationed there, because he wrote to the pharaoh: "Say to Ianhamu that (he) take money and clothing for the people of Gubla in Iarimuta."

The conflict over Iarimuta outlived Abdi-Ashirta (Ben-Hadad) and was still acute in the days of Aziru (Hazael).

The king of Sumur (Samaria) required of the pharaoh military assistance to compel the local chief, who had aligned himself with the king of Damascus, to give provisions to the peasants and soldiers.

LETTER 114: Formerly, my peasants have provided provisions from the land of Iarimuta. But, behold, Iapa-Addi did not permit them to go for provisions for the garrison. So, then, let the king send thy archers. ... Care for me. Who would be a friend if I should die? Is not Iapa-Addi with Aziru?

During his entire reign he insisted that Iarimuta's grain, as in former times, must belong to him and his people.

LETTER 125: Formerly, a royal garrison was with me, and the king gave grain from Iarimuta for their provisions. But, behold, now Aziru has again oppressed me.

There have been various conjectures about this place: that it was in Goshen in Egypt,[4] this according with the theory that Ianhama was the scriptural Joseph; that Iarimuta was the ancient name for Philistia and Sharon;[5] that it was in the plain of Antioch.[6] "Iarimuta's location is uncertain."[7]

If this place interested the king of Sumur (Samaria) so much that he involved himself in a prolonged conflict over it, claiming its produce, we may expect that the Scriptures will give the answer to the question: What place was it?

I KINGS 22:3 And the king of Israel said unto his servants: Know ye that Ramoth in Gilead is ours, and we be still, and take it not out of the hand of the king of Syria?

Ramoth in Gilead played a most important role in the wars of the king of Israel. He conferred with his prophets: "Shall we go against Ramoth-Gilead to battle, or shall we forbear?" His governor, Amon, supported him in this undertaking (I Kings 22:26). Jehoshaphat accompanied him to battle in Ramoth, as already narrated; there, according to one version, Ahab was killed, according to another he was only wounded.

At a later date the king of Israel "had kept Ramoth-Gilead, he and all Israel, because of Hazael king of Syria" (II Kings 9:14);[8] on this front Jehu was anointed by the messenger of Elisha (II Kings 9:4).

Josephus Flavius gave the name of the place for which Ahab battled with the Syrians as Aramatha in Galadene.[9]

Ramoth of the Bible and Aramatha of Josephus is Iarimuta or Rimuta of the el-Amarna letters.

The battles and wars waged for this place in the years when Samaria suffered from famine are well explained by the letters: the high land of Gilead was the breadbasket of the entire region, and famine had not touched it.

Ramoth in Gilead is prominently mentioned in the Scriptures during the time of Ahab-Jehoram and Jehoshaphat, and only during this time.[10]

Samaria (Sumur) under the Oligarchs

During the period of the el-Amarna letters Sumur (Samaria), though the center of Egyptian administration, was almost constantly beleaguered by the kings of Damascus. The two sieges of Samaria, and its liberation from the first by "the young men of the governors of the provinces" and from a long second siege as the result of a rumor of the arrival of an Egyptian army, with the parallels from the Scriptures and the el-Amarna letters, have been recorded in earlier pages.

Besides the king and the governor, the elders also exercised authority in the city. To their decision the king submitted the fate of the city when he received the ultimatum of the king of Damascus during the first siege (I Kings 20:7).

For the larger part of the period of the el-Amarna letters the king of Sumur lived in his second residence, and from there wrote most of his letters pleading for aid for his capital. At the beginning of that period the governor Aman-appa (the biblical governor Amon) relinquished his permanent post in Sumur, and for the greater part of the time was in Egypt. Ianhama's seat was in Damascus.

Because of the anarchical conditions that prevailed as a result of the many vicissitudes and frequent sieges, the elders of Sumur became more influential, and for some time were the only authority in the city. Aziru (Hazael), when his attempt to enter Sumur failed, wrote in a letter:

LETTER 157: But the chief men of Sumur have not admitted me.

To these chief men or "rulers and elders" of Samaria Jehu wrote, challenging them to choose one of the sons of Ahab to be king or yield to him (II Kings 10).

This time the oligarchs became frightened and admitted Jehu into the city.

The "King's City", Sumur

The city of Samaria was a "king's city", a residence built as the center of Egyptian administration of the Asiatic province. Pharaoh used to send silver to Samaria in the days of Omri, who built it.

LETTER 26: Formerly, however, there was sent to my father from the great palace silver. . . .

Egypt received meager, if any, tribute in return. The pharaoh also provided Samaria with chariots.

In the city was a palace, and the "king's house" of Sumur is often mentioned in the el-Amarna letters; nominally it was the residence of the pharaoh, but actually it was the residence of his regent—the vassal king—and probably also of the Egyptian deputy attached to the vassal. The "king's house" is referred to also in the stele of Mesha, who repaired it. The palace built by Ahab in Samaria is mentioned in the Scriptures; it was overlaid with ivory. The words of Mesha, "I have cut its cuttings", might refer to the ivory carvings on the palace. The ivory of the Samarian palace of Omri and Ahab was found in abundance during the excavations.[11] More will be said about it in a later section of this chapter.

The city of Samaria was surrounded by a wall. This wall is mentioned both in the Scriptures and in the letters of el-Amarna. It is also referred to in the Mesha stele.

The place before the gate was the scene of the two-king conference with the prophets; the thrones were placed "at the entering in of the gate of Samaria" (II Chronicles 18:9). The gate was also the station of the officer appointed "to have the charge of the gate" (II Kings 7:17). He had to collect silver in payment for the barley and flour abandoned by the Syrians who had fled from beneath the walls of Samaria. The officer was trodden to death. The custom of taking money at the gate is also mentioned in a letter by the king of Sumur: "All my gates have taken copper."[12] The gate of Samaria, which was an important place—a kind of forum for the people—is referred to in the Mesha inscription, as well as in the Scriptures and the el-Amarna letters.

In the city was a mound or acropolis called Ophel. It is mentioned in the Scriptures (II Kings 5:24) and in the stele of Mesha.

The excavations at Samaria disclosed two portions of the ancient wall, the place of the gate (the city wall had only one gate), and buildings of a palatial type.[13]

Since Mesh (Mesha) occupied Samaria in partnership with the king of Damascus, in his stele he ascribed the building activity in the city to himself, the work being done by the Israelite captives. The king of Damascus wrote similarly to a dignitary in Egypt, after all the devastation that the city had suffered in wars and sieges: "I built Sumur . . ."[14] meaning that he, with the work of the Israelite captives, repaired the walls of the palace and of other buildings. This privilege to build in Samaria was highly esteemed, and is mentioned in the treaty between Ben-Hadad and Ahab. After the defeat of the Syrians at Aphek, Ben-Hadad told the king of Samaria:

> I KINGS 20:34 Thou shalt make streets for thee in Damascus, as my father made in Samaria.

When, for a time, the city fell into his hands, he resumed—together with the rebel Mesh (King Mesha)—the building activity there.

The next king of Damascus, Aziru, the scriptural Hazael, during the time he held Sumur (Samaria), promised: "I will build it up Sumur now. . . . My lord, I will now build Sumur in haste,"[15] and again. "I have not built Sumur. But in one year I will build Zumur [Sumur]. . . ."[16]

Shalmaneser III Expels King Nikmed

According to the reconstruction of history presented here, the el-Amarna letters were written, not at the beginning of the fourteenth century, but during the three decades −870 to −840. In about the year −858, Shalmaneser III became king of Assyria. Later he became king of Babylonia, too, and made offerings in Babylon, Borsippa, and Kutha.[17] He undertook a number of devastating raids on the Phoenician shore and on northern Syria. Placing the el-Amarna letters in the time of Jehoshaphat, king of Jerusalem, and Ahab, king of Samaria, and Shalmaneser, their contemporary, challenges me to demonstrate agreement not only between the letters and the Books of Kings and Chronicles, but also between the letters and the Assyrian inscriptions. In his annals Shalmaneser gave an account of his wars, the larger part of which were carried on in Syria.

In a letter written by Abimilki, king of Tyre, and found in Tell el-Amarna, it is reported: "And fire has consumed Ugarit, the city of the king; half of it is consumed, and its other half is not; and the people of the army of Hatti are not there [any more]." In the section "End of Ugarit", in Chapter Five, we asked, Who was the invader? We quoted also a proclamation found in Ras Shamra-Ugarit in which the invading king decreed that "the Jaman [Ionians], the people of Didyme, the Khar [Carians], the Cypriotes, all foreigners, together with the king Nikmed," should be expelled from Ugarit. The opening portion of the proclamation, which might have revealed the name of the king who expelled Nikmed, is missing. Some intimation has been found that Nikmed was expelled by the Babylonians,"[18] who are called here Hatti. This suggestion is not far from the truth, since Babylon was incorporated by Shalmaneser into his empire.

We are curious to know whether Shalmaneser left any written record of his conquest of Ugarit. And in fact we find the following entry twice repeated in his annals:

> Year four. To the cities of Nikdime [and] Nikdiera I drew near. They became frightened at my mighty, awe-inspiring weapons and my grim warfare, [and] cast themselves upon the sea in wicker [?] boats——. I followed after them in boats of——, fought a great battle on the sea, defeated them, and with their blood I dyed the sea like wool.[19]

The city of Nikdime appears to be the city of Nikmed. Cities were named in honor of their kings, and in this case it is put clearly, "of Nikdime". The translator of this record also explained the words "city of Nikdime" by a gloss: "personal name". The inversion of two consonants, especially in personal names of foreign origin, is very common among oriental peoples; thus, not far from Ras Shamra is Iskanderun, a city named for Alexander. The city of Nikdime, like the city of Nikmed, was situated close to the sea. It may be that even Shalmaneser's simile of the sea looking like dyed wool was inspired by the trade of Ugarit-Ras Shamra. Factories for dyeing wool had been established in Ugarit, and stores of crushed shells prepared for the extraction of dye were found there. A written order to tint three loads of wool also came to light.

Some time before his expulsion Nikmed, together with Suppiluliuma, a contemporary city king, contributed to the goddess of the city of Arne. We shall come across these names in the annals of Shalmaneser, too.

Shalmaneser III is Opposed by a Syrian Coalition under Biridri (Biridia), the Commandant of Megiddo

Shalmaneser relates that in his sixth year, two years after he drove King Nikdem into the sea, a prince named Biridri, helped by a coalition of twelve princes, opposed him at Karkar.[20] Among Biridri's allies were Ahab, the prince of Israel ("with two thousand chariots and ten thousand warriors"), the city of Irqata (no prince is named), the prince of Arvad, Matinu-Bali, the prince of Usa (not named), and the prince of Siana, Adunu-Bali.

The inscription of Shalmaneser does not say that the allied princes, Ahab among them, participated personally in the battle of Karkar; it merely states: "These twelve kings he [Biridri] brought to his support." We meet some of the same princes in the el-Amarna letters. They wrote to the pharaoh that they were holding their garrisons in readiness to take a stand against the invading king of Hatti and some of them—from northern Syria, which was more immediately threatened than Palestine—might have taken part personally in the battle.

The city of Irqata wrote to the pharaoh:

> LETTER 100: Thus saith Irqata and the people of its inheritance ... Let the heart of the king, the lord, know that we protect Irqata for him. . . . Let the breath of the king not depart from us

We have closed the gate until the breath of the king comes to us. Powerful is the hostility against us, very powerful indeed.

In a letter from Rib-Addi[21] it is reported: "Aduna —— of Irqata, mercenaries have killed." In the same letter he wrote about the wars of the king of Hatti in the northeast.

Prince Aduni of Siana of Shalmaneser's inscriptions was probably Aduna "of —— and of Irqata" of the el-Amarna letters. But if Aduni of Shalmaneser's war annals was not the prince of Irqata, then the failure of Shalmaneser to name the prince of Irqata is explained by the el-Amarna letter: the prince of Irqata was killed, and the city defended itself without choosing a new prince.

Two letters of Mut-Balu are preserved. He wrote to the pharaoh:

LETTER 255: Let the king, my lord, send caravans to Karaduniash. I will bring them in so that they be completely safe.

Since Karaduniash is Babylon[22] and the city of Mut-Balu is on the way to Babylon, we may guess that Mut-Balu of the el-Amarna leters was Matinu-Bali, the king of Arvad, mentioned in the Assyrian inscription of Shalmaneser. Mut-Balu did not say from where he wrote his letter. The Assyrian inscription, by calling Matinu-Bali the king of Arvad, connects Mut-Balu of the letters with that city. From Josephus, who quoted Menander,[23] we have the information that Metten-Baal was a grandson of Ithobal [Ethbaal] and a nephew of Jezebel.

Uzu (Usa) was near Tyre. Tyre was on an island near the shore. It did not have enough drinking water to withstand a siege. In those rainless years the king of Tyre asked the pharaoh to transfer the city of Uzu to the domain of Tyre.

LETTER 150: Let the king direct his attention towards his servant and give him Uzu in order that he may live, and in order that he may drink water.

In both sources, in the inscription of Shalmaneser and in the letters, we have met Irqata, Arvada, and Uzu as small principalities opposing the invader coming from the north.

Megiddo was the strong military base behind these outposts. It had been fortified by Thutmose III, whose campaign and victory at Megiddo were emphasized in his annals. The city was a barrier in the path of the armies coming from the north. Its garrison also guarded the plain of Esdraelon (Jezreel).

At the time of the el-Amarna letters a captain by the name of Biridia was in command of Megiddo. It is obvious that a great responsibility was placed on him at this time of uneasiness. Eight of his letters have been found. Judged by them, he was a daring and faithful soldier.

> LETTER 243: And behold, I protect Makida, the city of the king, my lord, day and night. By day I guard from the fields. With chariots and soldiers I protect the walls of the king, my Lord.

It was to Megiddo that Ahaziah, the grandson of Jehoshaphat, tried to escape when the rebellion of Jehu struck him unawares in Jezreel. Megiddo, as the letters reveal, was a fortress, which Jehu would not have dared to enter in his pursuit of Ahaziah. But Ahaziah did not succeed in escaping: he was trapped and killed in his flight to Megiddo (II Kings 9:27).

The walls of Megiddo, once a formidable structure, when excavated were found to be very similar in construction and workmanship to the walls of Ahab's palace at Samaria, dating from the ninth century.[24]

Biridia, as commander of the most important fortress, was the man to lead the coalition of the vassal kings against the "king of Hatti".[25] The rank and position of Biridia of the el-Amarna letters correspond to those of Biridri of the Shalmaneser inscription. The slightly different spelling of the names is due to the fact that not only personal but even geographical names were spelled in the letters in different ways: the same Biridia (in one instance he wrote his name Biridi) announced to the pharaoh that he was defending Makida; another time he wrote that he was defending Magiidda. There are many similar examples in the letters.

Biridia, on watch at the stronghold of Megiddo, met the king of Assyria, the enemy of the pharaoh, and it was the chariots under his command that Shalmaneser took captive.

There were one thousand Musri soldiers in the army under Biridri, who joined battle with the army of Shalmaneser at Karkar, according to the record of Shalmaneser. Musri is the Assyrian name for Egypt (Mizraim in Hebrew). As it seemed unreasonable that the Egyptian king should have sent only one tenth the soldiers that Ahab sent, Musri was supposed to denote some realm other than Egypt. One theory[26] put it on the Sinai Peninsula; another[27] located it in northern Syria or in eastern Anatolia.

The letters of el-Amarna, when assigned to their proper time, make these theories superfluous. The presence of a small contingent of Egyptian soldiers in the allied army under Biridia is in accord with the contents of the el-Amarna letters. Biridia also sent regular reports to the pharaoh about his preparations to meet the king of Hatti on the battlefield.

In a number of letters the king of Sumur reported to the pharaoh the danger impending from the king of Hatti. In an early letter he wrote: "Let the king, my lord, know that the king of Hatti has overcome all lands, which belonged to the king of Mitta or the king of Nahma—the land of the great kings. Abdi-Ashirta, the slave, the dog, has gone with him."[28] The successes of the king of Hatti in Mitanni (Mitta) were only temporary, for King Tushratta of Mitanni, the father-in-law of the pharaoh, was able to send, among other numerous and rich presents, some of "the booty of the land of Hatti", and by this he let his son-in-law know that he was not defeated. The records of Shalmaneser, aside from his conquest of Babylon at the time of a dynastic war between two Babylonian princes, do not chronicle any decisive victory or permanent territorial acquisition in the east.

Almost every year during the first two decades of his reign Shalmaneser renewed his southern campaigns into Syria. In the el-Amarna correspondence the vassals of Egypt in northern Syria reported to the pharaoh the raids of the king of Hatti.

These raids were accompanied by holocausts. In one of his later letters Rib-Addi wrote: "I have heard from [about] the Hatti people that they burn the lands with fire. I have repeatedly written ... All lands of the king, my lord, are conquered ... they bring soldiers from the Hatti lands to conquer Gubla."[29] Rib-Addi in this letter expressed his fear that the attack would be directed against him. His information that the king of Hatti burned cities to the ground was true. Shalmaneser himself wrote: "I destroyed, I devastated, I burned with fire," a statement he often repeated; after his sixteenth march into Syria he perpetuated his deeds with these words: "Countless cities I destroyed, I devastated, I burned with fire."

The same was reported in many letters of Syrian city-princes to the pharaoh.

Rib-Addi in a still later letter[30] wrote about the exploits of the king of Hatti, but these lines are destroyed, and we can read among the defaced characters only the words "king of the Hatti lands".

Shalmaneser, the Assyrian king of the ninth century, was, according to this reconstruction of history, the king of Hatti of the el-Amarna letters. After he secured the domain he inherited from his father, Assurbanipal, and increased it by the conquest of Babylon and of other areas, he wrote:

> The land of Hatti to its farthest border I brought under my sway. From the source of the Tigris to the source of the Euphrates my hands conquered.[31]

Hatti is here apparently a broad geographical designation, not an ethnographical name.

When the Syrians under the walls of Samaria thought they heard an army approaching to relieve the city, they imagined that it was either the army of Egypt or of Hatti (II Kings 7:6). The latter means, apparently, the army of Shalmaneser. He put all other countries in a state of defense, and the pharaoh was regarded as his only real rival. Their rivalry in matters of Syria is perfectly reflected in the el-Amarna letters.

Shalmaneser III Invades Amuru Land and Is Opposed by the King of Damascus

In the tenth year of his reign Shalmaneser fought against another coalition, again under Biridri (Biridia). When Biridri died, Hazael, exploiting the absence of an Egyptian governor, killed his own father.

In Egypt, Aziru (Hazael) was suspected of being on the side of "the king of Hatti"—as his father at one time had been[32]—for he received messengers from the king of Hatti. At the same time he begged the pharaoh to confirm his status as king of Damascus, the throne of which he had usurped. He was summoned to Egypt, but he repeatedly delayed his departure with the excuse that the movements of the king of Hatti had to be watched: "If the king of Hatti comes for hostility against me, then, O king, my lord, give me soldiers and chariots for my assistance." But he was questioned: "Why hast thou taken care of the messengers of the king of Hatti? But thou hast not taken care of my messengers?" A long and interesting letter from the pharaoh to him is preserved. The royal ax would cut off his head and the heads of his brothers[33] if he took the side of the enemies of the pharaoh. But he would be in favor if faithful (Letter 162).

Aziru (Hazael) was "exceedingly glad" with "the pleasing and good" words of the pharaoh, but he could not comply with the invitation to come to Egypt. His double-dealing angered the king of Hatti too. Aziru (Hazael) wrote to his protector, Dudu: "But, O my lord, the king of Hatte is come to Nuhasse! Therefore I cannot come. May the king of Hatte depart." In a letter to the pharaoh he conceded that he had "arrived formerly to the king from [of] Hatte",[34] but now he looks only in the face of the sun, his lord (Akhnaton). "And the king of Hatte comes to Amurri, the land of the king, my lord ... for now he dwells in Nuhasse, two one-day's journey to Tunip,[35] and I am afraid that he will oppress Tunip. Let him depart!"[36] "I am afraid of him; I watch lest he come up to Amurri ... So I fear him."[37]

We must now look into the annals of Shalmaneser to see whether he really came to Amurri, and whether he threatened Hazael (Aziru).

In the letters of the kings of the Syrian cities we read about the awe Shalmaneser inspired in that country; it is also mentioned in his own annals: "The kings of the land of Amuru, all of them, became terrified at the approach of my mighty, awe-inspiring weapons."[38] He calls the land of Syria by the name applied to it in the el-Amarna letters—Amuru (Amurri). The annals of Shalmaneser also verify what was said in the el-Amarna letters concerning Hazael (Aziru): "In my eighteenth year of reign, I crossed the Euphrates for the sixteenth time. Hazael of Aram [Amuru] came forth to battle." He wrote that Hazael made his stand at Mount Senir (Anti-Lebanon), "Mount Saniru, a mountain peak which is in front of Mount Lebanon." This strategic position of Hazael protected Tunip (Baalbek). In the battle Shalmaneser captured 1121 chariots. "As far as Damascus, his royal city, I advanced. His orchards I cut down."

Four years later Shalmaneser, according to his annals, went

"against the cities of Hazael of Aram [Amuru]" and "captured four of his cities".

The Phoenicians Leave for a New Home

From the annals of Shalmaneser we know that he subjugated the seacoast of Syria as far as Tyre. The Assyrian terror and the heavy tribute forced the Phoenicians to look for a new home, and a number of them left Tyre and other cities, went to the North African shore halfway along the Mediterranean, and founded there the colony of Çarthage.[39]

As we look for further mention of the military activities of Shalmaneser in the letters of el-Amarna, it is only reasonable that we should direct our attention to the letters written from Tyre. Abimilki, king of Tyre, wrote several tablets to the pharaoh, reassuring him:

> LETTER 147: Behold, I protect Tyre, the great city, for the king, my lord.

And again,

> LETTER 149: The king, my lord, has appointed me to protect Tyre, the handmaid of the king.

The pharaoh requested Abimilki to give him information about the cities of Syria. Abimilki wrote that Aziru (Hazael) had conquered Sumura (Samaria); that the city of Danuna in Syria was quiet (Shalmaneser passed through this city in his second year—he called it Dihnunu); Abimilki also wrote that half the city of Ugarit was destroyed by fire, that the other half of the city was despoiled, and that the soldiers of the Hatti army had left there.

Shalmaneser captured Ugarit, the city of Nikdem, in the fourth year of his reign. With the fall of Danuna and Ugarit to Shalmaneser, the position of Tyre was very much endangered, especially because of the feud between that city and Sidon as reflected in the el-Amarna letters, and also because of the pillaging tribes that approached Tyre from across the Jordan. Tyre on the rock was dependent on the coast not only for water supply but also for the wood necessary in shipbuilding.

In his plight Abimilki wrote that he was a "serf of tears" and pleaded for help. And when finally he received word that the fleet of the pharaoh would arrive at Tyre he sent out his mariners to meet the Egyptian navy. He was encouraged for a while and wrote again that he would defend the city. But the help was insufficient and too late, if it came at all.

In the eighteenth year of his reign Shalmaneser wrote that he had received "the tribute of the men of Tyre, Sidon, and of Jehu, of the house of Omri".

In his last letter to the pharaoh Abimilki changed the manner he had used in writing his previous letters. He used to tell the pharaoh that he, the august overlord, "thundereth in the heavens like Adad". It should be noted here that this was the same attribute that Shalmaneser in his inscriptions applied to himself: "I thundered like Adad, the Storm-god." Shalmeneser also wrote, "Shalmeneser, the mighty king, the sun of all peoples." Akhnaton and Shalmaneser both claimed to be the sun, and before long Abimilki knew that there were two suns, a violent one in the Two-Stream Land and an inactive one in Egypt.

Abimilki still paid respect to the pharaoh and called him "the eternal sun", but in his long last letter he no longer wrote to the pharaoh as to the sole sovereign, and it was clearly made known to the pharaoh that Abimilki also bowed before another lord.

> LETTER 155: The king is sun for ever. The king has commanded that there be given to his servant and to the servant of Shalmaiati breath, and that there be given water for him to drink. But they have not done what the king, my lord, commanded. They have not given it. So, then, let the king care for the servant of Shalmaiati that water be given for the sake of his life. Further: let my lord, the king, as there is no wood, no water, no straw, no earth, no place for the dead, let the king, my lord, care for the servant of Shalmaiati that life be given to him.

Bizarre conjectures were published to explain the meaning of that perplexing name Shalmaiati. Without taking into consideration the contemporaneous account of Shalmaneser, who in his eighteenth year received tribute from Tyre and Sidon, the change in the overlordship of Tyre is easily misunderstood. Shalmaiati could not be another name of the pharaoh, because Abimilki called himself

Pharaoh's servant *and* the servant of Shalmaiati. It was also pro posed[40] that the word be read "Maya-ati" and to identify it wit Meryt-aten, a daughter of Akhnaton. It has been admitted that th evidence for this and similar explanations is defective.[41] Then it wa said that Shalmaiati was a god, and a hypothesis was constructe according to which ancient Jerusalem and Tyre both worshiped a deit Salem and this deity was called Shalmaiati by Abimilki. It was ob served, however, that a sign which usually accompanied the name of deity was missing every time Shalmaiati was named in the letter. Who then, could have been? And yet, in disregard of both facts—tha Shalmaiati and the pharaoh could not be the same person, and tha Shalmaiati could not be a deity—it was supposed that Shalmaiati wa the name of a god, equivalent to the sun or Aton of Akhnaton, but tha here this god is invoked as an *alter ego* of Akhanaton, which make the sign of divinity unnecessary, and yet is an indication of a secon personality.[42]

Abimilki's last letter is not yet ended. What should we like to rea in it? Of course, about the departure of the people of Tyre and othe Phonecian cities on their ships from the grievous yoke of Shal maneser and from the recurrent drought, to look for new have along the shores of the Mediterranean. We know that it was in th days of Shalmaneser and in fear of him that the refugees from Tyr and other Phoenician cities fled and laid the foundations of Car thage.

This last letter of Abimilki closes with the words:

> LETTER 155: Let the king set his face towards his servant an Tyre the city of Shalmaiati ... Behold, the man of Berut [Beirut] has gone in a ship, and the man of Zidon goes away i two ships, and I go away with all thy ships and my whole city.

The dramatic meaning of this letter becomes evident. Abimilk wrote that he would desert his rocky island and evacuate the popu lation of Tyre. He begged the pharaoh to take care of what woul remain in the deserted city, which was bound by tribute to a kin invader.

Who Is the Dreaded "King of Hatti" of the el-Amarna Correspondence?

The king of Hatti, always feared and often mentioned in the letters of the Syrian princes, might well have been one of the correspondents of the el-Amarna collection. Although in continued conflict with Egypt, he never made open war against the pharaoh; at least the pharaoh never sent a strong army to the assistance of his Syrian vassals. It is probable, therefore, that they exchanged letters. It is generally accepted that Suppululiuma, of whom only one very amiable letter is preserved, was the feared king of Hatti. A number of generations later another Suppululiuma was "a king of Hatti" and therefore it appeared reasonable that Suppululiuma of the el-Amarna period also had been a great king of Hatti.

Actually, in the time of Shalmaneser III (in the ninth century) there lived a prince called Suppululiuma (Sapalulme), to whom Shalmaneser referred in his annals.[43] He could have been the author of the letter in the el-Amarna collection signed with his name.[44]

In a short and broken text from Ugarit referring to donations made to the goddess of the city of Arne, Prince Nikmed of Ugarit-Ras Shamra as well as Suppululiuma are mentioned.[45] Apparently Nikmed and Suppululiuma, too, donated to the goddess of Arne. Arne was not far from Ugarit and was captured by Shalmaneser III in one of his campaigns. "Against the cities of Arame (personal name) I drew near. Arne his royal city I captured."[46]

Besides establishing the identity of Shalmaneser III and the king of Hatti, the invader from the north at the time of the el-Amarna correspondence, there is a basis for suggesting an identification of the king of Hatti as one of the el-Amarna correspondents. We have already shown that he is mentioned in the letters of the king of Tyre by the name of Shalmaiati; but no letter is signed with this name.

Inasmuch as Shalmaneser III was king of Assyria, who also became king of Babylon by intrigue and conquest, and Burraburiash (Burnaburiash) is the name of the king who wrote from Babylon and who referred in his letters to the Assyrians as his subjects,[47] most probably Burraburiash is the *alter ego* of Shalmaneser the Assyrian. It is well known from many instances that in Nineveh and in Babylon the king used various names.[48]

Shalmaneser wrote of himself: "Shalmaneser, the mighty king, of the universe, the king without a rival king the autocrat ... who shatters the might of the princes of the whole world, who has smashed all of his foes like pots, the mighty hero, the unsparing."

It also seems that Shalmaneser alone in his time would have written in the manner of Burraburiash. Burraburiash was very haughty and wrote letters often bordering on insult. Under the pretext of an indisposition he refused to grant an audience to the messenger of the pharaoh ("his messenger has never eaten food nor drunk date wine in my presence"); then, although the pharaoh was the injured party, Burraburiash said that he was angry—"I poured forth my wrath toward my brother"—because he had not received the message of solicitude to which an indisposed sovereign was entitled. But when he was informed, through the close questioning of his own messenger, that the road was long and that in the short period of time Pharaoh could not have responded to the news of his indisposition, he added to the same letter: "I no longer poured forth to thee my wrath." Then he found fault with the pharaoh's gifts: "The forty minas of gold which they brought—when I put them in the furnace, did not come out full weight." He called on him "to make good his loss" of a caravan plundered in Syria. "Khinshi is thy land and its kings are thy servants," therefore "bind them, and the money which they have stolen, make good." He also wrote: "Do not retain my messenger, let him come quickly." He ordered various art objects to be made and sent to him as presents: "Let experts, who are with thee, make animals, either of land or of river, as if they were alive; the skin, may it be made as if it were alive." He also sent presents, but remarked: "For the mistress of thy house I have sent only twenty seal rings of beautiful lapis lazuli because she has not done anything for me. ..."

This manner of writing to the pharaoh is unique in the el-Amarna collection. All clues point to the identification of Burraburiash with Shalmaneser of the Assyrian inscriptions, with Shalmaiati of the letter of the king of Tyre, and with the "king of Hatti".

There exists a very long register of gifts sent by Akhnaton to Burraburiash.[49] After an extensive enumeration of objects of gold, silver, precious stones, and ivory, the list mentions animals, but this part is defaced, and only "ibex" (*durah*) is discernible. The list gives the impression of being tribute rather than a gift, and the letters of

308

Burraburiash indicate that gifts delivered in excessive quantities were not made on an entirely voluntary basis. Treasures found in the tomb of Tutankhamen, the son-in-law of Akhnaton, are unequaled by any other archaeological discovery in Egypt or elsewhere, but they are trifles compared with the presents Akhnaton sent to Burraburiash. The list opens with the words:

> LETTER 14: These are the objects which Naphururia, the great king, king of Egypt [sar Miisrii] sent to his brother Burnaburiash the great king, king of Karaduniash [Babylon].

Shalmaneser pictured on an obelisk the presentation of the tribute from various lands. One of these lands was Musri (Egypt): "Tribute of the land of Musri." In addition to gold, and more appreciated than this metal, there were rare animals described as camels with two humps, a river ox, and other land and water creatures, some of which are depicted. It is conceivable that these were figures of animals "of land and river" in compliance with the demand expressed in the quoted letter.

Briefly summarizing the parallels between the annals of Shalmaneser and the el-Amarna letters concerning the king of Hatti, we find the following:

The record of the fourth-year campaign of Shalmaneser against Nikdem (Nikmed) has its counterpart in a letter of the king of Tyre and in a proclamation found in Ugarit-Ras Shamra. Shalmaneser's repeated campaigns in Syria—he himself mentioned sixteen marches during the first eighteen years of his reign—are told in many of the el-Amarna letters, especially in those written from the cities in northern Syria.

Shalmaneser wrote that "the land of Hatti to its farthest border I brought under my sway". He wrote also, "I received gifts from all the kings of Hatti."[50] In the letters of el-Amarna the invading king is called "the king of Hatti". His destruction of numerous cities by fire is mentioned in his inscriptions and also in the letters of el-Amarna; vestiges of the conflagration are visible in the ruins of Ugarit. "The chilling terror" of his arms, which, according to his annals, filled the kings of the land of Amuru, finds expression in many letters.

Cities and princes who wrote to the pharaoh about the approach

of the invader are also mentioned in the annals of Shalmaneser a
those who, with the assistance of Egypt (Musri), fought against him
in the sixth year of his reign. They fought again in his eleventh yea
and once more in his fourteenth year. From the commander of thi
coalition, Biridri of the Assyrian annals, or Biridia of the letter
(written also Biridi), there are military reports to the pharaoh abou
his preparations for defense against the aggressor.

The tributary status to which Shalmaneser, in his eighteenth yea
reduced the cities of Tyre and Sidon, and the departure of thei
population on ships for a new haven, are described in the last lette
of the king of Tyre. Shalmaneser's march against Hazael in the sam
year is related in the former's annals and is reported in the letters o
Hazael (Aziru). Hazael's stand at Anti-Lebanon (Senir) is also re
corded both in the annals and in the el-Amarna letters. The letter
likewise mention the king of Hatti's (Shalmaneser's) wars in th
region of Mesopotamia.

The el-Amarna letters describe, and the obelisk of Shalmanese
exhibits, the presents of gold and rare animals sent from Egypt.

Idioms of the el-Amarna Letters

The el-Amarna letters were written in cuneiform characters in Baby
lonian (Akkadian) with many "Syrian idioms". Having been writ
ten in the days of Jehoshaphat, king of Jerusalem, their manner o
expression, a characteristic of every land and every age, should b
expected to resemble the manner of expression in the earlier book
of the prophets, for only a hundred years separate Jehoshaphat fron
Amos, the prophet.

The similarity of expression in the el-Amarna letters from Pale
stine and in the prophets and psalmists did not pass unnoticed, an
identical turns of speech were detected and stressed. Here are som
examples.[51]

Loyalty is expressed by the metaphor, "to lay the neck to the yok
and bear it", in the letters of Yakhtiri (Iahtiri) and Baal-miir
and similarly in Jeremiah 27:11f.

The submission of the enemy is described in the words, "to ea
dust", in the letter of the men of Irqata as well as in Isaiah 49:23.

The king's "face" is against a man, or the king "casts down" man's face, or he throws the man out of his hand—thus, in letters of Rib-Addi (Ahab) and in Genesis 19:21 and I Samuel 25:29. Rib-Addi's "face is friendly toward the king"; "he has directed his face towards the glory of the king, and would see his gracious face". "Biblical ideas of 'face' and 'presence' will be at once recalled."

"Just as Ikhnaton's [Akhnaton] hymn reminds us of Psalm 104, so Psalm 139:7f is suggested by the words of Tagi (Letter 264): 'As for us, consider! My two eyes are upon thee. If we go up into heaven (shamema), or if we descend into the earth, yet is our head (rushunu) in thine hands.' "

"The footstool of his feet" is an expression found in a letter and in Psalm 110. Akhi-Yawi writes, "A brother art thou and love is in thy bowels and in thy heart," similar to Jeremiah 4:19. "The city weeps and its tears run down, and there is none taking hold of our hand" (i.e. helper)—these words, written by the men of Dunip (Tunip), remind us of Lamentations 1:2 and Isaiah 42:6.

Rib-Addi's appeals to the pharaoh's good name contain turns of speech used also in Deuteronomy 9:27f. and in Joshua 7:9. When he wishes to say that he confessed his sins, he uses the words, "opened his sins",[53] an expression also found in Proverbs 28:13. When he writes that he will die "if there is not another heart" in the king, he uses an expression also found in I Samuel 10:9 and Ezekiel 11:19.

The king of Jerusalem wrote to Pharaoh that "because he has his name upon Jerusalem for ever, the city cannot be neglected", words that remind us of a passage in Jeremiah 14:9. Again, he wrote: "See! the king, my lord, has set his name at the rising of the sun and at the setting of the sun," a parallel to which may be found in Malachi 1:11.

These and similar parallels moved the scholar who brought them together to write: "The repeated lyrical utterances of Rib-Addi and Abdi-Khiba are early examples of the unrestrained laments of the later Israelites who appeal, not to a divine king of Egypt, their overlord, but to Yahweh."[54] By establishing this, the author hit on only half the truth. He emphasized the similarity of the expressions and he was right; but, compelled by the conventional chronology, he regarded this fact as proof that the Canaanites, seven or eight hundred years before the Israelite prophets and scribes, had used the same peculiar expressions—and he was wrong.

The assumed heritage of the Canaanite culture in art and literature

was actually an Israelite creation. The proofs of the continuity o
culture in Palestine (before and after the conquest by the Israelites
are vanishing.

To this list of comparisons, taken from the research of another
may be added all those expressions that we have found to be ident
ical in the letters and in the dialogues and monologues of the Book
of Kings and Chronicles in the chapters dealing with the period o
Ahab and Jehoshaphat, and which we compared earlier in thes
pages. We compared even the forms of speech of Jehoshaphat witl
similar constructions in the letters of the king of Jerusalem; ex
pressions of the scriptural Ahab with expressions in the letters of th
king of Sumura; and we found every sentence of the biblical dialogu
of Hazael preserved in the letters of Aziru and in letters about him
It is easy to expand this collection, for there is not a monologue o
dialogue in the pertinent chapters of Kings and Chronicles that doe
not have turns of speech which are also in the el-Amarna letters. Th
conventional form of respectful address, such as "the king, my lord"
"my father", "thy son", "thy brother"; peculiar expressions, sucl
as, "But what, is thy servant a dog, that he should do this?"[55] o
"our eyes are upon thee";[56] and common idioms, such as "let go ou
of thine hand",[57] are found in el-Amarna letters and in the chapter
of Kings and Chronicles dealing with the period of Ahab and Je
hoshaphat.

The art of writing on clay was developed and exchange of letter
was common in the days of Jehoshaphat and Ahab. Jezebel sen
letters to the elders of Jezreel inviting them to bring false witnes
against Naboth. Ben-Hadad sent a letter to the king of Israel askin;
him to heal Naaman. Jehu, after the success of the conspiracy i
Jezreel, twice sent letters to the elders of Samaria.

The el-Amarna letters provide us with ample evidence that em
ployment of scribes and the writing of letters was rather common i
the Palestine of that time. They prove also that in the Palestine o
Ahab and Jehoshaphat scribes read and wrote cuneiform in additio
to Hebrew. This confirms the theory advanced at the beginning o
this century[58] that cuneiform was being written in Palestine at th
time when the annals, later used by the editor of the biblical texts
were composed.

The excavations in Samaria brought to light two tablets in cunei form. One of them contains these words: ". . . say Abiahi to the governor of the cities to deliver six oxen, twelve sheep"; the other tablet is almost obliterated. They have a Hebrew seal.[59] A Hebrew scribe must have written them. The exact date of these tablets from the palace at Samaria is not known. The palace was built early in the ninth century and destroyed late in the eighth century.

The Age of Ivory

The period of the el-Amarna correspondence might quite properly be called the "ivory age". Next to gold, the desideratum of all ages, objects of ivory were the most coveted, and with lapis lazuli they are the most frequently mentioned royal presents. Lapis lazuli was sent by the kings of Asia to the pharaohs Amenhotep III and Akhnaton; objects, and especially furniture, of ivory or inlaid with ivory were requested of the pharaohs by these kings.

Amenhotep III wrote to the king of Arzawa in Asia Minor: "Ten chairs of ebony, inlaid with ivory and lapis lazuli . . . I have dispatched."[60] And Tushratta, the king of Mitanni, wrote to Amenhotep III: "And let my brother give three statues of ivory."

Burraburiash, writing from Babylon, asked for objects of ivory:

LETTER 11: Let trees be made of ivory and colored! Let field plants be made of ivory and colored . . . and let them be brought!

The list of presents sent by Amenhotep IV (Akhnaton) to Burraburiash displays the "ivory age" before the eyes of the reader. Here are a few passages from this list:

LETTER 14: Eight *umninu* of ebony inlaid with ivory
Two *umninu* of ebony inlaid with ivory. . . .
—— Of ebony, inlaid with ivory.
Six beast-paws of ivory ——
nine plants of ivory,
ten —— which are —— of ivory ——
twenty-nine gherkin oil-vessels of ivory,
. . . forty-four oil vessels . . . of ivory,
three hundred and seventy-five oil vessels of ivory —

313

19 *gasu*, of ivory ——
19 breast ornaments, of ivory ——
13 *umninu* of ivory ——
3 for for the head, of ivory—
3 bowls of ivory ——
3 oil-containing oxen, of ivory,
3 oil —— *dushahu*, of ivory ——
of ivory . . . of ivory . . . of ivory . . .

The present research has established the fact that King Ahab was a contemporary of Amenhotep III and Akhnaton, and that Samaria was built with the help of the king of Egypt. In Samaria a house of ivory was built.

> I KINGS 22:39 Now the rest of the acts of Ahab, and all that he did, and the ivory house which he made, and all the cities that he built, are they not written in the book of the chronicles of the kings of Israel?

A few generations after Ahab, Amos, the prophet, prophesied about Israel and about Samaria, the capital, and the buildings therein:

> AMOS 3:15 And I will smite the winter house with the summer house; and the houses of ivory shall perish. . . .

And again he returned to prophesy evil "to them that are at ease in Zion, and trust in the mountain of Samaria".

> AMOS 6:4–5 That lie upon beds of ivory, and stretch themselves upon their couches . . . That chant to the sound of the viol . . .

The house of ivory and beds of ivory were supposed by earlier excavators to have been poetic inventions of the authors of the Scriptures.[61] But later excavations made on the site of ancient Samaria brought to light "hundreds of fragments of ivory".[62] They were identified as objects of jewelry, pieces of furniture, and ornamental work.[63] It can be established with exactitude that the time in which these objects originated was that of Ahab. Ivory fragments inscribed with Hebrew letters bear witness to this conclusion: a comparison of these letters with the characters of the Mesha inscription showed that they originated at one and the same time.[64]

The excavators felt justified in writing: "No other finds have told us so much about the art of the Israelite monarchy."[65]

On some of these objects there are Egyptian designs, and the Egyptian double crown, clearly cut, was found on several of the plates.[66] A student of the Bible would expect to find Assyrian motifs because of Assyrian domination in the north of Syria in the ninth century before this era, not a predominance of Egyptian influence in art at a time when, according to his table of chronology, arts were not cherished in the Egypt of the ignoble successors of Shoshenk (Sosenk) of the Twenty-second Dynasty, long after the brilliant Eighteenth Dynasty. The excavators of Samaria were surprised: "It is significant that in our ivories there is no sign of Assyrian influence", and "the influence of Egypt on the other hand is all-pervasive".[67] There are plaques which represent Egyptian gods; the subjects on furniture "are all Egyptian".[68]

By synchronizing the time of Akhnaton with that of Ahab, we realize that the part Egypt played in Samaria in the days of the el-Amarna letters makes the presence of Egyptian motifs and furniture styles in the ivory of Samaria very understandable. We are even in a position to compare the ivory finds in Samaria with those in Egypt at the time of Akhnaton. "Winged figures in human form" were found in Samaria. "The forms of winged figures on the ivories . . . are derived from Egyptian models. Tutelary goddesses of this type stand at the four corners of the shrine of Tutenkhamun."[69] Three winged sphinxes with human heads were also found in Samaria, and they, too, were recognized as similar to the human-headed lion from the tomb of Tutankhamen.[70]

Tutankhamen was a son-in-law of Akhnaton. The presence of similar figures on his sarcophagus and in the Samaria of Ahab is, from our point of view, not unexpected.

The excavators of Samaria's ivories, observing the style of the ornaments, recognize the influence of Egyptian art but think that in the days of Ahab a restoration of ancient forms in art took place, the "Egypt of yesterday" having been revived in Samaria after six hundred years.

In the same ornaments of Samarian ivory could be recognized motifs of the type described in the Scriptures, such as palm trees "between a cherub and a cherub" (Ezekiel 41:18). "Some of the figures [of the ivories of Samaria] . . . were like those which were carved on the House of the Lord in Jerusalem."[71] The conclusion arrived at was that the style of the ornaments in the Temple at Jerusalem represented an intermediate stage in the succession of influences. "The type goes back to the Egyptian art of the Eighteenth Dynasty."[72]

The place of origin of the style—Egypt or Palestine—can be disputed. Probably a reciprocal influence was the inspiration for the motifs on the ornaments. One fact, however, reverses all conclusions in the study of the comparative art of Egypt-Palestine: the Eighteenth Dynasty ruled from the days of Saul until Jehu, and the great temples of Luxor and Karnak, creations of Thutmose III and Amenhotep III, were built not before but after the Temple in Jerusalem was erected.

In the time of Solomon ivory was imported from distant countries, along with silver, apes, and peacocks. From Palestine it was exported to Egypt, and Hatshepsut brought ivory with her as the bas-reliefs of the Punt expedition show and the texts accompanying the bas-reliefs tell. Pharaoh Thutmose III, on plungering Megiddo in his campaign of conquest, took, according to his annals, "six large tables of ivory and six chairs of ivory" besides other spoils. The pharaoh brought from Jerusalem "the great throne of ivory, overlaid with pure gold"—this according to the Hebrew sources (II Chronicles 9:17; 12:9). In the eighth campaign Thutmose III took tribute including "vessels laden with ivory". In the records of the tribute he received in Punt, God's Land, or Rezenu (Palestine) there is frequent mention of ivory in tusks (thus, eighteen tusks from the chiefs of Rezenu in the sixteenth collection of tribute) and in furniture (tables of ivory in the thirteenth collection of tribute). After enlarging his fleet with the fleet of the Phoenicians, he sent his ships to collect tribute and, like Hatshepsut before him, used the maritime route for transport of ivory from Palestine to Egypt.

The art of working in ivory was transferred at that time to Egypt. In the tomb of Rekhmire, a vizier of Thutmose III, workers from Palestine are portrayed "making chests of ivory"—these were imported craftsmen.

Samaria, built a few decades later, became a manufacturing center of ivory, working mainly for Egypt. Among the ivories excavated in Samaria there are many unfinished pieces with Egyptian patterns.

Color effects were produced by staining the ivory with pigments.[73] Colored ivory was also found in the tomb of Tutankhamen. In an el-Amarna letter of Burraburiash we read about colored ivory.

The tablets of el-Amarna mention furniture and various objects of ivory delivered to Asia Minor, Cyprus, Assyria, and other countries of Western Asia. Objects similar to those found in Samaria were discovered in these lands.

In the last century inlaid plaques with Egyptian subjects were found in Mesopotamia. When the ivories of Samaria were unearthed they were acclaimed as closely related to those found previously in the palace of Nimrud and in other places: ivories which "may have come from the same workshop as the Samaria ivories, were found by Layard in the northwest palace at Nimrud; a few stray examples come from other sites".[74]

Ivories similar to those of Samaria have been found in a number of different places, sometimes together with Egyptian objects of the Eighteenth Dynasty. One of these places is Megiddo. Though the ivories of Samaria and Megiddo show the same patterns and the same workmanship, they were ascribed to two different periods.[75] Similarly, other discoveries of ivory were attributed by their finders either to the period of the Eighteenth Dynasty (the fifteenth and fourteenth centuries) or to the period of the kings of Samaria (the ninth and eighth centuries).[76] The second period is thought to have been the age of imitation of old Egyptian styles and of the renaissance of the old craft.

They were but one and the same period, and the ivories of the Samaria of Ahab and of the Thebes of Tutankhamen are products of one and the same golden age of ivory art.

Conclusions

If one is determined to keep to the traditional construction of history and insist that the letters of el-Amarna were written to and from archaic Canaanite princes, he is also bound to maintain that in Canaan events occurred which recurred half a millennium later in the time of Jehoshaphat and Ahab. This makes it necessary to hold that there already was a city of Sumur, of which not a relic remained; that this city, with a royal palace and fortified walls, was repeatedly besieged by a king of Damascus, who had a prolonged dispute and recurrent wars with the king of Sumur over a number of cities, in a conflict that endured for a number of decades; that on one occasion the king of Sumur captured the king of Damascus but released him; that on the occasion of a siege of Sumur by the king of Damascus the guard attached to the governors succeeded in driving away the Syrian host from the walls of Sumur; that on the occasion of another siege of Sumur the Syrian host, hearing rumors of the arrival of the Egyptian archers, left their camp and fled—every detail an exact image of what happened again half a millennium later at the walls of Samaria.

The traditional construction of history implies also that the king of Damascus, who was at the head of a coalition of many Arabian chieftains, succeeded in fomenting a revolt by a Trans-Jordan king named Mesh against the king of Sumur, whose vassal he was, and this rebellious vassal king captured cities of the king of Sumur and humiliated his people, as in the days of Mesha's rebellion against the king of Samaria. That Rimuta was the place in dispute between the king of Damascus and the king of Sumur, as Ramoth was in the second epoch; that the king of Sumur had a second residence where a deity was worshiped whose name, Baalith, was the same as that of the deity introduced by Jezebel, and the king of Sumur planted groves in his second residence, like Ahab in the field of Naboth; that the king of Damascus organized a number of ambuscades against the king of Sumur, and the king of Sumur each time managed to escape death, like the king of Samaria of the second period; that the king of Damascus became gravely ill, yet did not die from the illness but was put to a violent death on his sickbed, like the king of Damascus of the second period.

This hypothetical scholar would also be bound to admit that all these coincidences happened at a time when the land of Sumur was visited by a drought, and the springs dried up and a severe famine followed; that the drought lasted several years and caused starvation of the people and epidemics among the domestic animals; and that the inhabitants departed from the realm of the two residences—everything just as it happened in the second period.

He would have to maintain that the two periods do not differ in any respect whatsoever, and that each event of one period has its twin in the other. The land of Edom was ruled by a deputy of the king of Jerusalem—in both cases. Tribes from as far away as Mount Seir invaded Trans-Jordan—in both cases. In the first period as in the second, the invaders threatened Jerusalem and caused the population of the kingdom to flee from their homes. The king of Jerusalem, like Jehoshaphat centuries later, was afraid of being driven with his people from their inheritance and expressed his fear in similar terms, but everything turned out well when the tribes of Mount Seir and Trans-Jordan rose up one against the other, as they did five to six hundred years later.

This scholar would also have to admit that the military chiefs of the Canaanite king of Jerusalem signed their letters with the same names as the military chiefs of Jehoshaphat, king of Jerusalem, and that the names were as peculiar and unusual as, for example, Iahzibada (Jehozabad) and "son of Zuchru" ("son of Zichri"), or Addaia (Adaia), or Adadanu (Adadani, Adna), who was again the first among the chiefs; that the governor of Sumur had the same name as the governor of Samaria of the later period (Amon), and that the keeper of the palace in Sumur was named Arzaia like the chief steward, Arza, of the king of Israel.

Again, in the city of Shunem (Shunama) lived a "great lady", and already in the first period some miracle had happened to her so that she was called Baalat-Neše.

And again, the king of Damascus had a military governor (Naaman, Ianhama), by whose hand "deliverance was given to Syria", and who at first was feared by the king of Sumur but later on became the latter's friend, like his reincarnation six hundred years later.

Further, the successor of the murdered king of Damascus, by the name of Aziru or Azaru, acted like Hazael of the second period: he oppressed the land of Sumur; he conquered almost all the land of the realm; he burned with fire the strongholds and villages of the king of Sumur; he even spoke with the same peculiar expressions as Hazael did later on.

This scholar would also be faced with the fact that in the second period the city of Irqata again lost her king, and that King Matinu-Bali and King Adunu-Bali, under the leadership of Biridri, defied the mighty invader from the north, just as happened in the first period when a Biridia (Biridi) assumed the task of leading the kingless city of Irqata and King Mut-Balu and King Aduna against the invader from the north. In both cases this invader was the king of Assyria and the lord of Hatti. In both cases he was victorious over the coalition of Syrian and Palestinian princes helped by Egyptian battalions. In both cases he received placating presents from Musri (Egypt) in the form of rare animals or figures of such animals. Again, the king of Damascus, Hazael, battled with him between Lebanon and Hermon as did Azaru of the first period. Again, the kings of Tyre and Sidon, harassed by this invader, left their cities and departed in ships, as they did six hundred years earlier.

In both periods the art of ivory work flourished, and identical patterns were produced: designs and execution, characteristic of the earlier period, were repeated in the second period, and have been found to be so similar that they have been taken for copies of the art objects of the first period.

In both periods the same architecture and stone workmanship (Megiddo, Samaria) found expression.

In both periods the same idiomatic Hebrew was spoken.

Can one accept such a series of coincidences? And if it is accepted, is it only to have the old difficulties present themselves again? If the Habiru were the Israelites, why, then, in the Book of Joshua, which records the conquest of Canaan, and in the letters of el-Amarna are no common name and no common event preserved?

A Halfway Mark

At the beginning of this work I placed before the reader the unsolved problem of the correlation of Israelite and Egyptian histories. Of these two ancient nations, one professes to have had close ties with the other; actually the biblical story moves in the light and shadow of the great kingdom on the Nile. Egyptian history, on the other hand, in all its numerous inscriptions, on stone and on papyrus, denies any real contact with the neighboring kingdom on the Jordan. Even the glorious age of King Solomon, so exalted in the Old Testament, appears to have passed entirely unnoticed by the Egyptian kings and their scribes. And more than that, the great events of the Israelite past—their long bondage in Egypt and their departure from that land under unusual circumstances—appear to have been entirely unknown to the conventional history of Egypt. For that reason the time of the Exodus is debated and placed at almost every conceivable time point of the Egyptian past, from the beginning of the New Kingdom presumably in −1580 down many centuries. The uncertainty as to the time of the sojourn in Egypt and the departure is the direct result of the absence of references to the children of Israel in Egypt and to their leaving the country, and of the sterility of information concerning the relations as neighbors of these two peoples during the period covered by the Scriptures.

We have attempted to solve the problem of the synchronization of the histories of these two peoples of antiquity, both of whom occupy major places in the history of the ancient world. We made this attempt after we recognized that the biblical story of the Exodus contains frequently repeated references to some natural catastrophes. Logic thus required us to look in extant Egyptian sources for references to some disturbance in nature.

The search proved not to be fruitless. The Leiden Papyrus Ipuwer is a record of some natural catastrophe followed by a social upheaval; in the description of the catastrophe we recognized many details of the disturbances that accompanied the Exodus as narrated in the Scriptures. The inscription on the shrine from el-Arish contains another version of the cataclysm accompanied by a hurricane and nine days' darkness; and there we found also a description of the march of the pharaoh and his army toward the eastern frontier of his kingdom, where he was engulfed in a whirlpool. The name of the pharaoh is given in a royal cartouche which proves that the text was not regarded by its writer as mythical.

If we have in these documents the same story as found in the Book of Exodus, then a synchronical point between the histories of these two ancient nations is established. But here, where we expected to reach the solution of the problem of the date of the Exodus in Egyptian history, we were confronted with a problem that made the question of the date of the Exodus shrink into insignificance. Whatever theories have been offered concerning the time of the Exodus, not once has the thought occurred that the Israelites left Egypt on the eve of the arrival of the Hyksos. Consequently we found ourselves faced with a problem of very different magnitude. Either Egyptian history is much too long or biblical history is much too short. Must Egyptian history be shortened by some "ghost" centuries, or biblical history lengthened by the same number of "lost" centuries?

We could not know the answer to this problem until we traveled a long distance through the centuries of ancient history. We noticed a path on which to start this journey. If the Israelites left Egypt on the eve of its invasion by the Hyksos, who arrived from Asia, we might, perchance, find in the Scriptures a reference to a meeting outside the borders of Egypt of the children of Israel and the invaders. Actually, even before they reached Mount Sinai, the Israelites encountered the hordes of Amalekites. We turned to the old Arabian writers and found that the tradition of the Amalekites as the dominant tribe among the Arabs, who invaded Egypt and ruled it for four or five hundred years, is alive in the Arabian literary heritage from their early past.

As we compared point after point in the Egyptian hieroglyphic, Hebrew biblical and post-biblical sources, and Arab autochthonous traditions found in their medieval writings, we were forced to conclude that the time of the Hyksos domination of Egypt was the time of the Judges in scriptural history. The equation of Hyksos and Amalekites gave additional support to the synchronization of the fall of the Middle Kingdom and the Exodus. We then had to examine the historical moment of the collapse of Hyksos rule in Egypt and the end of Amalekite domination in the Near East. In the siege of Auaris, the Hyksos fortress, by Ahmose, some foreign troops played a decisive role. From parallels in the Book of Samuel it could be determined that it was King Saul, the first Jewish king, who was victorious over the Amalekites at el-Arish; and with the help of many proofs we could establish that el-Arish occupies the position of the ancient Auaris.

David was a contemporary of Ahmose, founder of the Eighteenth Dynasty, and of Amenhotep I; Solomon was a contemporary of Thutmose I and Hatshepsut. And we found that the celebrated journey to God's Land and Punt was the voyage to Palestine and Phoenicia described in the Scriptures as the visit of the Queen of Sheba.

We compared many details and always found that they coincided. But this drove us to the next station on the road. Five years after Solomon's death the Temple of Jerusalem and its palace were sacked by a pharaoh. Thutmose III succeeded Hatshepsut. If we were traveling on the right road, again we had to find a correspondence here: Thutmose III must have sacked Jerusalem of the treasures of its palace and Temple. This he actually did, and the pictures of his booty correspond very closely, in shape and number, with the description of the loot taken by a pharaoh in the fifth year after Solomon's death.

Under the next pharaoh Palestine was invaded again, according to scriptural and Egyptian sources. This time, however, the expedition was far from being victorious.

For three generations biblical scholars proved to the full satisfaction of all that many parts of the Scriptures were products of much later centuries than the Scriptures would indicate. Then, during the 1930s, with the discovery of the Ras Shamra texts, the estimate was revised in the diametrically opposite direction: the same biblical texts were now regarded as a heritage of Canaanite culture, six centuries older than the biblical texts. However, the collection of material from Hebrew literary sources, from Ras Shamra, and from Egypt convinced us that not only the earlier reduction of the age of biblical prose and verse but also its present increase is erroneous. In saying this we are actually ahead of what we may legitimately assert: we still do not know which of the two histories, Egyptian or Israelite, must be readjusted. At the same time we observed how the histories of other ancient countries and peoples accord with either the Israelite or the Egyptian chronology; and how the histories of Cyprus, Mycenae, and Crete, in correlating with one side or the other, create confusion in archaeology and chronology.

In three consecutive chapters we collated the historical evidence of three successive generations in Egypt (Hatshepsut, Thutmose III, Amenhotep II) and in Palestine (Solomon, Rehoboam, Asa), and found unfailing correspondence. It is possible that by sheer accident one age in Egypt bears a close resemblance to another age, and thus offers ground for a spurious coevality; but it is quite impossible that three consecutive generations in Egypt and in the neighboring Palestine of two different ages could produce consistent correspondence in so many details. What is even more striking, these three consecutive generations in Egypt as well as in Palestine were not selected at random, but were forced upon us by the deliberations and parallels of the earlier chapters, in which we scrutinized the time of the Exodus and the following centuries until Saul, and in Egypt the last days of the Middle Kingdom and the following centuries of Hyksos rule until the rise of the New Kingdom.

It would be a miracle, indeed, if all these coincidences were purely accidental. Anyone familiar with the theory of probabilities knows that with every additional coincidence the chances for another grow smaller, not in arithmetical or in geometric progression, but in a progression of a higher order; therefore the chance would be a trillion or quadrillion against one that all the parallels offered on previous pages are merely coincidences.

Following the three consecutive generations in Egypt and Palestine, there were Amenhotep III and Akhnaton in Egypt and Jehoshaphat in Judea and Ahab in Israel. It could not be by mere chance that the fourth generation again presented a picture in which the details fit together like the pieces of a jigsaw puzzle. The histories of two lands and the vicissitudes in the lives of their rulers and their peoples could not be in complete correspondence were there not exact synchronism. And so it happened that in this fourth generation the rulers and prominent personages in one country actually wrote letters to the rulers and prominent personages in the other country and received written answers from them. To what extent details and events correspond during those years of famine, sieges, invasion from Trans-Jordan, and military pressure from the north has been recounted at length. And this sequence of invariable correlation and conformity gives us a feeling of security that we are not on the wrong path.

However, we are not yet at the end of the journey. Notwithstanding all that has been said up to now concerning the numerous parallels, collations, coincidences, and conformities, as well as the theory of probabilities, we cannot regard the problem of ancient history as solved until we have covered the full distance to the point where the histories of the peoples of the ancient East no longer present a problem of synchronization.

We have before us the eighth and the following centuries, according to Israelite history. Where, then, shall we find room for the so-called Nineteenth Dynasty, that of Ramses II and other famous kings? And what about the Hittite king with whom Ramses II signed a treaty? And where is there room for the Twentieth and Twenty-first Dynasties, the Libyan and Ethiopian dominion in Egypt, and all the others up to and including the Thirtieth Dynasty, that which expired shortly before Alexander reached Egypt?

The identifications we have made will be of no avail if we are unable to reach safely this point in history—the end of the last native dynasty in Egypt. We must be able to disentangle the archaeological, historical, and chronological problems that we shall meet in the following centuries and, with the thread of Ariadne we took from the hands of Ipuwer, to proceed along the path to the point where the histories of the various nations of antiquity appear in harmony with one another. If we should not be able to do so the coincidences presented will necessarily be regarded as miraculous; for they are too many and too striking to be ascribed to accident. It is safer to assume, therefore, that by working diligently we shall arrive at our goal of a complete revision of ancient history.

NOTES

1. Letter 85.
2. Letter 85.
3. Letter 86.
4. Compare C. Niebuhr in *Mitteilungen der Vorderasiatisch-ägyptischen Gesellschaft*, I (1896), 208ff.; W. M. Müller, ibid., II (1897), 274f.; H. Ranke, *Keilinschriftliches Material zur Altaegyptischen Vokalisation* (*Berlin*, 1910), p. 22 and note 1.
5. Albright, *Journal of the Palestine Oriental Society*, II (Jerusalem, 1922), 112, note 2; ibid., IV (1924), 140.
6. Maisler, *Untersuchungen zur alten Geschichte und Ethnographie Syriens und Palästinas*, pp. 7ff.
7. Mercer, *Tell el-Amarna*, note to Letter 68.
8. See also II Kings 8:28.
9. *Jewish Antiquities*, VIII, 398.
10. Babylonian Talmud, Tractate Makkot 9b: "Sichem in the mountains, opposite to Ramoth in Gilead." See A. Neubauer, *La Géographie du Talmud* (Paris, 1868), p. 10.
11. J. W. Crowfoot and G. M. Crowfoot, *Early Ivories from Samaria* (London, 1938).
12. Letter 69.
13. G. A. Reisner, C. S. Fisher, and D. G. Lyon, *Harvard Excavations at Samaria, 1908–1910* (Cambridge, Mass., 1924).

14. Letter 62.

15. Letter 159.

16. Letter 160.

17. Luckenbill, *Records of Assyria*, I, Sec. 566.

18. Hrozný, "Les Ioniens à Ras-Shamra", *Archiv Orientální*, IV (1932), p. 178.

19. Luckenbill, *Records of Assyria*, I, Sec. 609.

20. In northern Syria.

21. Letter 75.

22. Knudtzon, *Die El-Amarna-Tafeln*, pp. 1013f.

23. *Against Apion*, I, 123–25.

24. See C. S. Fisher, *The Excavation of Armageddon* (Chicago, 1929), p. 16.

25. Since the el-Amarna letters were not considered as belonging to the time of Shalmaneser, the chief of the coalition, Biridri, was supposed to have been Ben-Hadad, the most powerful among the kings of Syria. (See Meyer, *Geschichte des Altertums*, II, Pt. 2, p. 274.) The identification of Ben-Hadad and Biridri gave rise to the question: Why did Ahab come to the help of Ben-Hadad, his enemy at Karkar? It was conjectured that Ben-Hadad conducted his war against Ahab to compel him to participate in the war against Shalmaneser.

26. By C. Beke. See his *Mount Sinai a Volcano*, p. 8.

27. By H. Winckler. The Encyclopaedia Biblica, ed. Cheyne and Black, by giving credence to this theory and all its consequences (relating to all contacts of the Israelites with Egypt), became worthless with respect to many important subjects.

28. Letter 75.

29. Letter 126.

30. Letter 129.

31. Luckenbill, *Records of Assyria*, I, Sec. 641.

32. Letter 75.

33. A similar punishment was meted out to the brothers of Ahab's heir by Jehu's order.

34. Letter 165.

35. Identified as Baalbek by Halévy and Winckler. Cf. Weber, in Knudtzon, pp. 1123ff.

36. Letter 165.

37. Letter 166.

38. Luckenbill, *Records of Assyria*, I, Sec. 601.

39. In the second part of the ninth century.

40. By W. F. Albright, *Journal of Egyptian Archaeology*, 23 (1937), 191f.; *Journal of Biblical Literature*, 61 (1942), 314.

41. ". . . nor is it clear what the etymology of the word is". Mercer, *Tell-el-Amarna Tablets*, pp. 504–505.

42. Weber, in Knudtzon, *Die El-Amarna-Tafeln*, pp. 1254f.

43. Luckenbill, *Records of Assyria*, I, Sec. 599.

44. Letter 41.

45. G. Virolleaud, "Suppiluliuma et Nigmad d'Ugarit". *Revue hittite et asianique*, V (1940), 173–74; C. H. Gordon, *Ugaritic Handbook* (Rome, 1948).

46. Luckenbill, *Records of Assyria*, I, Sec. 563.

47. Letter 9: "Assyrians, my subjects."

48. See article "Babylonia" in Encyclopaedia Biblica, ed. Cheyne and Black.

49. Letter 14.

50. Luckenbill, *Records of Assyria*, I, Sec. 563.

51. The following examples and quotations are from S. A. Cook, "Style and Ideas", in *Cambridge Ancient History*, Vol. II.

52. Letters 296, 257.

53. Letter 137.

54. S. A. Cook, in *Cambridge Ancient History*, II, 338.

55. "The dog" means also a male prostitute. Cf. Deuteronomy 23:18.
    The expression, "Is thy servant a dog that he shall not . . ." is also found in letters of Lachish, the modern Tell ed Duweir, in southern Palestine. These letters were written shortly before the destruction of the first Temple.

56. II Chronicles 20:2 (words of Jehoshaphat).

57. I Kings 20:42.

58. W. Winckler, "Der Gebrauch der Keilschrift bei den Juden", *Altorientalische Forschungen*, III (1902), Part I, 165f.; E. Naville, *Archaeology of the Old Testament* (London, 1913); Benzinger, *Hebräische Archaeology* (2nd ed., 1907), p. 176. Jeremias, *Das Alte Testament im Lichte des alten Orients*, p. 263.

59. Reisner, Fisher, and Lyon, *Harvard Excavations at Samaria*, I, 247.

60. Letter 31.

61. Cf. Reisner, Fisher, and Lyon, *Harvard Excavations at Samaria*, p. 61.

62. Crowfoot and Crowfoot, *Early Ivories*, p. 2.

63. Ibid., p. 55.

64. E. L. Sukenik, ibid.: "The result of this examination leads us to the conclusion that the Samaria ivories are, like those of Arslan Tash, of the ninth century and *earlier* than the Samaria ostraca."

65. Crowfoot and Crowfoot, *Early Ivories*, p. 49.

66. Ibid, p. 23

67. Ibid., p. 49.

68. Ibid., p. 9.

69. Ibid., p. 18.

70. H. Carter, *The Tomb of Tut.ankh.Amen* (London, 1923–33), Vol. II, Plate XIX.

71. Crowfoot and Crowfoot, *Early Ivories*, p. 53.

72. Ibid., p. 34.

73. Homer, in the *Iliad*, IV, 141–42, mentions the Carian woman who stains ivory red.
74. Crowfoot and Crowfoot, *Early Ivories*, p. 9.
75. See G. Loud, *The Megiddo Ivories* (Chicago, 1939).
76. Ibid.

Selective Name Index

(The names of modern authors, with a few exceptions, are not included)

Aaron, 17, 55

Abd-Alhakam, Ibn, 63

Abd' Astart, 226, 281

Abdi-Ashirta, 223, 225, 234, 235, 237–81 *passim*

Abdi-Hiba, 223, 225, 311

Abel-beth-Maachah, 235, 242

Abijah, 158, 193

Abimilki, 204, 297, 304–6

Abraham, 58, 124, 164, 197, 198, 217

Abu-el-Saud, 95

Abu-faid, 96

Abu'l Faradj, 58

Abulfeda, 58, 62

Abydos, 173

Abyssinia, Abyssinian, 47, 102, 125, 127, 195 *See also* Ethiopia

Adad, 281 *See also* Hadad, Adados

Adad (storm god), 305

Ada-danu, 229, 319

Adados, 279, 288 *See also* Adad, Hadad

Adaia, Addaia, 231, 319

Addudani, Addadani, 228–30

Adna, Adnah, 228, 229, 241, 319

Adonizedek, 218

Adoram, 141

Aduna, 278, 299, 319

Adunu-Bali, 298, 319

Afiru, 219, 270

Africa, African, 103–6, 117, 130

Agag, Agog, 67–8, 72, 74, 78, 81, 87

Ahab, 188, 220, 221–2, 225–7, 236, 243–9 *passim*, 256, 264, 285–8 *passim*, 295, 296, 312, 314, 315, 317, 318

Ahaziah of Israel, 243, 248, 252

Ahaziah of Judea, 188, 287, 300

Ahmose, 2, 3, 5, 73, 79, 80, 99, 135, 157

Ahmose (officer), 73, 74, 75, 79, 81

Ajalon, Ajaluna, 269, 277

Akhet-Aton, 50, 204, 214, 215, 217, 284

Akhi-Yawi, 311

Akhnaton, 2, 3, 5, 6, 7, 173, 204, 215, 216, 219, 223, 224, 247, 275, 284, 287, 303, 305, 306, 308, 309, 311, 313–15

Akkadian, 176, 177–8, 185, 215, 250, 310

Aksum, 126

Alasia (Cyprus), 176

Alexander the Great, 3, 91, 183, 298

Alexandria, 158

Alkan (Abou-Kabous), 62

Aloth, 133 *See also* Eloth

Al-Samhudi, 61, 95

Amalek, Amalekite, 55–92 *passim*, 107, 111, 140, 148, 195

329

330

105, 141, 159, 191, 193, 284, 295

Askelon, 165, 232

Assurbanipal, 302

Assyria, Assyrians, 3, 75, 87, 90, 91, 176, 185, 201, 203, 226, 227, 229, 236, 273, 297, 300, 304, 307–10, 315, 317, 320

Assyro-Babylonian, 215, 247

Astart, Astarte, 179, 200, 226

Ataroth, 259

Athaliah, 188, 248

Athenian, Athenians, 187, 207

Atlantic Ocean, 107

Atlantis, 191

Aton, 214, 215, 306

Auaris, Avaris, 44, 45, 62, 64, 68, 72–6, 78–9, 80–3, 87, 107, 148

Augustine, 5

Aye, 7

Azaelos, 288 *See also* Hazael

Azaru, Azira, Aziru, 224–6, 239, 281–310 *passim*, 319 *passim*

Baal, 179, 208

Baalat-Nese, 277–8, 319

Baalbek, 303

Baal-Meon, 259

Baal-miir, 310

Baasha, 235, 243, 266

Babylon, Babylonia, 2, 3, 90–1, 183, 201, 219, 252, 255, 297, 299, 301, 302, 307, 310

Baghdad, 64

Balaam, 67, 68

Barak, 69

Batruna, 221

Beer-sheba, 165, 232

Beirut, 104, 240, 285, 286, 287, 306

Beke, Charles, 13–15

Belias, 222, 223, 318

Belit, 222, 223, 318

Benaiah, 188

Ben-Hadad I, 234

Ben-Hadad II, 223–4, 225, 226, 234–81, 284

Ben-Hail, 232

Benjamin, 141, 155, 193, 228, 230, 231, 241

Beth-el, 8, 141, 150, 163

Bethlehem, 165

Beth-Ninib, 273

Beth-Shan, 79, 165

Beth-Zur, 141, 146

Bezaleel, 148, 151–2

Bilkis, 126

Biridia, 298, 300, 301, 302, 310, 319

Biridri, 225–6, 298, 300, 302, 310, 319

Bisitanu, 265

Bit-Arkha, 262

Bithynia, 207

Boghazkeui, 185

Borsippa, 297

Bosphorus, 207

Botrys, 221

Bronze Age, 172

Bubastis, arm of river, 81

Burkuna, 277, 278

Burraburiash, 307, 308, 309, 313, 317

Byblos, 104, 110, 208, 221

Canaan, Canaanites, *passim*

333

334

335

Maimonides, 135
Makeda, 125, 126
Makera, Makere, 101, 119, 125, 126 *See also* Hatshepsut
Makkedah, 290
Manasseh, 251
Manetho, 3, 4, 38, 45, 53, 54, 62, 63, 64, 65, 68, 70, 75–8, 81–2, 87–9
Maresha, Mareshah, 141, 193–4, 196, 200, 202, 203, 205
Mariette, A. E., 134
Masudi, 59–62, 83
Matinu-Bali, 298, 299, 319
Mayas, 124
Mecca, 58, 59, 61, 69, 87
Medeba, Mehedeba, 246, 259
Media, 87
Medina, 69, 95, 125
Mediterranean Sea, 12, 83, 105, 119, 129, 138, 157, 185, 186, 304, 306
Megasthenes, 88
Megiddo, 142–3, 144, 147, 155, 158, 160, 162, 165, 188, 226, 298–300, 317, 320
Melchizedek, 124
Memphis, 38, 50, 63, 192, 194
Menander, 221, 226, 281, 299
Menashe (Manasseh), 163
Menelik, 102, 125, 212
Menkheperre-Seneb, 159
Mercer, A. B., 230, 249
Meriba, 55
Merneptah, 2, 3, 7, 8, 44, 217, 218
Meryt-aten, 306
Mesha, Mesh, 68, 219, 224, 227, 244, 246, 247, 248, 255, 257–65

passim, 273, 278, 295, 296, 314, 318
Mesha stele, 219, 246, 247, 262, 263, 295
Mesopotamia, 80, 190, 218, 310, 317
Metropolitan Museum of Art, 159, 167
Metten-Baal, 299
Meyer, E., 167
Micaiah, 233
Middle Kingdom (in Egypt), 2, 3, 40–2, 43, 70–1, 124, 163, 171, 173
Midian, Midianites, 69, 70, 80
Milet, 177
Milkili, 277–8
Minet el Beida, 171–2, 175
Minoan (Early, Middle Late), 43, 171–4, 191, 205
Minos, 172, 186
Mitanni, Mitannian, 53, 185, 186, 190, 301
Mitta, 301
Mizraim (Egypt), 14, 300
Moab, Moabites, 67, 69, 80, 139, 219, 224, 227, 244–6, 257–70 *passim*
Mohammed, 125
Moresheth-Gath, 202
Moses, 5, 7, 11, 17, 27, 55–6, 60, 86–7
Mount Sinai, 13–15, 55, 111
Musri (Egypt), 300, 301, 309, 310, 320
Mut-Balu, 299, 319
Mycenae, Mycenaean, 91, 171–4, 182, 205
Mylasa, 190

337

Phaestus, 172
Philistia, Philistines, 8, 78–9, 85,
    99, 100, 107, 188, 227, 293
Philo of Alexandria, 35
Philo of Byblos, 132, 249
Phiops, 2
Phoenicia, Phoenicians, 78, 95,
    104, 105, 107, 118, 123, 124,
    129, 130, 155, 162, 179, 180,
    181, 187, 189, 191, 196, 197,
    198, 199, 201, 202, 204, 205,
    221, 226, 240, 297, 304, 316
Phoinikus, 187
Phoinix, 187
Pi-ha-Khiroth, 31, 33, 37, 38,
    59
Pi-Kharoti, 37
Pindar, 13
Piram, 218
Pirathon, 69
Pisoped, 38, 50
Pithom, 38, 46, 51, 52
"Place of the Whirlpool", 36
Plato, 88, 132
Pleti, 188, 202
Pliny the Younger, 14
Plutarch, 135
Pompeii, 14
Pontus, father of Sidon, 124
Poseidon, 104
Psammetich, 189
Ptolemaic dynasty, age, 3, 34, 53,
    76
Ptolemy, 3
Punic Wars, 124
Punt, 103–30 passim, 139, 161–2,
    166, 167, 316
Puti-Hiba, 224
Pythagoras, 88

Ra, 37–8, 44, 62, 103, 111, 126
Ra-Amon, 275
Rabbath-Ammon, 261, 269
Rachel, 218
Ra-Harakhti, 36
Ra-Harmachis, 36
Ramah, 235
Ramessides, 64, 164
Ramoth-Gilead, 240, 242–5 pas-
    sim, 248, 258, 293, 294, 318
Ramses II, 2, 3, 7, 44, 50, 134,
    216–18
Ramses III, 2, 3, 8
Ramses, City of, 46, 50, 51
Ras Shamra, 171–206 passim,
    297, 307
Rebecca, 217
Red Sea, 12, 56, 57, 59, 60, 80,
    100, 106, 107, 132, 134, 138,
    162, 180, 240
Rehoboam, 138, 140–1, 146,
    149, 158, 161, 165, 166, 167,
    169, 193, 224
Rekhmire, 159, 316
Rephidim, 55
Retenu, Rezenu, 100, 155, 159,
    161–3, 192, 195, 316
Rezon, 140, 234, 235
Rhinocolura, Rhinocorura, 82, 95
Rib-Addi, 223, 225, 233–4, 243,
    286, 287, 301, 302, 311
Rimmon, 275
Rome, Roman, 82, 89, 124, 154
Rubuda, Rubute, 261, 269

Saadia, 96
Saba (Sheba), land of, 59, 101–2,
    105, 125, 128

339

Sabi-ilu, 265
Safa, 59, 60
Salatis *See* Salitis
Salem (Jerusalem), 124, 143, 220, 306
Salitis, 63–4, 81, 82–3
Sallier Papyrus I, 72
Salt Lake, 12
Samaria, 3, 219–320 *passim*
Samuel, 73, 78
Sanchoniaton, 104
Sapasites, 200
Saqqara, 18
Sarah, 217
Sargon II, 3
Sarira, 201
Saul, 3, 8, 68, 73–4, 76, 78, 81, 82, 85, 87, 90, 99, 138, 148, 195, 316
Scythians, 53
Sea of Galilee, 240, 290
Sea of Passage, 14, 24, 35, 59
Sebastieh, 264
Sebekhotep I, 51
Sebeknofrure, 133
Seder Olam, 169
Sehlal, 265
Seir, Seeri, 267–9, 319
Sekhet-za, 96
Seknenre, 72
Seleucides, 183
Seleucus Nicator, 98
Semneh, 43
Senir, Mount, 310
Senmut, 115
Sennacherib, 87
Septuagint, 158, 159
Serbon, Lake, 12
Sesostris III, 163, 168

Seth, 64
Sethe, K., 41, 42, 167
Seti the Great, 2, 22, 216, 218
Seventeenth Dynasty, 2, 65, 81
Shalmaiati, 305, 306, 308
Shalmaneser III, 219, 225, 226, 227, 229, 247, 248, 281, 296–306 *passim*, 307–10
Shamash-Edom, 192
Shamshi-Ramman, 229
Sharon, 293
Sharuhen, 73, 76, 77–9, 84, 87
Sheba, Queen of, 99–130 *passim*, 139, 148, 161, 195, 224
Shechem, 141
Shem, 58
Shemaiah, 141, 146
Shemer, 220
Shiloh, 150
Shinar, 162, 164
Shishak, 100, 145, 146, 147, 149, 159, 165, 193, 194, 212, 224
Shou, 36
Shunem, Shunama, 277–8, 319
Shur, 74, 81, 95
Siana, 298, 299
Sidon, Sidonians, 104, 124, 139, 147, 155, 158, 179, 196, 200–2, 222, 236, 240, 247, 285, 304–6, 320
Sigata, 269–70
Sinai Peninsula, desert, 14, 74, 100, 107, 219, 301
Sirius, 71
Sisera, 69
Sixth Dynasty, 2, 110, 173
Socoh, 141, 146, 192
Sodom, Sodomites, 58, 270, 271